ELEMENTS OF POWER SYSTEM ANALYSIS

McGraw-Hill Series in Electrical Engineering

Consulting Editor
Stephen W. Director, Carnegie–Mellon University

Networks and Systems
Communications and Information Theory
Control Theory
Electronics and Electronic Circuits
Power and Energy
Electromagnetics
Computer Engineering and Switching Theory
Introductory and Survey
Radio, Television, Radar, and Antennas

Previous Consulting Editors

Ronald M. Bracewell, Colin Cherry, James F. Gibbons, Willis W. Harman, Hubert Heffner, Edward W. Herold, John G. Linvill, Simon Ramo, Ronald A. Rohrer, Anthony E. Siegman, Charles Susskind, Frederick E. Terman, John G. Truxal, Ernst Weber, and John R. Whinnery

Power and Energy

Consulting Editor
Stephen W. Director, Carnegie–Mellon University

Elgerd: *Electric Energy Systems Theory: An Introduction*
Fitzgerald, Kingsley, and Kusko: *Electric Machinery*
Meisel: *Principles of Electromechanical-Energy Conversion*
Odum and Odum: *Energy Basis for Man and Nature*
Stagg and El Abiad: *Advanced Computer Methods in Power System Analysis*
Stevenson: *Elements of Power System Analysis*
Sullivan: *Power System Planning*

ELEMENTS OF POWER SYSTEM ANALYSIS

Fourth Edition

William D. Stevenson, Jr.

Professor of Electrical Engineering, Emeritus
North Carolina State University

McGraw-Hill Book Company

New York St. Louis San Francisco Auckland Bogotá Hamburg
Johannesburg London Madrid Mexico Montreal New Delhi
Panama Paris São Paulo Singapore Sydney Tokyo Toronto

This book was set in Times Roman.
The editor was Frank J. Cerra;
the production supervisor was Diane Renda.
The cover was designed by Infield, D'Astolfo Associates.

ELEMENTS OF POWER SYSTEM ANALYSIS

4567890 HDHD 89876543

Library of Congress Cataloging in Publication Data

Stevenson, William D.
    Elements of power system analysis.

    (McGraw-Hill series in electrical engineering.
Power and energy)
    Includes index.
    1. Electric power distribution. 2. Electric
power systems. I. Title. II. Series.
TK3001.S85 1982   621.319   81-3741
ISBN 0-07-061278-1 (Text)   AACR2
ISBN 0-07-061279-X (Solutions manual)

CONTENTS

PREFACE

Each revision of this book has embodied many changes, this one more so than usual. Over the years, however, the objective has remained the same. The approach has always been to develop the thinking process of the student in reaching a sound understanding of a broad range of topics in the power-system area of electrical engineering. At the same time, another goal has been to promote the student's interest in learning more about the electric-power industry. The objective is not great depth, but the presentation is thorough enough to give the student the basic theory at a level that can be understood by the undergraduate. With this beginning, the student will have the foundation to continue his education while at work in the field or in graduate school. Footnotes throughout the book suggest sources of further information on most of the topics presented.

As in preparing previous revisions, I sent a questionnaire to a number of faculty members across the country, and I appreciate greatly the prompt and in many cases, the detailed responses to specific questions as well as the valuable additional comments. The most popular suggestion was for a chapter on system protection, and accordingly that subject has been added to the other four main topics of load flow, economic dispatch, fault calculations, and system stability. Perhaps surprisingly there were many requests to retain the material on transmission-line parameters. The per-unit system is first introduced in Chapter 2 and developed gradually to allow the student to become accustomed to scaled, or normalized, quantities. The need to review steady-state alternating-current circuits still exists and so the chapter on basic concepts is not altered. Direct formulation of the bus impedance matrix has been added. The Newton-Raphson method for load-flow calculations has been developed more fully. Increased attention has been given to developing the equivalent circuits of transformers and synchronous machines to help the many students who study power systems before they have had a course in machinery. Equations for transients on a lossless line have been developed to lead into the discussion of surge arresters. Other topics discussed briefly are direct-current transmission, reactive compensation, and underground cables. The subject of automatic load dispatch has been expanded.

I am particularly fortunate to have had two principal contributors to this edition. Arun G. Phadke, Consulting Engineer, American Electric Power Service Corporation, is the author of the new chapter on system protection. John J. Grainger, my colleague at N. C. State University, has completely rewritten the chapter on power-system stability. To both of these people who have added so much to this edition, I extend my sincere thanks. W. H. Kersting of New Mexico State University contributed to the section on reactive compensation. He as well as J. M. Feldman of Northeastern University, G. T. Heydt of Purdue University, and H. V. Poe of Clemson University provided new problems for a number of chapters. To all of them, I am extremely grateful.

I must thank three people who have always willingly given their time when I needed advice about this revision. Homer E. Brown with his long experience in power-system work as well as his secondary career in teaching has been a great help to me. A. J. Goetze who has often taught the course based on this book here at North Carolina State University has always been ready with suggestions when asked, and finally John Grainger should be mentioned again because he has provided me with much up-to-date information and counseling.

My thanks go also to the companies which have been so willing to provide information, photographs and even the review of some of the new descriptive material. These companies are: Carolina Power and Light Company, Duke Power Company, General Electric Company, Leeds and Northup Company, Utility Power Corporation, Virginia Electric and Power Company, and Westinghouse Electric Corporation.

As always, I have profited by the letters from users of past editions. I hope this correspondence will continue.

William D. Stevenson, Jr.

ONE

GENERAL BACKGROUND

Development of sources of energy to accomplish useful work is the key to the industrial progress which is essential to the continual improvement in the standard of living of people everywhere. To discover new sources of energy, to obtain an essentially inexhaustible supply of energy for the future, to make energy available wherever needed, and to convert energy from one form to another and use it without creating the pollution which will destroy our biosphere are among the greatest challenges facing this world today. The electric power system is one of the tools for converting and transporting energy which is playing an important role in meeting this challenge. The industry, by some standards, is the largest in the world. Highly trained engineers are needed to develop and implement the advances of science to solve the problems of the electric power industry and to assure a very high degree of system reliability along with the utmost regard for the protection of our ecology.

An electric power system consists of three principal divisions: the generating stations, the transmission lines, and the distribution systems. Utilization of the energy delivered to the customers of the operating companies is not within the responsibilities of the utility companies and will not be considered in this book. Transmission lines are the connecting links between the generating stations and the distribution systems and lead to other power systems over interconnections. A distribution system connects all the individual loads to the transmission lines at substations which perform voltage transformation and switching functions.

The objective of this book is to present methods of analysis, and we shall devote most of our attention to transmission lines and system operation. We shall not be concerned with distribution systems or any aspects of power plants other than the electrical characteristics of generators.

1.1 THE GROWTH OF ELECTRIC POWER SYSTEMS

The development of ac systems began in the United States in 1885, when George Westinghouse bought the American patents covering the ac transmission system developed by L. Gaulard and J. D. Gibbs of Paris. William Stanley, an early

1

associate of Westinghouse, tested transformers in his laboratory in Great Barrington, Massachusetts. There, in the winter of 1885–1886, Stanley installed the first experimental ac distribution system which supplied 150 lamps in the town. The first ac transmission line in the United States was put into operation in 1890 to carry electric energy generated by water power a distance of 13 mi from Willamette Falls to Portland, Oregon.

The first transmission lines were single-phase, and the energy was usually consumed for lighting only. Even the first motors were single-phase, but on May 16, 1888, Nikola Tesla presented a paper describing two-phase induction and synchronous motors. The advantages of polyphase motors were apparent immediately, and a two-phase ac distribution system was demonstrated to the public at the Columbian Exposition in Chicago in 1893. Thereafter, the transmission of electric energy by alternating current, especially three-phase alternating current, gradually replaced dc systems. In January 1894, there were five polyphase generating plants in the United States, of which one was two-phase and the others three-phase. Transmission of electric energy in the United States is almost entirely by means of alternating current. One reason for the early acceptance of ac systems was the transformer, which makes possible the transmission of electric energy at a voltage higher than the voltage of generation or utilization with the advantage of greater transmission capability.

In a dc transmission system ac generators supply the dc line through a transformer and electronic rectifier. An electronic inverter changes the direct current to alternating current at the end of the line so that the voltage can be reduced by a transformer. By providing both rectification and inversion at each end of the line power can be transferred in either direction. Economic studies have shown that dc overhead transmission is not economical in the United States for distances of less than 350 mi. In Europe, where transmission lines are generally much longer than in the United States, dc transmission lines are in operation in several locations for both underground and overhead installations. In California large amounts of hydro power are transferred from the Pacific Northwest to the southern part of California over 500-kV ac lines along the coast and farther inland through Nevada by direct current at 800 kV line to line.

Statistics reported from 1920 until early in the 1970–1980 decade showed an almost constant rate of increase of both installed generating capacity and annual energy production which amounted quite closely to doubling each of these values every 10 years. Growth then became more erratic and unpredictable, but in general somewhat slower.

In the early days of ac power transmission in the United States, the operating voltage increased rapidly. In 1890 the Willamette-Portland line was operated at 3300 V. In 1907 a line was operating at 100 kV. Voltages rose to 150 kV in 1913, 220 kV in 1923, 244 kV in 1926, and 287 kV on the line from Hoover Dam to Los Angeles, which began service in 1936. In 1953 came the first 345-kV line. The first 500-kV line was operating in 1965. Four years later in 1969 the first 765-kV line was placed in operation.

Until 1917, electric systems were usually operated as individual units because they started as isolated systems and spread out only gradually to cover the whole country. The demand for large blocks of power and increased reliability suggested the interconnection of neighboring systems. Interconnection is advantageous economically because fewer machines are required as a reserve for operation at peak loads (reserve capacity) and fewer machines running without load are required to take care of sudden, unexpected jumps in load (spinning reserve). The reduction in machines is possible because one company can usually call on neighboring companies for additional power. Interconnection also allows a company to take advantage of the most economical sources of power, and a company may find it cheaper to buy some power than to use only its own generation during some periods. Interconnection has increased to the point where power is exchanged between the systems of different companies as a matter of routine. The continued service of systems depending on water power for a large part of their generation is possible in times of unusual and extreme water shortage only because of the power obtained from other systems through interconnections.

Interconnection of systems brought many new problems, most of which have been solved satisfactorily. Interconnection increases the amount of current which flows when a short circuit occurs on a system and requires the installation of breakers able to interrupt a larger current. The disturbance caused by a short circuit on one system may spread to interconnected systems unless proper relays and circuit breakers are provided at the point of interconnection. Not only must the interconnected systems have the same nominal frequency, but also the synchronous generators of one system must remain in step with the synchronous generators of all the interconnected systems.

Planning the operation, improvement, and expansion of a power system requires load studies, fault calculations, the design of means of protecting the system against lightning and switching surges and against short circuits, and studies of the stability of the system. An important problem in efficient system operation is that of determining how the total generation required at any time should be distributed among the various plants and among the units within each plant. In this chapter we shall consider the general nature of these types of problems after a brief discussion of energy production and of transmission and distribution. We shall see the great contributions made by computers to the planning and operation of power systems.

1.2 ENERGY PRODUCTION

Most of the electric power in the United States is generated in steam-turbine plants. Water power accounts for less than 20% of the total and that percentage will drop because most of the available sources of water power have been developed. Gas turbines are used to a minor extent for short periods when a system is carrying peak load.

Coal is the most widely used fuel for the steam plants. Nuclear plants fueled by uranium account for a continually increasing share of the load, but their construction is slow and uncertain because of the difficulty of raising capital to meet the sharply rising costs of construction, constantly increasing safety requirements which cause redesign, public opposition to the operation of the plants, and delays in licensing.

Many plants converted to oil between 1970 and 1972, but in the face of the continuous escalation in the price of oil and the necessity of reducing dependence on foreign oil reconversion from oil to coal has taken place wherever possible.

The supply of uranium is limited, but the fast breeder reactors now prohibited in the United States, have greatly extended the total energy available from uranium in Europe. Nuclear fusion is the great hope for the future, but a controllable fusion process on a commercial scale is not expected to become feasible until well after the year 2000, if ever. That year, however, is now the target date for demonstrating the first pilot model of a controlled fusion reactor. As this comes to pass, electric power systems must continue to grow and take over the direct fuel applications. For instance, the electric car will probably be widely used in order to reserve the fossil fuels (including petroleum and gas synthesized from coal) for aircraft and long-distance trucking.

There is some use of geothermal energy in the form of live steam issuing from the ground in the United States and foreign countries. Solar energy, now chiefly in the form of direct heating of water for residential use, should eventually become practical through research on photovoltaic cells which convert sunlight to electricity directly. Great progress has been made in increasing the efficiency and reducing the cost of these cells, but the distance still to go is extremely large. Windmills driving generators are operating in a number of locations to provide small amounts of power to power systems. Efforts to extract power from the changing tides and from waves are under way. An indirect form of solar energy is alcohol grown from grain and mixed with gasoline to make an acceptable fuel for automobiles. Synthetic gas from garbage and sewage is another indirect form of solar energy.

Finally, in producing energy by any means, protection of our environment is extremely important. Atmospheric pollution is all too apparent to residents of industrialized countries. Thermal pollution is less obvious, but cooling water for nuclear reactors is very important and adds greatly to construction costs. Too great a rise in the temperature of rivers is harmful to fish, and artificial lakes for cooling water often use up too much productive land. Here cooling towers, though expensive, seem to be the answer for cooling at nuclear plants.

1.3 TRANSMISSION AND DISTRIBUTION

The voltage of large generators usually is in the range of 13.8 kV to 24 kV. Large modern generators, however, are built for voltages ranging from 18 to 24 kV. No standard for generator voltages has been adopted.

Generator voltage is stepped up to transmission levels in the range of 115 to 765 kV. The standard high voltages (HV) are 115, 138, and 230 kV. Extra-high voltages (EHV) are 345, 500, and 765 kV. Research is being conducted on lines in the ultra-high-voltage (UHV) levels of 1000 to 1500 kV. The advantage of higher levels of transmission-line voltage is apparent when consideration is given to the transmission capability in megavoltamperes (MVA) of a line. Roughly, the capability of lines of the same length varies at a rate somewhat greater than the square of the voltage. No definite capability can be specified for a line of any given voltage, however, because capability is dependent on the thermal limits of the conductor, allowable voltage drop, reliability, and requirements for maintaining synchronism between the machines of the system, which is known as stability. Most of these factors are dependent on line length.

Undergound transmission cables for a particular voltage seem to be developed about 10 years after operation is initiated at that voltage on open-wire lines. Underground transmission is negligible in terms of mileage but is increasing significantly. It is mostly confined to heavily populated urban areas, or used under wide bodies of water.

The first step-down of voltage from transmission levels is at the bulk-power substation, where the reduction is to a range of 34.5 to 138 kV, depending, of course, upon transmission-line voltage. Some industrial customers may be supplied at these voltage levels. The next step-down in voltage is at the distribution substation, where the voltage on lines leaving the substation ranges from 4 to 34.5 kV and is commonly between 11 and 15 kV. This is the primary distribution system. A very popular voltage at this level is 12,470 V line to line, which means 7200 V from line to ground, or neutral. This voltage is usually described as 12,470Y/7200 V. A lower primary-system voltage which is less widely used is 4160Y/2400 V. Most industrial loads are fed from the primary system, which also supplies the distribution transformers providing secondary voltages over single-phase three-wire circuits for residential use. Here the voltage is 240 V between two wires and 120 V between each of these and the third wire, which is grounded. Other secondary circuits are three-phase four-wire systems rated 208Y/120 V, or 480Y/277 V.

1.4 LOAD STUDIES

A load study is the determination of the voltage, current, power, and power factor or reactive power at various points in an electric network under existing or contemplated conditions of normal operation. Load studies are essential in planning the future development of the system because satisfactory operation of the system depends on knowing the effects of interconnections with other power systems, of new loads, new generating stations, and new transmission lines before they are installed.

Before the development of large digital computers load-flow studies were made on ac calculating boards, which provided small-scale single-phase replicas of actual systems by interconnecting circuit elements and voltage sources. Setting up the connections, making adjustments, and reading the data was tedious and time-consuming. Digital computers now provide the solutions of load-flow studies on complex systems. For instance, the computer program may handle more than 1500 buses, 2500 lines, 500 transformers with tap changing under load, and 25 phase-shifting transformers. Complete results are printed quickly and economically.

System planners are interested in studying a power system as it will exist 10 or 20 years in the future. More than 10 years elapse between initiating the plans for a new nuclear plant and bringing it on the line. A power company must know far in advance the problems associated with the location of the plant and the best arrangement of lines to transmit the power to load centers which do not exist when the planning must be done.

We shall see in Chap. 8 how load-flow studies are made on the computer. Figure 8.2 shows the computer printout of the load flow of a small system which we shall be studying.

1.5 ECONOMIC LOAD DISPATCH

The power industry may seem to lack competition. This idea arises because each power company operates in a geographic area not served by other companies. Competition is present, however, in attracting new industries to an area. Favorable electric rates are a compelling factor in the location of an industry, although this factor is much less important in times when costs are rising rapidly and rates charged for power are uncertain than in periods of stable economic conditions. Regulation of rates by state utility commissions, however, places constant pressure on companies to achieve maximum economy and earn a reasonable profit in the face of advancing costs of production.

Economic dispatch is the name given to the process of apportioning the total load on a system between the various generating plants to achieve the greatest economy of operation. We shall see that all the plants on a system are controlled continuously by a computer as load changes occur so that generation is allocated for the most economical operation.

1.6 FAULT CALCULATIONS

A fault in a circuit is any failure which interferes with the normal flow of current. Most faults on transmission lines of 115 kV and higher are caused by lightning, which results in the flashover of insulators. The high voltage between a conductor and the grounded supporting tower causes ionization, which provides a path to ground for the charge induced by the lightning stroke. Once the ionized path to ground is established, the resultant low impedance to ground allows the flow of power current from the conductor to ground and through the ground to the grounded neutral of a transformer or generator, thus completing the circuit. Line-to-line faults not involving ground are less common. Opening circuit breakers to isolate the faulted portion of the line from the rest of the system interrupts the flow of current in the ionized path and allows deionization to take place. After an interval of about 20 cycles to allow deionization, breakers can usually be reclosed without reestablishing the arc. Experience in the operation of transmission lines has shown that ultra-high-speed reclosing breakers successfully reclose after most faults. Of those cases where reclosure is not successful, an appreciable number are caused by permanent faults where reclosure would be impossible regardless of the interval between opening and reclosing. Permanent faults are caused by lines being on the ground, by insulator strings breaking because of ice loads, by permanent damage to towers, and by lightning-arrester failures. Experience has shown that between 70 and 80% of transmission-line faults are single line-to-ground faults, which arise from the flashover of only one line to the tower and ground. The smallest number of faults, roughly 5%, involve all three phases and are called three-phase faults. Other types of transmission-line faults are line-to-line faults, which do not involve ground, and double line-to-ground faults. All the above faults except the three-phase type are unsymmetrical and cause an imbalance between the phases.

The current which flows in different parts of a power system immediately after the occurrence of a fault differs from that flowing a few cycles later just before circuit breakers are called upon to open the line on both sides of the fault, and both these currents differ widely from the current which would flow under steady-state conditions if the fault were not isolated from the rest of the system by the operation of circuit breakers. Two of the factors upon which the proper selection of circuit breakers depends are the current flowing immediately after the fault occurs and the current which the breaker must interrupt. Fault calculations consist in determining these currents for various types of faults at various locations in the system. The data obtained from fault calculations also serve to determine the settings of relays which control the circuit breakers.

Analysis by symmetrical components is a powerful tool which we shall study later and which makes the calculation of unsymmetrical faults almost as easy as the calculation of three-phase faults. Again it is the digital computer which is invaluable in making fault calculations. We shall examine the fundamental operations called for by the computer programs.

1.7 SYSTEM PROTECTION

Faults can be very destructive to power systems. A great deal of study, development of devices, and design of protection schemes have resulted in continual improvement in the prevention of damage to transmission lines and equipment and interruptions in generation following the occurrence of a fault.

We shall be discussing the problem of transients on a transmission line for a very simplified case. This study will lead us to a discussion of how surge arresters protect apparatus such as transformers at plant buses and substations against the very high voltage surges caused by lightning and, in the case of EHV and UHV lines, by switching.

Faults caused by surges are usually of such short duration that any circuit breakers which may open will reclose automatically after a few cycles to restore normal operation. If arresters are not involved or faults are permanent the faulted sections of the system must be isolated to maintain normal operation of the rest of the system.

Operation of circuit breakers is controlled by relays which sense the fault. In the application of relays zones of protection are specified to define the parts of the system for which various relays are responsible. One relay will also back up another relay in an adjacent zone or zones where the fault occurs in case the relay in the adjacent zone fails to respond. In Chap. 13 we shall discuss the characteristics of the basic types of relays and look at some numerical examples of relay applications and coordination.

1.8 STABILITY STUDIES

The current which flows in an ac generator or synchronous motor depends on the magnitude of its generated (or internal) voltage, on the phase angle of its internal voltage with respect to the phase angle of the internal voltage of every other machine in the system, and on the characteristics of the network and loads. For example, two ac generators operating in parallel but without any external circuit connections other than the paralleling circuit will carry no current if their internal voltages are equal in magnitude and in phase. If their internal voltages are equal in magnitude but different in phase, the voltage of one subtracted from the voltage of the other will not be zero, and a current will flow, as determined by the difference in voltages and the impedance of the circuit. One machine will supply power to the other, which will run as a motor rather than a generator.

The phase angles of the internal voltages depend upon the relative positions of the rotors of the machines. If synchronism were not maintained between the generators of a power system, the phase angles of their internal voltages would be changing constantly with respect to each other and satisfactory operation would be impossible.

The phase angles of the internal voltages of synchronous machines remain constant only as long as the speeds of the various machines remain constant at the speed which corresponds to the frequency of the reference phasor. When the load on any one generator or on the system as a whole changes, the current in the generator or throughout the system changes. If the change in current does not result in a change in magnitude of the internal voltages of the machines, the phase angles of the internal voltages must change. Thus, momentary changes in speed are necessary to obtain adjustment of the phase angles of the voltages with respect to each other, since the phase angles are determined by the relative positions of the rotors. When the machines have adjusted themselves to the new phase angles, or when some disturbance causing a momentary change in speed has been removed, the machines must operate again at synchronous speed. If any machine does not remain in synchronism with the rest of the system, large circulating currents result; in a properly designed system, the operation of relays and circuit breakers removes the machine from the system. The problem of stability is the problem of maintaining the synchronous operation of the generators and motors of the system.

Stability studies are classified according to whether they involve steady-state or transient conditions. There is a definite limit to the amount of power an ac generator is capable of delivering and to the load which a synchronous motor can carry. Instability results from attempting to increase the mechanical input to a generator or the mechanical load on a motor beyond this definite amount of power, called the *stability limit*. A limiting value of power is reached even if the change is made gradually. Disturbances on a system, caused by suddenly applied loads, by the occurrence of faults, by the loss of excitation in the field of a generator, and by switching, may cause loss of synchronism even when the change in the system caused by the disturbance would not exceed the stability limit if the change were made gradually. The limiting value of power is called the *transient stability limit* or the *steady-state stability limit*, according to whether the point of instability is reached by a sudden or a gradual change in conditions of the system.

Fortunately, engineers have found methods of improving stability and of predicting the limits of stable operation under both steady-state and transient conditions. The stability studies we shall investigate for a two-machine system are less complex than studies of multimachine systems, but many of the methods of improving stability can be seen by the analysis of a two-machine system. Digital computers are used to advantage in predicting the stability limits of a complex system.

1.9 THE POWER-SYSTEM ENGINEER

This chapter has attempted to sketch some of the history of the basic developments of electric power systems and to describe some of the analytic studies important in planning the operation, improvement, and expansion of a modern power system. The power-system engineer should know the methods of making load studies, fault analyses, and stability studies and the principles of economic dispatch because such studies affect the design and operation of the system and the selection of apparatus for its control. Before we can consider these problems in more detail, we must study some fundamental concepts relating to power systems in order to understand how these fundamental concepts affect the larger problems.

1.10 ADDITIONAL READING

Footnotes throughout the book provide sources of further information about many of the topics which we will be discussing. The reader is also referred to the books listed below which treat most of the same subjects as this text although some include other topics or the same ones in greater depth.

Elgerd, O. I., "Electric Energy Systems Theory: An Introduction," 2d ed., McGraw-Hill Book Company, New York, 1982.

Gross, C. A., "Power System Analysis," John Wiley & Sons, New York, 1979.

Neuenswander, J. R., "Modern Power Systems," Intext Educational Publishers, New York, 1971.

Weedy, B. M., "Electric Power Systems," 3d ed., John Wiley & Sons Ltd., London, 1979.

TWO

BASIC CONCEPTS

The power-system engineer is just as concerned with the normal operation of the system as he is with the abnormal conditions which may occur. Therefore, he must be very familiar with steady-state ac circuits, particularly three-phase circuits. It is the purpose of this chapter to review a few of the fundamental ideas of such circuits, establish the notation which will be used throughout the book, and introduce the expression of values of voltage, current, impedance, and power in per unit.

2.1 INTRODUCTION

The waveform of voltage at the buses of a power system can be assumed to be purely sinusoidal and of constant frequency. In developing most of the theory in this book we shall be concerned with the phasor representations of sinusoidal voltages and currents and shall use the capital letters V and I to indicate these phasors (with appropriate subscripts where necessary). Vertical bars enclosing V and I, that is, $|V|$ and $|I|$, will designate the magnitude of the phasors. Lower-case letters will indicate instantaneous values. Where a generated voltage (electromotive force) is specified, the letter E rather than V will be used for voltage to emphasize the fact that an emf rather than a general potential difference between two points is being considered.

If a voltage and a current are expressed as functions of time, such as

$$v = 141.4 \cos (\omega t + 30°)$$

and

$$i = 7.07 \cos \omega t$$

their maximum values are obviously $V_{max} = 141.4$ V and $I_{max} = 7.07$ A, respectively. Vertical bars are not needed when the subscript max with V and I is used to indicate maximum value. The term magnitude refers to root-mean-square or (rms) values, which equal the maximum values divided by $\sqrt{2}$. Thus, for the above expressions for v and i,

$$|V| = 100 \text{ V} \qquad \text{and} \qquad |I| = 5 \text{ A}$$

These are the values read by the ordinary types of voltmeters and ammeters. Another name for the rms value is the *effective value*. The average power expended in a resistor is $|I|^2 R$.

To express these quantities as phasors a reference must be chosen. If the current is the reference phasor

$$I = 5\underline{/0^\circ} = 5 + j0 \text{ A}$$

the voltage which leads the reference phasor by 30° is

$$V = 100\underline{/30^\circ} = 86.6 + j50 \text{ V}$$

Of course, we might not choose as the reference phasor either the voltage or current whose instantaneous expressions are v and i, in which case their phasor expressions would involve other angles.

In circuit diagrams it is often most convenient to use polarity marks in the form of plus and minus signs to indicate the terminal assumed positive when specifying voltage. An arrow on the diagram specifies the direction assumed positive for the flow of current. In the single-phase equivalent of a three-phase circuit single-subscript notation is usually sufficient, but double-subscript notation is usually simpler when dealing with all three phases.

2.2 SINGLE-SUBSCRIPT NOTATION

Figure 2.1 shows an ac circuit with an emf represented by a circle. The emf is E_g, and the voltage between nodes a and o is identified as V_t. The current in the circuit is I_L and the voltage across Z_L is V_L. To specify these voltages as phasors, however, the $+$ and $-$ markings, called polarity marks, on the diagram and an arrow for current direction are necessary.

In an ac circuit the terminal marked $+$ is positive with respect to the terminal marked $-$ for half a cycle of voltage and is negative with respect to the

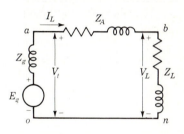

Figure 2.1 An ac circuit with emf E_g and load impedance Z_L.

other terminal during the next half cycle. We mark the terminals to enable us to say that the voltage between the terminals is positive at any instant when the terminal marked plus is actually at a higher potential than the terminal marked minus. For instance, in Fig. 2.1 the instantaneous voltage v_t is positive when the terminal marked plus is actually at a higher potential than the terminal marked with a negative sign. During the next half cycle the positively marked terminal is actually negative, and v_t is negative. Some authors use an arrow but must specify whether the arrow points toward the terminal which would be labeled plus or toward the terminal which would be labeled minus in the convention described above.

The current arrow performs a similar function. The subscript, in this case L, is not necessary unless other currents are present. Obviously the actual direction of current flow in an ac circuit reverses each half cycle. The arrow points in the direction which is to be called positive for current. When the current is actually flowing in the direction opposite that of the arrow, the current is negative. The phasor current is

$$I_L = \frac{V_t - V_L}{Z_A} \tag{2.1}$$

and

$$V_t = E_g - I_L Z_g \tag{2.2}$$

Since certain nodes in the circuit have been assigned letters, the voltages may be designated by the single-letter subscripts identifying the node whose voltages are expressed with respect to a reference node. In Fig. 2.1 the instantaneous voltage v_a and the phasor voltage V_a express the voltage of node a with respect to the reference node o, and v_a is positive when a is at a higher potential than o. Thus

$$v_a = v_t \qquad v_b = v_L$$
$$V_a = V_t \qquad V_b = V_L$$

2.3 DOUBLE-SUBSCRIPT NOTATION

The use of polarity marks for voltages and direction arrows for currents can be avoided by double-subscript notation. Understanding of three-phase circuits is considerably clarified by adopting a system of double subscripts. The convention to be followed is quite simple.

In denoting a current the order of the subscripts assigned to the symbol for current defines the direction of flow of current when the current is considered to be positive. In Fig. 2.1 the arrow pointing from a to b defines the positive direction for the current I_L associated with the arrow. The instantaneous current i_L is positive when the current is actually in the direction from a to b, and in double-subscript notation this current is i_{ab}. The current i_{ab} is equal to $-i_{ba}$.

In double-subscript notation the letter subscripts on a voltage indicate the nodes of the circuit between which the voltage exists. We shall follow the convention which says that the first subscript denotes the voltage of that node with respect to the node identified by the second subscript. This means that the instantaneous voltage v_{ab} across Z_A of the circuit of Fig. 2.1 is the voltage of node a with respect to node b and that v_{ab} is positive during that half cycle when a is at a higher potential than b. The corresponding phasor voltage is V_{ab}, and

$$V_{ab} = I_{ab} Z_A \tag{2.3}$$

where Z_A is the complex impedance through which I_{ab} flows between nodes a and b, which may also be called Z_{ab}.

Reversing the order of the subscripts of either a current or voltage gives a current or voltage 180° out of phase with the original; that is,

$$V_{ab} = V_{ba} \underline{/180°} = -V_{ba}$$

The relation of single- and double-subscript notation for the circuit of Fig. 2.1 is summarized as follows:

$$V_t = V_a = V_{ao} \qquad V_L = V_b = V_{bo}$$
$$I_L = I_{ab}$$

In writing Kirchhoff's voltage law the order of the subscripts is the order of tracing a closed path around the circuit. For Fig. 2.1,

$$V_{oa} + V_{ab} + V_{bn} = 0 \tag{2.4}$$

Nodes n and o are the same in this circuit, and n has been introduced to identify the path more precisely. Replacing V_{oa} by $-V_{ao}$ and noting that $V_{ab} = I_{ab} Z_A$ yields

$$-V_{ao} + I_{ab} Z_A + V_{bn} = 0 \tag{2.5}$$

and so

$$I_{ab} = \frac{V_{ao} - V_{bn}}{Z_A} \tag{2.6}$$

2.4 POWER IN SINGLE-PHASE AC CIRCUITS

Although the fundamental theory of the transmission of energy describes the travel of energy in terms of the interaction of electric and magnetic fields, the power-system engineer is almost always more concerned with describing the rate of change of energy with respect to time (which is the definition of power) in terms of voltage and current. The unit of power is a watt. The power in watts being absorbed by a load at any instant is the product of the instantaneous voltage drop across the load in volts and the instantaneous current into the load in amperes. If the terminals of the load are designated a and n, and if the voltage

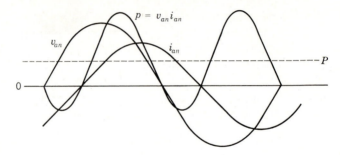

Figure 2.2 Current, voltage, and power plotted versus time.

and current are expressed by

$$v_{an} = V_{\max} \cos \omega t \quad \text{and} \quad i_{an} = I_{\max} \cos(\omega t - \theta)$$

the instantaneous power is

$$p = v_{an} i_{an} = V_{\max} I_{\max} \cos \omega t \cos(\omega t - \theta) \tag{2.7}$$

The angle θ in these equations is positive for current lagging the voltage and negative for leading current. A positive value of p expresses the rate at which energy is being absorbed by the part of the system between the points a and n. The instantaneous power is obviously positive when both v_{an} and i_{an} are positive but will become negative when v_{an} and i_{an} are opposite in sign. Figure 2.2 illustrates this point. Positive power calculated as $v_{an} i_{an}$ results when current is flowing in the direction of a voltage drop and is the rate of transfer of energy to the load. Conversely, negative power calculated as $v_{an} i_{an}$ results when current is flowing in the direction of a voltage rise and means energy is being transferred from the load into the system to which the load is connected. If v_{an} and i_{an} are in phase, as they are in a purely resistive load, the instantaneous power will never become negative. If the current and voltage are out of phase by 90°, as in a purely inductive or purely capacitive ideal circuit element, the instantaneous power will have equal positive and negative half cycles and its average value will be zero.

By using trigonometric identities the expression of Eq. (2.7) is reduced to

$$p = \frac{V_{\max} I_{\max}}{2} \cos \theta (1 + \cos 2\omega t) + \frac{V_{\max} I_{\max}}{2} \sin \theta \sin 2\omega t \tag{2.8}$$

where $V_{\max} I_{\max}/2$ may be replaced by the product of the rms voltage and current $|V_{an}| \cdot |I_{an}|$ or $|V| \cdot |I|$.

Another way of looking at the expression for instantaneous power is to consider the component of the current in phase with v_{an} and the component 90° out of phase with v_{an}. Figure 2.3a shows a parallel circuit for which Fig. 2.3b is the phasor diagram. The component of i_{an} in phase with v_{an} is i_R, and from Fig. 2.3b, $|I_R| = |I_{an}| \cos \theta$. If the maximum value of i_{an} is $I_{\max}$, the maximum value of i_R is $I_{\max} \cos \theta$. The instantaneous current i_R must be in phase with v_{an}.

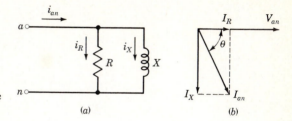

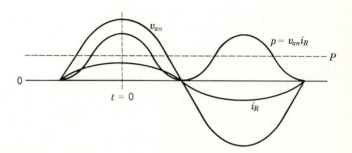

Figure 2.3 Parallel RL circuit and the corresponding phasor diagram.

(a)

(b)

For $v_{an} = V_{max} \cos \omega t$,

$$i_R = \underbrace{I_{max} \cos \theta}_{\text{max } i_R} \cos \omega t \qquad (2.9)$$

Similarly the component of i_{an} lagging v_{an} by 90° is i_X, whose maximum value is $I_{max} \sin \theta$. Since i_X must lag v_{an} by 90°,

$$i_X = \underbrace{I_{max} \sin \theta}_{\text{max } i_X} \sin \omega t \qquad (2.10)$$

Then

$$v_{an} i_R = V_{max} I_{max} \cos \theta \cos^2 \omega t$$

$$= \frac{V_{max} I_{max}}{2} \cos \theta \, (1 + \cos 2\omega t) \qquad (2.11)$$

which is the instantaneous power in the resistance and is the first term in Eq. (2.8). Figure 2.4 shows $v_{an} i_R$ plotted versus t.

Similarly,

$$v_{an} i_X = V_{max} I_{max} \sin \theta \sin \omega t \cos \omega t$$

$$= \frac{V_{max} I_{max}}{2} \sin \theta \sin 2\omega t \qquad (2.12)$$

which is the instantaneous power in the inductance and is the second term in Eq. (2.8). Figure 2.5 shows v_{an}, i_X, and their product plotted versus t.

Figure 2.4 Voltage, current in phase with the voltage, and the resulting power plotted versus time.

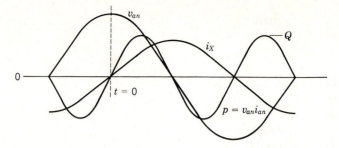

Figure 2.5 Voltage, current lagging the voltage by 90°, and the resulting power plotted versus time.

Examination of Eq. (2.8) shows that the first term, the term which contains $\cos \theta$, is always positive and has an average value of

$$P = \frac{V_{\max} I_{\max}}{2} \cos \theta \qquad (2.13)$$

or, when rms values of voltage and current are substituted,

$$P = |V| \cdot |I| \cos \theta \qquad (2.14)$$

P is the quantity to which the word power refers when not modified by an adjective identifying it otherwise. P, the average power, is also called the real power. The fundamental unit for both instantaneous and average power is the watt, but a watt is such a small unit in relation to power-system quantities that P is usually measured in kilowatts or megawatts.

The cosine of the phase angle θ between the voltage and the current is called the *power factor*. An inductive circuit is said to have a lagging power factor, and a capacitive circuit is said to have a leading power factor. In other words, the terms lagging power factor and leading power factor indicate, respectively, whether the current is lagging or leading the applied voltage.

The second term of Eq. (2.8), the term containing $\sin \theta$, is alternately positive and negative and has an average value of zero. This component of the instantaneous power p is called the *instantaneous reactive power* and expresses the flow of energy alternately toward the load and away from the load. The maximum value of this pulsating power, designated Q, is called reactive power or reactive voltamperes and is very useful in describing the operation of a power system, as will become increasingly evident in further discussion. The reactive power is

$$Q = \frac{V_{\max} I_{\max}}{2} \sin \theta \qquad (2.15)$$

or

$$Q = |V| \cdot |I| \sin \theta \qquad (2.16)$$

The square root of the sum of the squares of P and Q is equal to the product of $|V|$ and $|I|$, for

$$\sqrt{P^2 + Q^2} = \sqrt{(|V| \cdot |I| \cos \theta)^2 + (|V| \cdot |I| \sin \theta)^2} = |V| \cdot |I| \quad (2.17)$$

Of course P and Q have the same dimensional units, but it is usual to designate the units for Q as vars (for voltamperes reactive). The more practical units for Q are kilovars or megavars.

In a simple series circuit where Z is equal to $R + jX$, we can substitute $|I| \cdot |Z|$ for $|V|$ in Eqs. (2.14) and (2.16) to obtain

$$P = |I|^2 \cdot |Z| \cos \theta \qquad (2.18)$$

and

$$Q = |I|^2 \cdot |Z| \sin \theta \qquad (2.19)$$

Then recognizing that $R = |Z| \cos \theta$ and $X = |Z| \sin \theta$, we find

$$P = |I|^2 R \qquad \text{and} \qquad Q = |I|^2 X \qquad (2.20)$$

as expected.

Equations (2.14) and (2.16) provide another method of computing the power factor since we see that $Q/P = \tan \theta$. The power factor is therefore

$$\cos \theta = \cos \tan^{-1} \frac{Q}{P}$$

or from Eqs. (2.14) and (2.17)

$$\cos \theta = \frac{P}{\sqrt{P^2 + Q^2}}$$

If the instantaneous power expressed by Eq. (2.8) is the power in a predominantly capacitive circuit with the same impressed voltage, θ would be negative, making $\sin \theta$ and Q negative. If capacitive and inductive circuits are in parallel, the instantaneous reactive power for the RL circuit would be 180° out of phase with the instantaneous reactive power of the RC circuit. The net reactive power is the difference between Q for the RL circuit and Q for the RC circuit. A positive value is assigned to Q for an inductive load and a negative sign to Q for a capacitive load.

Power-system engineers usually think of a capacitor as a generator of positive reactive power rather than a load requiring negative reactive power. This concept is very logical, for a capacitor drawing negative Q in parallel with an inductive load reduces the Q which would otherwise have to be supplied by the system to the inductive load. In other words, the capacitor supplies the Q required by the inductive load. This is the same as considering a capacitor as a device that delivers a lagging current rather than as a device which draws a leading current, as shown in Fig. 2.6. An adjustable capacitor in parallel with an inductive load, for instance, can be adjusted so that the leading current to the capacitor is exactly equal in magnitude to the component of current in the

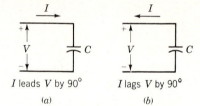

Figure 2.6 Capacitor considered (*a*) as a passive circuit element drawing leading current and (*b*) as a generator supplying lagging current.

I leads V by 90°

(*a*)

I lags V by 90°

(*b*)

inductive load which is lagging the voltage by 90°. Thus, the resultant current is in phase with the voltage. The inductive circuit still requires positive reactive power, but the net reactive power is zero. It is for this reason that the power-system engineer finds it convenient to consider the capacitor to be supplying this reactive power to the inductive load. When the words positive and negative are not used, positive reactive power is assumed.

2.5 COMPLEX POWER

If the phasor expressions for voltage and current are known, the calculation of real and reactive power is accomplished conveniently in complex form. If the voltage across and the current into a certain load or part of a circuit are expressed by $V = |V|\underline{/\alpha}$ and $I = |I|\underline{/\beta}$, the product of voltage times the conjugate of the current is

$$VI^* = V\underline{/\alpha} \times I\underline{/-\beta} = |V| \cdot |I|\underline{/\alpha - \beta} \tag{2.21}$$

This quantity, called the *complex power*, is usually designated by S. In rectangular form

$$S = |V| \cdot |I| \cos (\alpha - \beta) + j|V| \cdot |I| \sin (\alpha - \beta) \tag{2.22}$$

Since $\alpha - \beta$, the phase angle between voltage and current, is θ in the previous equations,

$$S = P + jQ \tag{2.23}$$

Reactive power Q will be positive when the phase angle $\alpha - \beta$ between voltage and current is positive, that is, when $\alpha > \beta$, which means that current is lagging the voltage. Conversely, Q will be negative for $\beta > \alpha$, which indicates current leading the voltage. This agrees with the selection of a positive sign for the reactive power of an inductive circuit and a negative sign for the reactive power of a capacitive circuit. To obtain the proper sign for Q, it is necessary to calculate S as VI^*, rather than V^*I, which would reverse the sign for Q.

2.6 THE POWER TRIANGLE

Equation (2.23) suggests a graphical method of obtaining the overall P, Q, and phase angle for several loads in parallel since $\cos \theta$ is $P/|S|$. A power triangle can be drawn for an inductive load, as shown in Fig. 2.7. For several loads in

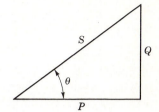

Figure 2.7 Power triangle for an inductive load.

parallel, the total P will be the sum of the average powers of the individual loads, which should be plotted along the horizontal axis for a graphical analysis. For an inductive load, Q will be drawn vertically upward since it is positive. A capacitive load will have negative reactive power, and Q will be vertically downward. Figure 2.8 illustrates the power triangle composed of P_1, Q_1, and S_1 for a lagging load having a phase angle θ_1 combined with the power triangle composed of P_2, Q_2, and S_2, which is for a capacitive load with a negative θ_2. These two loads in parallel result in the triangle having sides $P_1 + P_2$, $Q_1 + Q_2$ and hypotenuse S_R. The phase angle between voltage and current supplied to the combined loads is θ_R.

2.7 DIRECTION OF POWER FLOW

The relation between P, Q, and bus voltage V, or generated voltage E, with respect to the signs of P and Q is important when the flow of power in a system is considered. The question involves the direction of flow of power, that is, whether power is being generated or absorbed when a voltage and current are specified.

The question of delivering power to a circuit or absorbing power from a circuit is rather obvious for a dc system. Consider the current and voltage relationship shown in Fig. 2.9, where dc current I is flowing through a battery. If $I = 10$ A and $E = 100$ V, the battery is being charged (absorbing energy) at the rate of 1000 W. On the other hand, with the arrow still in the direction shown, the current might be $I = -10$ A. Then the conventional direction of current is

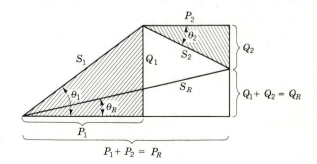

Figure 2.8 Power triangle for combined loads. Note that Q_2 is negative.

Figure 2.9 A dc representation of charging a battery if E and I are both positive or both negative.

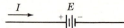

opposite to the direction of the arrow, the battery is discharging (delivering energy), and the product of E and I is -1000 W. By drawing Fig. 2.9 with I flowing through the battery from the positive to the negative terminal, charging of the battery seems to be indicated, but this is the case only if E and I are positive so that the power calculated as the product of E and I is positive. With this relationship between E and I the positive sign for power is obviously assigned to charging the battery.

If the direction of the arrow for I in Fig. 2.9 had been reversed, discharging of the battery would be indicated by a positive sign for I and for power. Thus the circuit diagram determines whether a positive sign for power is associated with charging or discharging the battery. This explanation seems unnecessary, but it provides the background for interpreting the ac circuit relationships.

For an ac system Fig. 2.10 shows an ideal voltage source (constant magnitude, constant frequency, zero impedance) with polarity marks which, as usual, indicate the terminal which is positive during the half cycle of positive instantaneous voltage. Of course, the positively marked terminal is actually the negative terminal during the negative half cycle of the instantaneous voltage. Similarly the arrow indicates the direction of current during the positive half cycle of current.

In Fig. 2.10a a generator is expected since the current is positive when flowing away from the positively marked terminal. However, the positively marked terminal may be negative when the current is flowing away from it. The approach to understanding the problem is to resolve the phasor I into a component along the axis of the phasor E and a component 90° out of phase with E. The product of $|E|$ and the magnitude of the component of I along the E axis is P. The product of $|E|$ and the magnitude of the component of I which is 90° out of phase with E is Q. If the component of I along the axis of E is in phase with E, the power is *generated* power which is being *delivered* to the system, for this component of current is always flowing away from the positively marked terminal when that terminal is actually positive (and toward that terminal when the terminal is negative). P, the real part of EI^*, is positive.

If the component of current along the axis of E is negative (180° out of phase with E), power is being absorbed and the situation is that of a motor. P, the real part of EI^*, would be negative.

The voltage and current relationship might be as shown in Fig. 2.10b, and a motor would be expected. However, an average power *absorbed* would occur only if the component of the phasor I along the axis of the phasor E was found

Figure 2.10 An ac circuit representation of an emf and current to illustrate polarity marks.

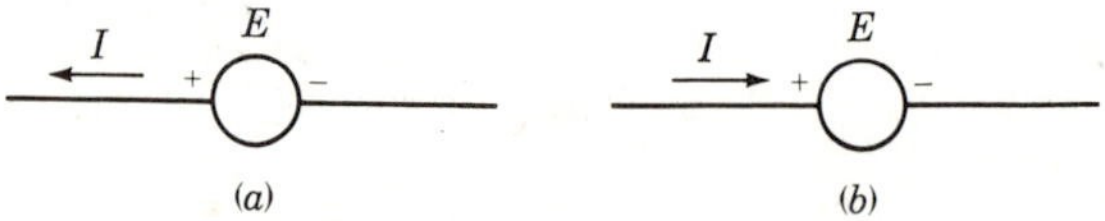

(a) (b)

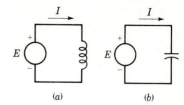

Figure 2.11 Alternating emf applied (*a*) to a purely inductive element and (*b*) to a purely capacitive element.

(*a*)   (*b*)

Table 2.1

Circuit diagram	Calculated from EI^*
E I Generator action assumed	If P is +, emf supplies power If P is −, emf absorbs power If Q is +, emf supplies reactive power (I lags E) If Q is −, emf absorbs reactive power (I leads E)
E I Motor action assumed	If P is +, emf absorbs power If P is −, emf supplies power If Q is +, emf absorbs reactive power (I lags E) If Q is −, emf supplies reactive power (I leads E)

to be in phase rather than 180° out of phase with E, so that this component of current would be always in the direction of the drop in potential. In this case P, the real part of EI^* would be positive. Negative P here would indicate generated power.

To consider the sign of Q, Fig. 2.11 is helpful. In Fig. 2.11*a* positive reactive power equal to $|I|^2X$ is supplied to the inductance since inductance draws positive Q. Then I lags E by 90°, and Q, the imaginary part of EI^*, is positive. In Fig. 2.11*b* negative Q must be supplied to the capacitance of the circuit, or the source with emf E is receiving positive Q from the capacitor. I leads E by 90°.

If the direction of the arrow in Fig. 2.11*a* is reversed, I will lead E by 90° and the imaginary part of EI^* would be negative. The inductance could be viewed as supplying negative Q rather than absorbing positive Q. Table 2.1 summarizes these relationships.

Example 2.1 Two ideal voltage sources designated as machines 1 and 2 are connected as shown in Fig. 2.12. If $E_1 = 100\underline{/0°}$ V, $E_2 = 100\underline{/30°}$ V, and $Z = 0 + j5$ Ω, determine (*a*) whether each machine is generating or consuming power and the amount, (*b*) whether each machine is receiving or supplying reactive power and the amount, and (*c*) the P and Q absorbed by the impedance.

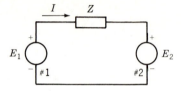

Figure 2.12 Ideal voltage sources connected through impedance Z.

SOLUTION

$$I = \frac{E_1 - E_2}{Z} = \frac{100 + j0 - (86.6 + j50)}{j5}$$

$$\frac{51.76\,\angle 75^\circ}{5\,\angle 90}$$

$$= \frac{13.4 - j50}{j5} = -10 - j2.68 = 10.35\underline{/195^\circ}$$

$$E_1 I^* = 100(-10 + j2.68) = -1000 + j268$$

$$E_2 I^* = (86.6 + j50)(-10 + j2.68)$$

$$= -866 + j232 - j500 - 134 = -1000 - j268$$

$$|I|^2 X = 10.35^2 \times 5 = 536 \text{ var}$$

Machine 1 may be expected to be a generator because of the current direction and polarity markings. Since P is negative and Q is positive, the machine consumes energy at the rate of 1000 W and supplies reactive power of 268 var. The machine is actually a motor.

Machine 2, expected to be a motor, has negative P and negative Q. Therefore, this machine generates energy at the rate of 1000 W and supplies reactive power of 268 var. The machine is actually a generator.

Note that the supplied reactive power of $268 + 268$ is equal to 536 var, which is required by the inductive reactance of 5 Ω. Since the impedance is purely reactive, no P is consumed by the impedance, and all the P generated by machine 2 is transferred to machine 1.

2.8 VOLTAGE AND CURRENT IN BALANCED THREE-PHASE CIRCUITS

Electric power systems are supplied by three-phase generators. Usually the generators are supplying balanced three-phase loads, which means loads with identical impedances in all three phases. Lighting loads and small motors are, of course, single-phase, but distribution systems are designed so that overall the phases are essentially balanced. Figure 2.13 shows a Y-connected generator with neutral marked o supplying a balanced-Y load with neutral marked n. In discussing this circuit we shall assume that the impedances of the connections between the terminals of the generator and the load, as well as the impedance of the direct connection between o and n, are negligible.

The equivalent circuit of the three-phase generator consists of an emf in each of the three phases, as indicated by circles on the diagram. Each emf is in series

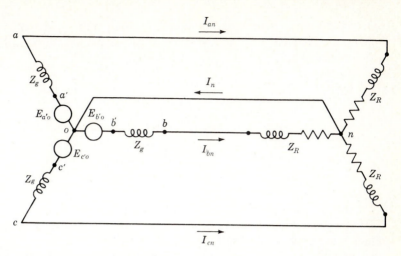

Figure 2.13 Circuit diagram of a Y-connected generator connected to a balanced-Y load.

with a resistance and inductive reactance composing the impedance Z_g. Points a', b', and c' are fictitious since the generated emf cannot be separated from the impedance of each phase. The terminals of the machine are the points a, b, and c. Some attention will be given to this equivalent circuit in a later chapter. In the generator the emfs $E_{a'o}$, $E_{b'o}$, $E_{c'o}$ are equal in magnitude and displaced from each other 120° in phase. If the magnitude of each is 100 V with $E_{a'o}$ as reference,

$$E_{a'o} = 100\underline{/0°} \text{ V} \qquad E_{b'o} = 100\underline{/240°} \text{ V} \qquad E_{c'o} = 100\underline{/120°} \text{ V}$$

provided the phase sequence is abc, which means that $E_{a'o}$ leads $E_{b'o}$ by 120° and $E_{b'o}$ in turn leads $E_{c'o}$ by 120°. The circuit diagram gives no indication of phase sequence, but Fig. 2.14 shows these emfs with phase sequence abc.

At the generator terminals (and at the load in this case) the terminal voltages to neutral are

$$V_{ao} = E_{a'o} - I_{an} Z_g$$

$$V_{bo} = E_{b'o} - I_{bn} Z_g \qquad\qquad (2.24)$$

$$V_{co} = E_{c'o} - I_{cn} Z_g$$

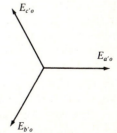

Figure 2.14 Phasor diagram of the emfs of the circuit shown in Fig. 2.13.

Since o and n are at the same potential, V_{ao}, V_{bo}, and V_{co} are equal to V_{an}, V_{bn}, and V_{cn}, respectively, and the line currents (which are also the phase currents for a Y connection) are

$$I_{an} = \frac{E_{a'o}}{Z_g + Z_R} = \frac{V_{an}}{Z_R}$$

$$I_{bn} = \frac{E_{b'o}}{Z_g + Z_R} = \frac{V_{bn}}{Z_R} \qquad (2.25)$$

$$I_{cn} = \frac{E_{c'o}}{Z_g + Z_R} = \frac{V_{cn}}{Z_R}$$

Since $E_{a'o}$, $E_{b'o}$, and $E_{c'o}$ are equal in magnitude and 120° apart in phase and the impedances seen by each of these emfs are identical, the currents will also be equal in magnitude and displaced 120° from each other in phase. The same must also be true of V_{an}, V_{bn}, and V_{cn}. In this case we describe the voltages and currents as balanced. Figure 2.15a shows three line currents of a balanced system. In Fig. 2.15b the sum of these currents is shown to be a closed triangle. It is obvious that their sum is zero. Therefore, I_n in the connection shown in Fig. 2.13 between the neutrals of the generator and load must be zero. Then the connection between n and o may have any impedance, or even be open, and n and o will remain at the same potential.

If the load is *not balanced*, the sum of the currents will not be zero and a current will flow between o and n. For the unbalanced condition, in the absence of a connection of zero impedance, o and n will not be at the same potential.

The line-to-line voltages are V_{ab}, V_{bc}, and V_{ca}. Tracing a path from a to b through n in the circuit of Fig. 2.13 yields

$$V_{ab} = V_{an} + V_{nb} = V_{an} - V_{bn} \qquad (2.26)$$

Although $E_{a'o}$ and V_{an} are not in phase, we could decide to use V_{an} rather than $E_{a'o}$ as reference in defining the voltages. Then Fig. 2.16a is the phasor diagram of voltages to neutral, and Fig. 2.16b shows how V_{ab} is found. The magnitude of V_{ab} is

$$|V_{ab}| = 2|V_{an}| \cos 30°$$

$$= \sqrt{3}|V_{an}| \qquad (2.27)$$

Figure 2.15 Phasor diagram of currents in a balanced three-phase load: (a) phasors drawn from a common point; (b) addition of the phasors forming a closed triangle.

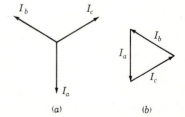

(a) (b)

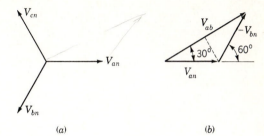

Figure 2.16 Voltages in a balanced three-phase circuit: (a) voltages to neutral; (b) relation between a line voltage and voltages to neutral.

(a)   (b)

As a phasor, V_{ab} leads V_{an} by 30°, and so

$$V_{ab} = \sqrt{3}\, V_{an} \underline{/30°} \tag{2.28}$$

The other line-to-line voltages are found in a similar manner, and Fig. 2.17 shows all the line-to-line and line-to-neutral voltages. The fact that the magnitude of line-to-line voltages of a balanced three-phase circuit is always equal to $\sqrt{3}$ times the magnitude of the line-to-neutral voltages is very important.

Figure 2.18 is another way of displaying the line-to-line and line-to-neutral voltages. The line-to-line voltage phasors are drawn to form a closed triangle oriented to agree with the chosen reference, in this case V_{an}. The vertices of the triangle are labeled so that each phasor begins and ends at the vertices corresponding to the order of the subscripts of that phasor voltage. Line-to-neutral

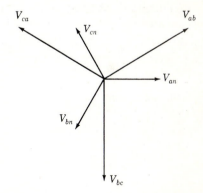

Figure 2.17 Phasor diagram of voltages in a balanced three-phase circuit.

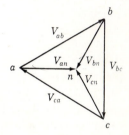

Figure 2.18 Alternative method of drawing the phasors of Fig. 2.17.

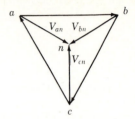

Figure 2.19 Phasor diagram of voltages for Example 2.2.

voltage phasors are drawn to the center of the triangle. Once this phasor diagram is understood, it will be found to be the simplest way to determine the various voltages.

The order in which the vertices a, b, and c of the triangle follow each other when the triangle is rotated counterclockwise about n indicates the phase sequence. We shall see later an example of the importance of phase sequence when we discuss symmetrical components as a means of analyzing unbalanced faults on power systems.

A separate current diagram can be drawn to relate each current properly with respect to its phase voltage.

Example 2.2 In a balanced three-phase circuit the voltage V_{ab} is $173.2\underline{/0°}$ V. Determine all the voltages and the currents in a Y-connected load having $Z_L = 10\underline{/20°}$ Ω. Assume that the phase sequence is abc.

SOLUTION The phasor diagram of voltages is drawn as shown in Fig. 2.19, from which it is determined that

$$V_{ab} = 173.2\underline{/0°} \text{ V} \qquad V_{an} = 100\underline{/-30°} \text{ V}$$

$$V_{bc} = 173.2\underline{/240°} \text{ V} \qquad V_{bn} = 100\underline{/210°} \text{ V}$$

$$V_{ca} = 173.2\underline{/120°} \text{ V} \qquad V_{cn} = 100\underline{/90°} \text{ V}$$

Each current lags the voltage across its load impedance by 20°, and each current magnitude is 10 A. Figure 2.20 is the phasor diagram of the currents

$$I_{an} = 10\underline{/-50°} \text{ A} \qquad I_{bn} = 10\underline{/190°} \text{ A} \qquad I_{cn} = 10\underline{/70°} \text{ A}$$

Balanced loads are often connected in Δ, as shown in Fig. 2.21. Here it is left to the reader to show that the magnitude of a line current such as I_a is equal to $\sqrt{3}$ times the magnitude of a phase current such as I_{ab} and that I_a lags I_{ab} by 30° when the phase sequence is abc.

When solving balanced three-phase circuits it is never necessary to work with the entire three-phase circuit diagram of Fig. 2.13. To solve the circuit a neutral connection of zero impedance is assumed to be present and to carry the sum of the three phase currents, which is zero, however, for balanced conditions. The circuit is solved by applying Kirchhoff's voltage law around a closed path which includes one phase and neutral. Such a closed path is shown in Fig. 2.22. This circuit is

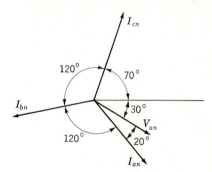

Figure 2.20 Phasor diagram of currents for
Example 2.2.

the single-phase equivalent of the circuit of Fig. 2.13. Calculations made for this
path are extended to the whole three-phase circuit by recalling that the currents in
the other two phases are equal in magnitude to the current of the phase calculated
and are displaced 120 and 240° in phase. It is immaterial whether the balanced
load, specified by its line-to-line voltage, total power, and power factor, is Δ- or
Y-connected since the Δ can always be replaced for purposes of calculaton by its
equivalent Y. The impedance of each phase of the equivalent Y will be one-third
the impedance of each phase of the Δ which it replaces.

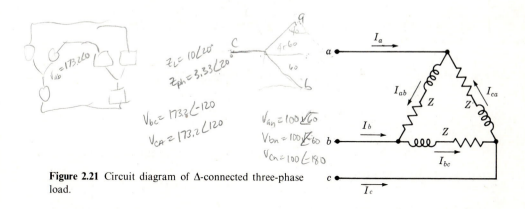

Figure 2.21 Circuit diagram of Δ-connected three-phase
load.

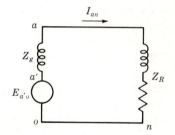

Figure 2.22 One phase of the circuit of Fig. 2.13.

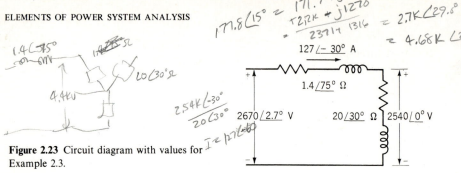

Figure 2.23 Circuit diagram with values for Example 2.3.

Example 2.3 The terminal voltage of a Y-connected load consisting of three equal impedances of $20\underline{/30°}$ Ω is 4.4 kV line to line. The impedance in each of the three lines connecting the load to a bus at a substation is $Z_L = 1.4\underline{/75°}$ Ω. Find the line-to-line voltage at the substation bus.

SOLUTION The magnitude of the voltage to neutral at the load is $4400/\sqrt{3} = 2540$ V. If V_{an}, the voltage across the load, is chosen as reference,

$$V_{an} = 2540\underline{/0°} \text{ V} \qquad \text{and} \qquad I_{an} = \frac{2540\underline{/0°}}{20\underline{/30°}} = 127.0\underline{/-30°} \text{ A}$$

The line-to-neutral voltage at the substation is

$$V_{an} + I_{an}Z_L = 2540\underline{/0°} + 127\underline{/-30°} \times 1.4\underline{/75°}$$
$$= 2540\underline{/0°} + 177.8\underline{/45°}$$
$$= 2666 + j125.7 = 2670\underline{/2.70°} \text{ V}$$

and the magnitude of the voltage at the substation bus is

$$\sqrt{3} \times 2.67 = 4.62 \text{ kV}$$

Figure 2.23 shows the circuit and quantities involved.

2.9 POWER IN BALANCED THREE-PHASE CIRCUITS

The total power delivered by a three-phase generator or absorbed by a three-phase load is found simply by adding the power in each of the three phases. In a balanced circuit this is the same as multiplying the power in any one phase by 3, since the power is the same in all phases.

If the magnitude of the voltages to neutral V_p for a Y-connected load is

$$V_p = |V_{an}| = |V_{bn}| = |V_{cn}| \tag{2.29}$$

and if the magnitude of the phase current I_p for a Y-connected load is

$$I_p = |I_{an}| = |I_{bn}| = |I_{cn}| \tag{2.30}$$

the total three-phase power is

$$P = 3V_p I_p \cos \theta_p \tag{2.31}$$

where θ_p is the angle by which phase current lags the phase voltage, that is, the angle of the impedance in each phase. If V_L and I_L are the magnitudes of line-to-line voltage and line current, respectively,

$$V_p = \frac{V_L}{\sqrt{3}} \quad \text{and} \quad I_p = I_L \tag{2.32}$$

and substituting in Eq. (2.31) yields

$$P = \sqrt{3} \, V_L I_L \cos \theta_p \tag{2.33}$$

The total vars are

$$Q = 3V_p I_p \sin \theta_p \tag{2.34}$$

$$Q = \sqrt{3} \, V_L I_L \sin \theta_p \tag{2.35}$$

and the voltamperes of the load are

$$|S| = \sqrt{P^2 + Q^2} = \sqrt{3} \, V_L I_L \tag{2.36}$$

Equations (2.33), (2.35), and (2.36) are the usual ones for calculating P, Q, and $|S|$ in balanced three-phase networks since the quantities usually known are line-to-line voltage, line current, and the power factor, $\cos \theta_p$. In speaking of a three-phase system, balanced conditions are assumed unless described otherwise; and the terms voltage, current, and power, unless identified otherwise, are understood to mean line-to-line voltage, line current, and total power of all three phases.

If the load is connected Δ, the voltage across each impedance is the line-to-line voltage and the current through each impedance is the magnitude of the line current divided by $\sqrt{3}$, or

$$V_p = V_L \quad \text{and} \quad I_p = \frac{I_L}{\sqrt{3}} \tag{2.37}$$

The total three-phase power is

$$P = 3V_p I_p \cos \theta_p \tag{2.38}$$

and substituting in this equation the values of V_p and I_p in Eq. (2.37) gives

$$P = \sqrt{3} \, V_L I_L \cos \theta_p \tag{2.39}$$

which is identical to Eq. (2.33). It follows that Eqs. (2.35) and (2.36) are also valid regardless of whether a particular load is connected Δ or Y.

2.10 PER-UNIT QUANTITIES

Power transmission lines are operated at voltage levels where the kilovolt is the most convenient unit to express voltage. Because of the large amount of power transmitted kilowatts or megawatts and kilovoltamperes or megavoltamperes are the common terms. However, these quantities as well as amperes and ohms are often expressed as a percent or per unit of a base or reference value specified for each. For instance, if a base voltage of 120 kV is chosen, voltages of 108, 120, and 126 kV become 0.90, 1.00, and 1.05 per unit, or 90, 100, and 105%, respectively. The per-unit value of any quantity is defined as the ratio of the quantity to its base value expressed as a decimal. The ratio in percent is 100 times the value in per unit. Both the percent and per-unit methods of calculation are simpler than the use of actual amperes, ohms, and volts. The per-unit method has an advantage over the percent method because the product of two quantities expressed in per unit is expressed in per unit itself, but the product of two quantities expressed in percent must be divided by 100 to obtain the result in percent.

Voltage, current, kilovoltamperes, and impedance are so related that selection of base values for any two of them determines the base values of the remaining two. If we specify the base values of current and voltage, base impedance and base kilovoltamperes can be determined. The base impedance is that impedance which will have a voltage drop across it equal to the base voltage when the current flowing in the impedance is equal to the base value of the current. The base kilovoltamperes in single-phase systems is the product of base voltage in kilovolts and base current in amperes. Usually base megavoltamperes and base voltage in kilovolts are the quantities selected to specify the base. For single-phase systems, or three-phase systems where the term current refers to line current, where the term voltage refers to voltage to neutral, and where the term kilovoltamperes refers to kilovoltamperes per phase, the following formulas relate the various quantities:

$$\text{Base current, A} = \frac{\text{base kVA}_{1\phi}}{\text{base voltage, kV}_{LN}} \tag{2.40}$$

$$\text{Base impedance} = \frac{\text{base voltage, V}_{LN}}{\text{base current, A}} \tag{2.41}$$

$$\text{Base impedance} = \frac{(\text{base voltage, kV}_{LN})^2 \times 1000}{\text{base kVA}_{1\phi}} \tag{2.42}$$

$$\text{Base impedance} = \frac{(\text{base voltage, kV}_{LN})^2}{\text{base MVA}_{1\phi}} \tag{2.43}$$

$$\text{Base power, kW}_{1\phi} = \text{base kVA}_{1\phi} \tag{2.44}$$

$$\text{Base power, MW}_{1\phi} = \text{base MVA}_{1\phi} \tag{2.45}$$

$$\frac{\text{Per-unit impedance}}{\text{of a circuit element}} = \frac{\text{actual impedance, } \Omega}{\text{base impedance, } \Omega} \tag{2.46}$$

In these equations the subscripts 1ϕ and LN denote "per phase" and "line-to-neutral," respectively, where the equations apply to three-phase circuits. If the equations are used for a single-phase circuit, kV_{LN} means the voltage across the single-phase line, or line-to-ground voltage if one side is grounded.

Since three-phase circuits are solved as a single line with a neutral return, the bases for quantities in the impedance diagram are kilovoltamperes per phase and kilovolts from line to neutral. Data are usually given as total three-phase kilovoltamperes or megavoltamperes and line-to-line kilovolts. Because of this custom of specifying line-to-line voltage and total kilovoltamperes or megavoltamperes, confusion may arise regarding the relation between the per-unit value of line voltage and the per-unit value of phase voltage. Although a line voltage may be specified as base, the voltage in the single-phase circuit required for the solution is still the voltage to neutral. The base voltage to neutral is the base voltage from line to line divided by $\sqrt{3}$. Since this is also the ratio between line-to-line and line-to-neutral voltages of a balanced three-phase system, *the per-unit value of a line-to-neutral voltage on the line-to-neutral voltage base is equal to the per-unit value of the line-to-line voltage at the same point on the line-to-line voltage base if the system is balanced.* Similarly, the three-phase kilovoltamperes is three times the kilovoltamperes per phase, and the three-phase kilovoltamperes base is three times the base kilovoltamperes per phase. Therefore, *the per-unit value of the three-phase kilovoltamperes on the three-phase kilovoltampere base is identical to the per-unit value of the kilovoltamperes per phase on the kilovoltampere-per-phase base.*

A numerical example may serve to clarify the relationships discussed. For instance, if

$$\text{Base } kVA_{3\phi} = 30{,}000 \text{ kVA}$$

and

$$\text{Base } kV_{LL} = 120 \text{ kV}$$

where the subscripts 3ϕ and LL mean "three-phase" and "line-to-line," respectively,

$$\text{Base } kVA_{1\phi} = \frac{30{,}000}{3} = 10{,}000 \text{ kVA}$$

and

$$\text{Base } kV_{LN} = \frac{120}{\sqrt{3}} = 69.2 \text{ kV}$$

For an actual line-to-line voltage of 108 kV, the line-to-neutral voltage is $108/\sqrt{3} = 62.3$ kV, and

$$\text{Per-unit voltage} = \frac{108}{120} = \frac{62.3}{69.2} = 0.90$$

For total three-phase power of 18,000 kW, the power per phase is 6000 kW, and

$$\text{Per-unit power} = \frac{18,000}{30,000} = \frac{6000}{10,000} = 0.6$$

Of course, megawatt and megavoltampere values may be substituted for kilowatt and kilovoltampere values throughout the above discussion. Unless otherwise specified, a given value of base voltage in a three-phase system is a line-to-line voltage, and a given value of base kilovoltamperes or base megavoltamperes is the total three-phase base.

Base impedance and base current can be computed directly from three-phase values of base kilovolts and base kilovoltamperes. If we interpret *base kilovoltamperes* and *base voltage in kilovolts* to mean *base kilovoltamperes for the total of the three phases* and *base voltage from line to line*, we find

$$\text{Base current, A} = \frac{\text{base kVA}_{3\phi}}{\sqrt{3} \times \text{base voltage, kV}_{LL}} \tag{2.47}$$

and from Eq. (2.42)

$$\text{Base impedance} = \frac{(\text{base voltage, kV}_{LL}/\sqrt{3})^2 \times 1000}{\text{base kVA}_{3\phi}/3} \tag{2.48}$$

$$\text{Base impedance} = \frac{(\text{base voltage, kV}_{LL})^2 \times 1000}{\text{base kVA}_{3\phi}} \tag{2.49}$$

$$\text{Base impedance} = \frac{(\text{base voltage, kV}_{LL})^2}{\text{base MVA}_{3\phi}} \tag{2.50}$$

Except for the subscripts, Eqs. (2.42) and (2.43) are identical to Eqs. (2.49) and (2.50), respectively. Subscripts have been used in expressing these relations in order to emphasize the distinction between working with three-phase quantities and quantities per phase. We shall use these equations without the subscripts, but we must (1) use line-to-line kilovolts with three-phase kilovoltamperes or megavoltamperes and (2) use line-to-neutral kilovolts with kilovoltamperes or megavoltamperes per phase. Equation (2.40) determines the base current for single-phase systems or for three-phase systems where the bases are specified in kilovoltamperes per phase and kilovolts to neutral. Equation (2.47) determines the base current for three-phase systems where the bases are specified in total kilovoltamperes for the three phases and in kilovolts from line to line.

Example 2.4 Find the solution of Example 2.3 by working in per unit on a base of 4.4 kV, 127 A so that both voltage and current magnitudes will be 1.0 per unit. Current rather than kilovoltamperes is specified here since the latter quantity does not enter the problem.

SOLUTION Base impedance is

$$\frac{4400/\sqrt{3}}{127} = 20.0 \ \Omega$$

and therefore the magnitude of the load impedance is also 1.0 per unit. The line impedance is

$$Z = \frac{1.4\underline{/75^\circ}}{20} = 0.07\underline{/75^\circ} \text{ per unit}$$

$$V_{an} = 1.0\underline{/0^\circ} + 1.0\underline{/-30^\circ} \times 0.07\underline{/75^\circ}$$

$$= 1.0\underline{/0^\circ} + 0.07\underline{/45^\circ}$$

$$= 1.0495 + j0.0495 = 1.051\underline{/2.70^\circ} \text{ per unit}$$

$$V_{LN} = 1.051 \times \frac{4400}{\sqrt{3}} = 2670 \text{ V, or } 2.67 \text{ kV}$$

$$V_{LL} = 1.051 \times 4.4 = 4.62 \text{ kV}$$

When the problems to be solved are more complex and particularly when transformers are involved the advantages of calculations in per unit will be more apparent.

2.11 CHANGING THE BASE OF PER-UNIT QUANTITIES

Sometimes the per-unit impedance of a component of a system is expressed on a base other than the one selected as base for the part of the system in which the component is located. Since all impedances in any one part of a system must be expressed on the same impedance base when making computations, it is necessary to have a means of converting per-unit impedances from one base to another. Substituting the expression for base impedance given by Eq. (2.42) or (2.49) for base impedance in Eq. (2.46) gives

Per-unit impedance of a circuit element   15.49 X.

$$= \frac{(\text{actual impedance, } \Omega) \times (\text{base kVA})}{(\text{base voltage, kV})^2 \times 1{,}000} \quad (2.51)$$

which shows that per-unit impedance is directly proportional to base kilovolt-amperes and inversely proportional to the square of the base voltage. Therefore,

$$0.8 \left(\frac{440}{440}\right)^2 \left(\frac{100kV}{10kVA}\right)^2$$

to change from per-unit impedance on a given base to per-unit impedance on a new base, the following equation applies:

$$\text{Per-unit } Z_{\text{new}} = \text{per-unit } Z_{\text{given}} \left(\frac{\text{base kV}_{\text{given}}}{\text{base kV}_{\text{new}}}\right)^2 \left(\frac{\text{base kVA}_{\text{new}}}{\text{base kVA}_{\text{given}}}\right) \quad (2.52)$$

This equation has nothing to do with transferring the ohmic value of impedance from one side of the transformer to another. The great value of the equation is in changing the per-unit impedance given on a particular base to a new base.

Rather than using Eq. 2.52, however, the change in base may also be accomplished by converting the per-unit value on the given base to ohms and dividing by the new base impedance.

Example 2.5 The reactance of a generator designated X'' is given as 0.25 per unit based on the generator's nameplate rating of 18 kV, 500 MVA. The base for calculations is 20 kV, 100 MVA. Find X'' on the new base.

SOLUTION By Eq. (2.52)

$$X'' = 0.25 \left(\frac{18}{20}\right)^2 \left(\frac{100}{500}\right) = 0.0405 \text{ per unit}$$

or by converting the given value to ohms and dividing by the new base impedance

$$X'' = \frac{0.25(18^2/500)}{20^2/100} = 0.0405 \text{ per unit}$$

Resistance and reactance of a device in percent or per unit are usually available from the manufacturer. The base is understood to be the rated kilovoltamperes and kilovolts of the device. Tables A.4 and A.5 in the Appendix list some representative values of reactance for generators and transformers. We shall discuss per-unit quantities further in Chap. 6 in connection with our study of transformers.

PROBLEMS

2.1 If $v = 141.4 \sin(\omega t + 30°)$ V and $i = 11.31 \cos(\omega t - 30°)$ A, find for each (a) the maximum value, (b) the rms value, and (c) the phasor expression in polar and rectangular form if voltage is the reference. Is the circuit inductive or capacitive?

2.2 If the circuit of Prob. 2.1 consists of a purely resistive and a purely reactive element, find R and X, (a) if the elements are in series and (b) if the elements are in parallel.

2.3 In a single-phase circuit $V_a = 120\underline{/45°}$ V and $V_b = 100\underline{/-15°}$ V with respect to a reference node o. Find V_{ba} in polar form.

2.4 A single-phase ac voltage of 240 V is applied to a series circuit whose impedance is $10\underline{/60°}$ Ω. Find R, X, P, Q, and the power factor of the circuit.

2.5 If a capacitor is connected in parallel with the circuit of Prob. 2.4 and if this capacitor supplies 1250 var, find the P and Q supplied by the 240-V source, and find the resultant power factor.

2.6 A single-phase inductive load draws 10 MW at 0.6 power factor lagging. Draw the power triangle and determine the reactive power of a capacitor to be connected in parallel with the load to raise the power factor to 0.85.

2.7 A single-phase induction motor is operating at a very light load during a large part of every day and draws 10 A from the supply. A device is proposed to "increase the efficiency" of the motor. During a demonstration the device is placed in parallel with the unloaded motor and the current drawn from the supply drops to 8 A. When two of the devices are placed in parallel the current drops to 6 A. What simple device will cause this drop in current? Discuss the advantages of the device. Is the efficiency of the motor increased by the device? (Recall that an induction motor draws lagging current).

2.8 If the impedance between machines 1 and 2 of Example 2.1 is $Z = 0 - j5$ Ω determine (a) whether each machine is generating or consuming power, (b) whether each machine is receiving or supplying positive reactor power and the amount, and (c) the value of P and Q absorbed by the impedance.

2.9 Repeat Prob. 2.8 if $Z = 5 + j0$ Ω.

2.10 A voltage source $E_{an} = -120\underline{/210°}$ V and the current through the source is given by $I_{na} = 10\underline{/60°}$ A. Find the values of P and Q and state whether the source is delivering or receiving each.

2.11 Solve Example 2.1 if $E_1 = 100\underline{/0°}$ V and $E_2 = 120\underline{/30°}$ V. Compare the results with Example 2.1 and form some conclusions about the effect of variation of the magnitude of E_2 in this circuit.

2.12 Three identical impedances of $10\underline{/-15°}$ Ω are Y-connected to balanced three-phase line voltages of 208 V. Specify all the line and phase voltages and the currents as phasors in polar form with V_{ca} as reference for a phase sequence of abc.

2.13 In a balanced three-phase system the Y-connected impedances are $10\underline{/30°}$ Ω. If $V_{bc} = 416\underline{/90°}$ V, specify I_{cn} in polar form.

2.14 The terminals of a three-phase supply are labeled a, b, and c. Between any pair a voltmeter measures 115 V. A resistor of 100 Ω and a capacitor of 100 Ω at the frequency of the supply are connected in series from a to b with the resistor connected to a. The point of connection of the elements to each other is labeled n. Determine graphically the voltmeter reading between c and n if phase sequence is abc and if phase sequence is acb.

2.15 Determine the current drawn from a three-phase 440-V line by a three-phase 15-hp motor operating at full load, 90% efficiency, and 80% power factor lagging. Find the values of P and Q drawn from the line.

2.16 If the impedance of each of the three lines connecting the motor of Prob. 2.15 to a bus is $0.3 + j1.0$ Ω, find the line-to-line voltage at the bus which supplies 440 V at the motor.

2.17 A balanced-Δ load consisting of pure resistances of 15 Ω per phase is in parallel with a balanced-Y load having phase impedances of $8 + j6$ Ω. Identical impedances of $2 + j5$ Ω are in each of the three lines connecting the combined loads to a 110-V three-phase supply. Find the current drawn from the supply and the line voltage at the combined loads.

2.18 A three-phase load draws 250 kW at a power factor of 0.707 lagging from a 440-V line. In parallel with this load is a three-phase capacitor bank which draws 60 kVA. Find the total current and resultant power factor.

2.19 A three-phase motor draws 20 kVA at 0.707 power factor lagging from a 220-V source. Determine the kilovoltampere rating of capacitors to make the combined power factor 0.90 lagging, and determine the line current before and after the capacitors are added.

2.20 A coal mining "drag line" machine in an open-pit mine consumes 0.92 MVA at 0.8 power factor lagging when it digs coal, and it generates (delivers to the electric system) 0.10 MVA at 0.5 power factor leading when the loaded shovel swings away from the pit wall. At the end of the "dig" period, the change in supply current magnitude can cause tripping of a protective relay which is constructed of solid-state circuitry. Therefore it is desired to minimize the change in current magnitude. Consider the placement of capacitors at the machine terminals and find the amount of capacitive correction (in kvar) to eliminate the change in steady-state current magnitude. The machine is energized from a 36.5 kV, three-phase supply. Start the solution by letting Q be the total three-phase megavars of the capacitors connected across the machine terminals, and write an expression for the magnitude of the line current *drawn by* the machine in terms of Q for both the digging and generating operations.

2.21 A generator (which may be represented by an emf in series with an inductive reactance) is rated 500 MVA, 22 kV. Its Y-connected windings have a reactance of 1.1 per unit. Find the ohmic value of the reactance of the windings.

2.22 The generator of Prob. 2.21 is in a circuit for which the bases are specified as 100 MVA, 20 kV. Starting with the per-unit value given in Prob. 2.21, find the per-unit value of reactance of the generator windings on the specified base.

2.23 Draw the single-phase equivalent circuit for the motor (an emf in series with inductive reactance labeled Z_m) and its connection to the voltage supply described in Probs. 2.15 and 2.16. Show on the diagram the per-unit values of the line impedance and the voltage at the motor terminals on a base of 20 kVA, 440 V. Then using per-unit values find the supply voltage in per unit and convert the per-unit value of the supply voltage to volts.

THREE

SERIES IMPEDANCE OF TRANSMISSION LINES

$cos\ x = cos\ y$

sin

csc

An electric transmission line has four parameters which affect its ability to fulfill its function as part of a power system: resistance, inductance, capacitance, and conductance. In this chapter we discuss the first two of these parameters, and we shall consider capacitance in the next chapter.

Conductance between conductors or between conductors and the ground accounts for the leakage current at the insulators of overhead lines and through the insulation of cables. Since leakage at insulators of overhead lines is negligible, the conductance between conductors of an overhead line is assumed to be zero.

When current flows in an electric circuit, we explain some of the properties of the circuit by the magnetic and electric fields present. Figure 3.1 shows a single-phase line and its associated magnetic and electric fields. The lines of magnetic flux form closed loops linking the circuit, and the lines of electric flux originate on the positive charges on one conductor and terminate on the negative charges on the other conductor. Variation of the current in the conductors causes a change in the number of lines of magnetic flux linking the circuit. Any change in the flux linking a circuit induces a voltage in the circuit which is proportional to the rate of change of flux. Inductance is the property of the circuit that relates the voltage induced by changing flux to the rate of change of current.

Capacitance exists between the conductors and is the charge on the conductors per unit of potential difference between them.

37

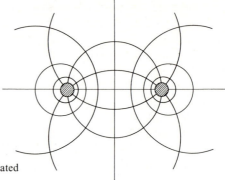

Figure 3.1 Magnetic and electric fields associated with a two-wire line.

The resistance and inductance uniformly distributed along the line form the series impedance. The conductance and capacitance existing between conductors of a single-phase line or from a conductor to neutral of a three-phase line form the shunt admittance. Although the resistance, inductance, and capacitance are distributed, the equivalent circuit of a line is made up of lumped parameters, as we shall see when we discuss them.

3.1 TYPES OF CONDUCTORS

In the early days of the transmission of electric power, conductors were usually copper, but aluminum conductors have completely replaced copper because of the much lower cost and lighter weight of an aluminum conductor compared with a copper conductor of the same resistance. The fact that an aluminum conductor has a larger diameter than a copper conductor of the same resistance is also an advantage. With a larger diameter the lines of electric flux originating on the conductor will be farther apart at the conductor surface for the same voltage. This means a lower voltage gradient at the conductor surface and less tendency to ionize the air around the conductor. Ionization produces the undesirable effect called *corona*.

Symbols identifying different types of aluminum conductors are as follows:

AAC   all-aluminum conductors
AAAC   all-aluminum-alloy conductors
ACSR   aluminum conductor, steel-reinforced
ACAR   aluminum conductor, alloy-reinforced

Aluminum-alloy conductors have higher tensile strength than the ordinary electrical-conductor grade of aluminum. ACSR consists of a central core of steel strands surrounded by layers of aluminum strands. ACAR has a central core of higher-strength aluminum surrounded by layers of electrical-conductor-grade aluminum.

Alternate layers of wire of a stranded conductor are spiraled in opposite directions to prevent unwinding and make the outer radius of one layer coincide with the inner radius of the next. Stranding provides flexibility for a large cross-sectional area. The number of strands depends on the number of layers and on whether all the strands are the same diameter. The total number of strands in concentrically stranded cables, where the total annular space is filled with strands of uniform diameter, is 7, 19, 37, 61, 91, or more.

Figure 3.2 shows the cross section of a typical steel-reinforced aluminum cable (ACSR). The conductor shown has 7 steel strands forming a central core, around which are two layers of aluminum strands. There are 24 aluminum strands in the two outer layers. The conductor stranding is specified as 24 Al/7 St, or simply 24/7. Various tensile strengths, current capacities, and conductor sizes are obtained by using different combinations of steel and aluminum.

Appendix Table A.1 gives some electrical characteristics of ACSR. Code names, uniform throughout the aluminum industry, have been assigned to each conductor for easy reference.

A type of conductor known as *expanded* ACSR has a filler such as paper separating the inner steel strands from the outer aluminum strands. The paper gives a larger diameter (and hence, lower corona) for a given conductivity and tensile strength. Expanded ACSR is used for some extra-high-voltage (EHV) lines.

Cables for underground transmission are usually made with stranded copper conductors rather than aluminum. The conductors are insulated with oil-impregnated paper. Up to voltages of 46 kV the cables are of the solid type which means that the only insulating oil in the cable is that which is impregnated during manufacture. The voltage rating of this type of cable is limited by the tendency of voids to develop between the layers of insulation. Voids cause early breakdown of the insulation. A lead sheath surrounds the cable which may consist of a single conductor or three conductors.

At voltages from 46 to 345 kV low-pressure oil-filled cables are available. Oil reservoirs at intervals along the length of the cable supply the oil to ducts in the center of single-conductor cables or to the spaces between the insulated conductors of the three-phase type. These conductors are also enclosed in a lead sheath.

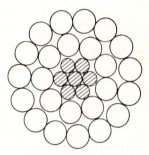

Figure 3.2 Cross section of a steel-reinforced conductor, 7 steel strands, 24 aluminum strands.

High-pressure pipe-type cables are the most widely used cables for underground transmission at voltages from 69 to 550 kV. The paper-insulated cables lie in a steel pipe of diameter somewhat larger than necessary to contain the insulated conductors which lie together along the bottom of the pipe.

Gas-insulated cables are available at voltages up to 138 kV. Research is constantly conducted on other types of cables especially for voltage levels of 765 and 1100 kV. Manufacturers furnish details of construction for the various cables.

In this book we shall devote our attention to overhead lines almost exclusively since underground transmission is usually restricted to large cities or transmission under wide rivers, lakes, and bays. Underground lines cost at least eight times as much as overhead lines and 20 times as much at the highest voltage.

3.2 RESISTANCE

The resistance of transmission-line conductors is the most important cause of power loss in a transmission line. The term resistance, unless specifically qualified, means effective resistance. The effective resistance of a conductor is

$$R = \frac{\text{power loss in conductor}}{|I|^2} \quad \Omega \tag{3.1}$$

where the power is in watts and I is the rms current in the conductor in amperes. The effective resistance is equal to the dc resistance of the conductor only if the distribution of current throughout the conductor is uniform. We shall discuss nonuniformity of current distribution briefly after reviewing some fundamental concepts of dc resistance.

Direct-current resistance is given by the formula

$$R_0 = \frac{\rho l}{A} \quad \Omega \tag{3.2}$$

where ρ = resistivity of conductor
  l = length
  A = cross-sectional area

Any consistent set of units may be used. In power work in the United States, l is usually given in feet, A in circular mils (cmil), and ρ in ohm–circular mils per foot, sometimes called ohms per circular mil–foot. In SI units l is in meters, A in square meters, and ρ in ohm-meters.†

A circular mil is the area of a circle having a diameter of 1 mil. A mil is equal to 10^{-3} in. The cross-sectional area of a solid cylindrical conductor in circular mils is equal to the square of the diameter of the conductor expressed in mils. The number of circular mils multiplied by $\pi/4$ equals the number of square

---

† SI is the official designation for the International System of Units.

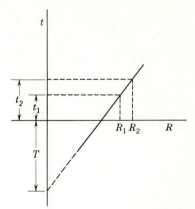

Figure 3.3 Resistance of a metallic conductor as a function of temperature.

mils. Since manufacturers in the United States identify conductors by their cross-sectional area in circular mils we must use this unit occasionally. The area in square millimeters equals the area in circular mils multiplied by 5.067×10^{-4}.

The international standard of conductivity is that of annealed copper. Commercial hard-drawn copper wire has 97.3% and aluminum 61% of the conductivity of standard annealed copper. At 20°C for hard-drawn copper ρ is $1.77 \times 10^{-8} \ \Omega \cdot m$ (10.66 $\Omega \cdot$cmil/ft). For aluminum at 20°C ρ is $2.83 \times 10^{-8} \ \Omega \cdot m$ (17.00 $\Omega \cdot$cmil/ft).

The dc resistance of stranded conductors is greater than the value computed by Eq. (3.2) because spiraling of the strands makes them longer than the conductor itself. For each mile of conductor the current in all strands except the one in the center flows in more than a mile of wire. The increased resistance due to spiraling is estimated as 1% for three-strand conductors and 2% for concentrically stranded conductors.

The variation of resistance of metallic conductors with temperature is practically linear over the normal range of operation. If temperature is plotted on the vertical axis and resistance on the horizontal axis, as in Fig. 3.3, extension of the straight-line portion of the graph provides a convenient method of correcting resistance for changes in temperature. The point of intersection of the extended line with the temperature axis at zero resistance is a constant of the material. From the geometry of Fig. 3.3

$$\frac{R_2}{R_1} = \frac{T + t_2}{T + t_1} \tag{3.3}$$

where R_1 and R_2 are the resistances of the conductor at temperatures t_1 and t_2, respectively, in degrees Celsius and T is the constant determined from the graph. Values of the constant T are as follows:

$$T = \begin{cases} 234.5 & \text{for annealed copper of 100\% conductivity} \\ 241 & \text{for hard-drawn copper of 97.3\% conductivity} \\ 228 & \text{for hard-drawn aluminum of 61\% conductivity} \end{cases}$$

Uniform distribution of current throughout the cross section of a conductor exists only for direct current. As the frequency of alternating current increases, the nonuniformity of distribution becomes more pronounced. An increase in frequency causes nonuniform current density. This phenomenon is called *skin effect*. In a circular conductor the current density *usually* increases from the interior toward the surface. For conductors of sufficiently large radius, however, a current density oscillatory with respect to radial distance from the center may result.

As we shall see when discussing inductance, some lines of magnetic flux exist inside a conductor. Filaments on the surface of a conductor are not linked by internal flux, and the flux linking a filament near the surface is less than the flux linking a filament in the interior. The alternating flux induces higher voltages acting on the interior filaments than are induced on filaments near the surface of the conductor. By Lenz's law the induced voltage opposes the change of current producing it, and the higher induced voltages acting on the inner filaments cause the higher current density in filaments nearer the surface and therefore higher effective resistance. Even at power frequencies skin effect is a significant factor in large conductors.

3.3 TABULATED RESISTANCE VALUES

The dc resistance of various types of conductors is easily found by Eq. (3.2), and the increased resistance due to spiraling can be estimated. Temperature corrections are determined by Eq. (3.3). The increase in resistance caused by skin effect can be calculated for round wires and tubes of solid material, and curves of R/R_0 are available for these simple conductors.† This information is not necessary, however, since manufacturers supply tables of electrical characteristics of their conductors. Table A.1 is an example of some of the data available.

Example 3.1 Tables of electrical characteristics of all-aluminum *Marigold* stranded conductor list a dc resistance of 0.01558 Ω per 1000 ft at 20°C and an ac resistance of 0.0956 Ω/mi at 50°C. The conductor has 61 strands and its size is 1,113,000 cmil. Verify the dc resistance and find the ratio of ac to dc resistance.

SOLUTION At 20°C from Eq. (3.2) with an increase of 2% for spiraling

$$R_0 = \frac{17.0 \times 1000}{1{,}113 \times 10^3} \times 1.02 = 0.01558 \ \Omega \text{ per 1000 ft}$$

---

† See The Aluminum Association, "Aluminum Electrical Conductor Handbook," New York, 1971.

At a temperature of 50°C from Eq. (3.3)

$$R_0 = 0.01558 \frac{228 + 50}{228 + 20} = 0.01746 \ \Omega \text{ per } 1000 \text{ ft}$$

$$\frac{R}{R_0} = \frac{0.0956}{0.01746 \times 5.280} = 1.037$$

Skin effect causes a 3.7% increase in resistance.

3.4 DEFINITION OF INDUCTANCE

Two fundamental equations serve to explain and define inductance. The first equation relates induced voltage to the rate of change of flux linking a circuit. The induced voltage is

$$e = \frac{d\tau}{dt} \tag{3.4}$$

where e is the induced voltage in volts and τ is the number of *flux linkages* of the circuit in weber-turns (Wbt). The number of weber-turns is the product of each weber of flux and the number of turns of the circuit linked. For the two-wire line of Fig. 3.1 each line of flux external to the conductors links the circuit only once. If we had been considering a coil instead of the circuit of Fig. 3.1, most of the lines of flux produced would have linked more than one turn of the coil. If some of the flux links less than all the turns of a coil, the total flux linkages are reduced. In terms of lines of flux, each line is multiplied by the number of turns it links, and these products are added to obtain the total flux linkages.

When the current in a circuit is changing, its associated magnetic field (which is described by the flux linkages) must be changing. If constant permeability is assumed for the medium in which the magnetic field is set up, the number of flux linkages is directly proportional to the current, and therefore the induced voltage is proportional to the rate of change of current. Thus our second fundamental equation is

$$e = L \frac{di}{dt} \quad \text{V} \tag{3.5}$$

where L = constant of proportionality
$\quad L$ = inductance of circuit, H
$\quad e$ = induced voltage, V
di/dt = rate of change of current, A/s

Equation (3.5) may be used where the permeability is not constant, but in such a case the inductance is not a constant.

When Eqs. (3.4) and (3.5) are solved for L, the result is

$$L = \frac{d\tau}{di} \quad \text{H} \tag{3.6}$$

If the flux linkages of the circuit vary linearly with current, which means that the magnetic circuit has a constant permeability,

$$L = \frac{\tau}{i} \quad \text{H} \tag{3.7}$$

from which arises the definition of the self-inductance of an electric circuit as the flux linkages of the circuit per unit of current. In terms of inductance the flux linkages are

$$\tau = Li \quad \text{Wbt} \tag{3.8}$$

In Eq. (3.8), since i is instantaneous current, τ represents instantaneous flux linkages. For sinusoidal alternating current, flux linkages are sinusoidal. Where ψ is the phasor expression for the flux linkages,

$$\psi = LI \quad \text{Wbt} \tag{3.9}$$

Since ψ and I are in phase, L is real, as is consistent with Eqs. (3.7) and (3.8). The phasor voltage drop due to the flux linkages is

$$V = j\omega LI \quad \text{V} \tag{3.10}$$

$$V = j\omega\psi \quad \text{V} \tag{3.11}$$

Mutual inductance between two circuits is defined as the flux linkages of one circuit due to the current in the second circuit per ampere of current in the second circuit. If the current I_2 produces ψ_{12} flux linkages with circuit 1, the mutual inductance is

$$M_{12} = \frac{\psi_{12}}{I_2} \quad \text{H}$$

The phasor voltage drop in circuit 1 caused by the flux linkages of circuit 2 is

$$V_1 = j\omega M_{12} I_2 = j\omega\psi_{12} \quad \text{V}$$

Mutual inductance is important in considering the influence of power lines on telephone lines and the coupling between parallel power lines.

3.5 INDUCTANCE OF A CONDUCTOR DUE TO INTERNAL FLUX

Only flux lines external to the conductors are shown in Fig. 3.1. Some of the magnetic field, however, exists inside the conductors, as we mentioned when considering skin effect. The changing lines of flux inside the conductors also contribute to the induced voltage of the circuit and therefore to the inductance. The correct value of inductance due to internal flux can be computed as the ratio of flux linkages to current by taking into account the fact that each line of internal flux links only a fraction of the total current.

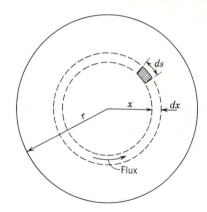

Figure 3.4 Cross section of a cylindrical conductor.

To obtain an accurate value for the inductance of a transmission line, it is necessary to consider the flux inside each conductor as well as the external flux. Let us consider the long cylindrical conductor whose cross section is shown in Fig. 3.4. We shall assume that the return path for the current in this conductor is so far away that it does not appreciably affect the magnetic field of the conductor shown. Then the lines of flux are concentric with the conductor.

The magnetomotive force (mmf) in ampere-turns around any closed path is equal to the current in amperes enclosed by the path. The mmf is also equal to the integral of the tangential component of the magnetic field intensity around the path. Thus

$$\text{mmf} = \oint H \cdot ds = I \qquad \text{At} \qquad (3.12)$$

where H = magnetic field intensity, At/m
$\quad\quad\; s$ = distance along path, m
$\quad\quad\; I$ = current, A, enclosed†

The dot between H and ds indicates that the value of H is the component of the field intensity tangent to ds.

Let the field intensity at a distance x meters from the center of the conductor be designated H_x. Since the field is symmetrical, H_x is constant at all points equidistant from the center of the conductor. If the integration indicated in Eq. (3.12) is performed around a circular path concentric with the conductor at x meters from the center, H_x is constant over the path and tangent to it. Equation (3.12) becomes

$$\oint H_x \, ds = I_x \qquad (3.13)$$

and

$$2\pi x H_x = I_x \qquad (3.14)$$

† Our work here applies equally to alternating and direct current. As shown, H and I are phasors and represent sinusoidally alternating quantities. For simplicity the current I could be interpreted as a direct current and H as a real number.

where I_x is the current enclosed. Then, assuming uniform current density,

$$I_x = \frac{\pi x^2}{\pi r^2} I \tag{3.15}$$

where I is the total current in the conductor. Then substituting Eq. (3.15) in Eq. (3.14) and solving for H_x, we obtain

$$H_x = \frac{x}{2\pi r^2} I \qquad \text{At/m} \tag{3.16}$$

The flux density x meters from the center of the conductor is

$$B_x = \mu H_x = \frac{\mu x I}{2\pi r^2} \qquad \text{Wb/m}^2 \tag{3.17}$$

where μ is the permeability of the conductor.†

In the tubular element of thickness dx, the flux $d\phi$ is B_x times the cross-sectional area of the element normal to the flux lines, the area being dx times the axial length. The flux per meter of length is

$$d\phi = \frac{\mu x I}{2\pi r^2} dx \qquad \text{Wb/m} \tag{3.18}$$

The flux linkages $d\psi$ per meter of length, which are caused by the flux in the tubular element, are the product of the flux per meter of length and the fraction of the current linked. Thus

$$d\psi = \frac{\pi x^2}{\pi r^2} d\phi = \frac{\mu I x^3}{2\pi r^4} dx \qquad \text{Wbt/m} \tag{3.19}$$

Integrating from the center of the conductor to its outside edge to find ψ_{int}, the total flux linkages inside the conductor, we obtain

$$\psi_{\text{int}} = \int_0^r \frac{\mu I x^3}{2\pi r^4} dx$$

$$\psi_{\text{int}} = \frac{\mu I}{8\pi} \qquad \text{Wbt/m} \tag{3.20}$$

For a relative permeability of 1, $\mu = 4\pi \times 10^{-7}$ H/m, and

$$\psi_{\text{int}} = \frac{I}{2} \times 10^{-7} \qquad \text{Wbt/m} \tag{3.21}$$

$$L_{\text{int}} = \frac{1}{2} \times 10^{-7} \qquad \text{H/m} \tag{3.22}$$

---

† In SI units the permeability of free space is $\mu_0 = 4\pi \times 10^{-7}$ H/m, and the relative permeability is $\mu_r = \mu/\mu_0$.

We have computed the inductance per unit length (henrys per meter) of a round conductor attributed only to the flux inside the conductor. Hereafter, for convenience, we shall refer to *inductance per unit length* simply as *inductance*, but we must be careful to use the correct dimensional units.

The validity of computing the internal inductance of a solid round wire by the method of partial flux linkages can be demonstrated by deriving the internal inductance in an entirely different manner. Equating energy stored in the magnetic field within the conductor per unit length at any instant to $L_{int} i^2/2$ and solving for L_{int} will yield Eq. (3.22).

3.6 FLUX LINKAGES BETWEEN TWO POINTS EXTERNAL TO AN ISOLATED CONDUCTOR

As a step in computing inductance due to flux external to a conductor, let us derive an expression for the flux linkages of an isolated conductor due only to that portion of the external flux which lies between two points distant D_1 and D_2 meters from the center of the conductor. In Fig. 3.5, P_1 and P_2 are two such points. The conductor carries a current of I A. Since the flux paths are concentric circles around the conductor, all the flux between P_1 and P_2 lies within the concentric cylindrical surfaces (indicated by solid circular lines) which pass through P_1 and P_2. At the tubular element which is x meters from the center of the conductor the field intensity is H_x. The mmf around the element is

$$2\pi x H_x = I \tag{3.23}$$

Solving for H_x and multiplying by μ yields the flux density B_x in the element, so that

$$B_x = \frac{\mu I}{2\pi x} \qquad \text{Wb/m}^2 \tag{3.24}$$

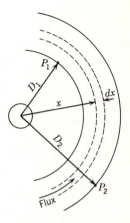

Figure 3.5 A conductor and external points P_1 and P_2.

The flux $d\phi$ in the tubular element of thickness dx is

$$d\phi = \frac{\mu I}{2\pi x}\, dx \qquad \text{Wb/m} \tag{3.25}$$

The flux linkages $d\psi$ per meter are numerically equal to the flux $d\phi$, since flux external to the conductor links all the current in the conductor once and only once. The total flux linkages between P_1 and P_2 are obtained by integrating $d\psi$ from $x = D_1$ to $x = D_2$. We obtain

$$\psi_{12} = \int_{D_1}^{D_2} \frac{\mu I}{2\pi x}\, dx = \frac{\mu I}{2\pi} \ln \frac{D_2}{D_1} \qquad \text{Wbt/m} \tag{3.26}$$

or, for a relative permeability of 1,

$$\psi_{12} = 2 \times 10^{-7} I \ln \frac{D_2}{D_1} \qquad \text{Wbt/m} \tag{3.27}$$

The inductance due only to the flux included between P_1 and P_2 is

$$L_{12} = 2 \times 10^{-7} \ln \frac{D_2}{D_1} \qquad \text{H/m} \tag{3.28}$$

3.7 INDUCTANCE OF A SINGLE-PHASE TWO-WIRE LINE

Before proceeding to the more general case of multiconductor lines and three-phase lines, let us consider a simple two-wire line composed of solid round conductors. Figure 3.6 shows a circuit having two conductors of radii r_1 and r_2. One conductor is the return circuit for the other. First consider only the flux linkages of the circuit caused by the current in conductor 1. A line of flux set up

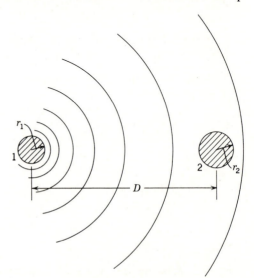

Figure 3.6 Conductors of different radii and the magnetic field due to current in conductor 1 only.

by current in conductor 1 at a distance equal to or greater than $D + r_2$ from the center of conductor 1 does not link the circuit and cannot induce a voltage in the circuit. Stated in another manner, such a line of flux links a net current of zero, since the current in conductor 2 is equal in value and opposite in direction to the current in conductor 1. The fraction of the total current linked by a line of flux external to conductor 1 at a distance equal to or less than $D - r_2$ is 1. Between $D - r_2$ and $D + r_2$ (that is, over the surface of conductor 2), the fraction of the total current in the circuit linked by a line of flux set up by current in conductor 1 varies from 1 to 0. Therefore, it is logical to simplify the problem, when D is much greater than r_1 and r_2 and the flux density through the conductor is nearly uniform, by assuming that all the external flux set up by current in conductor 1 extending to the center of conductor 2 links all the current I and that flux beyond the center of conductor 2 links none of the current. In fact, it can be shown that calculations made on this assumption are correct even when D is small.

The inductance of the circuit due to current in conductor 1 is determined by Eq. (3.28), with the distance D between conductors 1 and 2 substituted for D_2 and the radius r_1 of conductor 1 substituted for D_1. For external flux only

$$L_{1,\,\text{ext}} = 2 \times 10^{-7} \ln \frac{D}{r_1} \qquad \text{H/m} \tag{3.29}$$

For internal flux only

$$L_{1,\,\text{int}} = \frac{1}{2} \times 10^{-7} \qquad \text{H/m} \tag{3.30}$$

The total inductance of the circuit due to the current in conductor 1 only is

$$L_1 = \left(\frac{1}{2} + 2 \ln \frac{D}{r_1} \right) \times 10^{-7} \qquad \text{H/m} \tag{3.31}$$

The expression for inductance may be put in a more concise form by factoring Eq. (3.31) and by noting that $\ln \varepsilon^{1/4} = 1/4$, whence

$$L_1 = 2 \times 10^{-7} \left(\ln \varepsilon^{1/4} + \ln \frac{D}{r_1} \right) \tag{3.32}$$

Upon combining terms, we obtain

$$L_1 = 2 \times 10^{-7} \ln \frac{D}{r_1 \varepsilon^{-1/4}} \tag{3.33}$$

If we substitute r_1' for $r_1 \varepsilon^{-1/4}$,

$$L_1 = 2 \times 10^{-7} \ln \frac{D}{r_1'} \qquad \text{H/m} \tag{3.34}$$

The radius r_1' is that of a fictitious conductor assumed to have no internal flux but with the same inductance as the actual conductor of radius r_1. The quantity

$\varepsilon^{-1/4}$ is equal to 0.7788. Equation (3.34) gives the same value for inductance as Eq. (3.31). The difference is that Eq. (3.34) omits the term to account for internal flux but compensates for it by using an adjusted value for the radius of the conductor. We should note carefully that Eq. (3.31) was derived for a solid round conductor and that Eq. (3.34) was found by algebraic manipulation of Eq. (3.31). Therefore, the multiplying factor of 0.7788 to adjust the radius in order to account for internal flux applies only to solid round conductors. We shall consider other conductors later.

Since the current in conductor 2 flows in the direction opposite to that in conductor 1 (or is 180° out of phase with it), the flux linkages produced by current in conductor 2 considered alone are in the same direction through the circuit as those produced by current in conductor 1. The resulting flux for the two conductors is determined by the sum of the mmfs of both conductors. For constant permeability, however, the flux linkages (and likewise the inductances) of the two conductors considered separately may be added.

By comparison with Eq. (3.34) the inductance due to current in conductor 2 is

$$L_2 = 2 \times 10^{-7} \ln \frac{D}{r_2'} \qquad \text{H/m} \tag{3.35}$$

and for the complete circuit

$$L = L_1 + L_2 = 4 \times 10^{-7} \ln \frac{D}{\sqrt{r_1' r_2'}} \qquad \text{H/m} \tag{3.36}$$

If $r_1' = r_2' = r'$, the total inductance reduces to

$$L = 4 \times 10^{-7} \ln \frac{D}{r'} \qquad \text{H/m} \tag{3.37}$$

Equation (3.37) is the inductance of the two-wire line taking into account the flux linkages caused by current in both conductors, one of which is the return path for current in the other. This value of inductance is sometimes called the inductance per loop meter or per loop mile to distinguish it from that component of the inductance of the circuit attributed to the current in one conductor only. The latter, as given by Eq. (3.34), is one-half the total inductance of a single-phase line and is called the inductance per conductor.

3.8 FLUX LINKAGES OF ONE CONDUCTOR IN A GROUP

A more general problem than that of the two-wire line is presented by one conductor in a group of conductors where the sum of the currents in all the conductors is zero. Such a group of conductors is shown in Fig. 3.7. Conductors 1, 2, 3, ..., n carry the phasor currents $I_1, I_2, I_3, ..., I_n$. The distances of these conductors from a remote point P are indicated on the figure as $D_{1P}, D_{2P}, D_{3P}, ..., D_{nP}$. Let us determine ψ_{1P1}, the flux linkages of conductor 1 due to I_1 including internal flux linkages but excluding all the flux beyond the point P.

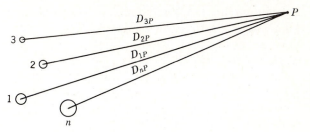

Figure 3.7 Cross-sectional view of a group of n conductors carrying currents whose sum is zero. Point P is remote from the conductors.

By Eqs. (3.21) and (3.27),

$$\psi_{1P1} = \left(\frac{I_1}{2} + 2I_1 \ln \frac{D_{1P}}{r_1}\right) 10^{-7} \tag{3.38}$$

$$\psi_{1P1} = 2 \times 10^{-7} I_1 \ln \frac{D_{1P}}{r_1'} \qquad \text{Wbt/m} \tag{3.39}$$

The flux linkages ψ_{1P2} with conductor 1 *due to I_2*, but excluding flux beyond point P, is equal to the flux produced by I_2 between the point P and conductor 1 (that is, within the limiting distances D_{2P} and D_{12} from conductor 2), and so

$$\psi_{1P2} = 2 \times 10^{-7} I_2 \ln \frac{D_{2P}}{D_{12}} \tag{3.40}$$

The flux linkages ψ_{1P} with conductor 1 *due to all the conductors* in the group, but excluding flux beyond point P, is

$$\psi_{1P} = 2 \times 10^{-7} \left(I_1 \ln \frac{D_{1P}}{r_1'} + I_2 \ln \frac{D_{2P}}{D_{12}} + I_3 \ln \frac{D_{3P}}{D_{13}} + \cdots + I_n \ln \frac{D_{nP}}{D_{1n}}\right) \tag{3.41}$$

which becomes, by expanding the logarithmic terms and regrouping,

$$\psi_{1P} = 2 \times 10^{-7} \left(I_1 \ln \frac{1}{r_1'} + I_2 \ln \frac{1}{D_{12}} + I_3 \ln \frac{1}{D_{13}} + \cdots + I_n \ln \frac{1}{D_{1n}}\right.$$

$$\left. + I_1 \ln D_{1P} + I_2 \ln D_{2P} + I_3 \ln D_{3P} + \cdots + I_n \ln D_{nP}\right) \tag{3.42}$$

Since the sum of all the currents in the group is zero,

$$I_1 + I_2 + I_3 + \cdots + I_n = 0$$

and, solving for I_n, we obtain

$$I_n = -(I_1 + I_2 + I_3 + \cdots + I_{n-1}) \tag{3.43}$$

Substituting Eq. (3.43) in the second term containing I_n in Eq. (3.42) and recombining some logarithmic terms, we have

$$\psi_{1P} = 2 \times 10^{-7}\left(I_1 \ln \frac{1}{r_1'} + I_2 \ln \frac{1}{D_{12}} + I_3 \ln \frac{1}{D_{13}} + \cdots + I_n \ln \frac{1}{D_{1n}}\right.$$

$$\left. + I_1 \ln \frac{D_{1P}}{D_{nP}} + I_2 \ln \frac{D_{2P}}{D_{nP}} + I_3 \ln \frac{D_{3P}}{D_{nP}} + \cdots + I_{n-1} \ln \frac{D_{(n-1)P}}{D_{nP}}\right) \quad (3.44)$$

Now letting the point P move infinitely far away so that the set of terms containing logarithms of ratios of distances from P becomes infinitesimal, since the ratios of the distances approach 1, we obtain

$$\psi_1 = 2 \times 10^{-7}\left(I_1 \ln \frac{1}{r_1'} + I_2 \ln \frac{1}{D_{12}} + I_3 \ln \frac{1}{D_{13}} + \cdots + I_n \ln \frac{1}{D_{1n}}\right) \quad \text{Wbt/m}$$

$$(3.45)$$

By letting point P move infinitely far away we have included all the flux linkages of conductor 1 in our derivation. Therefore, Eq. (3.45) expresses all the flux linkages of conductor 1 in a group of conductors, provided the sum of all the currents is zero. If the currents are alternating, they must be expressed as instantaneous currents to obtain instantaneous flux linkages or as complex rms values to obtain the rms value of flux linkages as a complex number.

3.9 INDUCTANCE OF COMPOSITE-CONDUCTOR LINES

Stranded conductors come under the general classification of *composite* conductors, which means conductors composed of two or more elements or strands electrically in parallel. We are now ready to study the inductance of a transmission line composed of composite conductors, but we shall limit ourselves to the case where all the strands are identical and share the current equally. The method can be expanded to apply to all types of conductors containing strands of different sizes and conductivities, but this will not be done here since the values of internal inductance of specific conductors are generally available from the various manufacturers and can be found in handbooks. The method to be developed indicates the approach to the more complicated problems of nonhomogeneous conductors and unequal division of current between strands. The method is applicable to the determination of inductance of lines consisting of circuits electrically in parallel since two conductors in parallel can be treated as strands of a single composite conductor.

Figure 3.8 shows a single-phase line composed of two conductors. In order to be more general, each conductor forming one side of the line is shown as an arbitrary arrangement of an indefinite number of conductors. The only restrictions are that the parallel filaments are cylindrical and share the current equally.

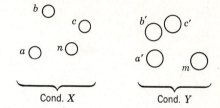

Figure 3.8 Single-phase line consisting of two composite conductors.

Conductor X is composed of n identical, parallel filaments, each of which carries the current I/n. Conductor Y, which is the return circuit for the current in conductor X, is composed of m identical, parallel filaments, each of which carries the current $-I/m$. Distances between the elements will be designated by the letter D with appropriate subscripts. Applying Eq. (3.45) to filament a of conductor X, we obtain for flux linkages of filament a

$$\psi_a = 2 \times 10^{-7} \frac{I}{n}\left(\ln \frac{1}{r_a'} + \ln \frac{1}{D_{ab}} + \ln \frac{1}{D_{ac}} + \cdots + \ln \frac{1}{D_{an}}\right)$$

$$- 2 \times 10^{-7} \frac{I}{m}\left(\ln \frac{1}{D_{aa'}} + \ln \frac{1}{D_{ab'}} + \ln \frac{1}{D_{ac'}} + \cdots + \ln \frac{1}{D_{am}}\right) \quad (3.46)$$

from which

$$\psi_a = 2 \times 10^{-7} I \ln \frac{\sqrt[m]{D_{aa'} D_{ab'} D_{ac'} \cdots D_{am}}}{\sqrt[n]{r_a' D_{ab} D_{ac} \cdots D_{an}}} \qquad \text{Wbt/m} \qquad (3.47)$$

Dividing Eq. (3.47) by the current I/n, we find that the inductance of filament a is

$$L_a = \frac{\psi_a}{I/n} = 2n \times 10^{-7} \ln \frac{\sqrt[m]{D_{aa'} D_{ab'} D_{ac'} \cdots D_{am}}}{\sqrt[n]{r_a' D_{ab} D_{ac} \cdots D_{an}}} \qquad \text{H/m} \qquad (3.48)$$

Similarly, the inductance of filament b is

$$L_b = \frac{\psi_b}{I/n} = 2n \times 10^{-7} \ln \frac{\sqrt[m]{D_{ba'} D_{bb'} D_{bc'} \cdots D_{bm}}}{\sqrt[n]{D_{ba} r_b' D_{bc} \cdots D_{bn}}} \qquad \text{H/m} \qquad (3.49)$$

The average inductance of the filaments of conductor X is

$$L_{av} = \frac{L_a + L_b + L_c + \cdots + L_n}{n} \qquad (3.50)$$

Conductor X is composed of n filaments electrically in parallel. If all the filaments had the same inductance, the inductance of the conductor would be $1/n$ times the inductance of one filament. Here all the filaments have different inductances, but the inductance of all of them in parallel is $1/n$ times the average inductance. Thus the inductance of conductor X is

$$L_X = \frac{L_{av}}{n} = \frac{L_a + L_b + L_c + \cdots + L_n}{n^2} \qquad (3.51)$$

Substituting the logarithmic expression for inductance of each filament in

Eq. (3.51) and combining terms, we obtain

$$L_X = 2 \times 10^{-7}$$

$$\times \ln \frac{\sqrt[mn]{(D_{aa'}D_{ab'}D_{ac'} \cdots D_{am})(D_{ba'}D_{bb'}D_{bc'} \cdots D_{bm}) \cdots (D_{na'}D_{nb'}D_{nc'} \cdots D_{nm})}}{\sqrt[n^2]{(D_{aa}D_{ab}D_{ac} \cdots D_{an})(D_{ba}D_{bb}D_{bc} \cdots D_{bn}) \cdots (D_{na}D_{nb}D_{nc} \cdots D_{nn})}}$$

$$\text{H/m} \quad (3.52)$$

where r_a', r_b', and r_n' have been replaced by D_{aa}, D_{bb}, and D_{nn}, respectively, to make the expression appear more symmetrical.

Note that the numerator of the argument of the logarithm in Eq. (3.52) is the mnth root of mn terms, which are the products of the distances from all the n filaments of conductor X to all the m filaments of conductor Y. For each filament in conductor X, there are m distances to filaments in conductor Y, and there are n filaments in conductor X. The product of m distances for each of n filaments results in mn terms. The mnth root of the product of the mn distances is called the *geometric mean distance* between conductor X and conductor Y. It is abbreviated D_m or GMD and is also called the *mutual* GMD between the two conductors.

The denominator of the argument of the logarithm in Eq. (3.52) is the n^2 root of n^2 terms. There are n filaments, and for each filament there are n terms consisting of r' for that filament times the distances from that filament to every other filament in conductor X. Thus we account for n^2 terms. Sometimes r_a' is called the distance from filament a to itself, especially when it is designated as D_{aa}. With this in mind the terms under the radical in the denominator may be described as the product of the distances from every filament in the conductor to itself and to every other filament. The n^2 root of these terms is called the *self* GMD of conductor X, and the r' of a separate filament is called the self GMD of the filament. Self GMD is also called *geometric mean radius*, or GMR. The correct mathematical expression is self GMD, but common practice has made GMR more prevalent. We shall use GMR in order to conform to this practice and identify it by D_s.

In terms of D_m and D_s, Eq. (3.52) becomes

$$L_X = 2 \times 10^{-7} \ln \frac{D_m}{D_s} \quad \text{H/m} \quad (3.53)$$

If we compare Eq. (3.53) with Eq. (3.34), the similarity between them is apparent. The equation for the inductance of one conductor of a composite-conductor line is obtained by substituting in Eq. (3.34) the GMD between conductors of the composite-conductor line for the distance between the solid conductors of the single-conductor line and by substituting the GMR of the composite conductor for the GMR (r') of the single conductor. Equation (3.53) gives the inductance of one conductor of a single-phase line. The conductor is composed of all the strands which are electrically in parallel. The inductance is the total number of flux linkages of the composite conductor per unit of line current. Equation (3.34) gives the inductance of one conductor of a single-phase line for the special case where the conductor is a solid round wire.

The inductance of conductor Y is determined in a similar manner, and the inductance of the line is

$$L = L_X + L_Y$$

Example 3.2 One circuit of a single-phase transmission line is composed of three solid 0.25-cm-radius wires. The return circuit is composed of two 0.5-cm-radius wires. The arrangement of conductors is shown in Fig. 3.9. Find the inductance due to the current in each side of the line and the inductance of the complete line in henrys per meter (and in millihenrys per mile).

SOLUTION Find the GMD between sides X and Y:

$$D_m = \sqrt[6]{D_{ad} D_{ae} D_{bd} D_{be} D_{cd} D_{ce}}$$

$$D_{ad} = D_{be} = 9 \text{ m}$$

$$D_{ae} = D_{bd} = D_{ce} = \sqrt{6^2 + 9^2} = \sqrt{177}$$

$$D_{cd} = \sqrt{9^2 + 12^2} = 15 \text{ m}$$

$$D_m = \sqrt[6]{9^2 \times 15 \times 177^{3/2}} = 10.743 \text{ m}$$

Then find the GMR for side X

$$D_s = \sqrt[9]{D_{aa} D_{ab} D_{ac} D_{ba} D_{bb} D_{bc} D_{ca} D_{cb} D_{cc}}$$
$$= \sqrt[9]{(0.25 \times 0.7788 \times 10^{-2})^3 \times 6^4 \times 12^2} = 0.481 \text{ m}$$

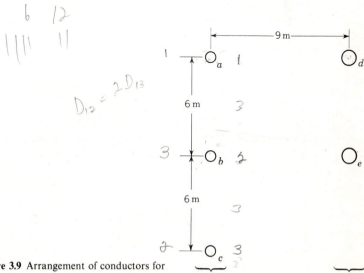

Figure 3.9 Arrangement of conductors for Example 3.2.

Side X Side Y

and for side Y

$$D_s = \sqrt[4]{(0.5 \times 0.7788 \times 10^{-2})^2 \times 6^2} = 0.153 \text{ m}$$

$$L_X = 2 \times 10^{-7} \ln \frac{10.743}{0.481} = 6.212 \times 10^{-7} \text{ H/m}$$

$$L_Y = 2 \times 10^{-7} \ln \frac{10.743}{0.153} = 8.503 \times 10^{-7} \text{ H/m}$$

$$L = L_X + L_Y = 14.715 \times 10^{-7} \text{ H/m}$$

$$(L = 14.715 \times 10^{-7} \times 1609 \times 10^3 = 2.37 \text{ mH/mi})$$

If a single-phase line consists of two stranded cables it is seldom necessary to calculate the GMD between strands of the two sides, for the GMD would be almost equal to the distance between centers of the cables. The calculation of mutual GMD is important only where the various strands (or conductors) electrically in parallel are separated from each other by distances more nearly approaching the distance between the two sides of the circuit. For instance, in Example 3.2 the conductors in parallel on one side of the line are separated by 6 m, and the distance between the two sides of the line is 9 m. Here the calculation of mutual GMD is important. For stranded conductors the distance between sides of a line composed of one conductor per side is usually so great that the mutual GMD can be taken as equal to the center-to-center distance with negligible error.

If the effect of the steel core of ACSR is neglected in calculating inductance, a high degree of accuracy results, provided the aluminum strands are in an even number of layers. The effect of the core is more apparent for an odd number of layers of aluminum strands, but the accuracy is good when the calculations are based on the aluminum strands alone.

3.10 THE USE OF TABLES

Tables listing values of GMR are generally available for standard conductors and provide other information for calculating inductive reactance as well as shunt capacitive reactance and resistance. Since industry in the United States continues to use units of inches, feet, and miles so do these tables. Therefore some of our examples will use feet and miles, but others will use meters and kilometers.

Inductive reactance rather than inductance is usually desired. The inductive reactance of one conductor of a single-phase two-conductor line is

$$X_L = 2\pi f L = 2\pi f \times 2 \times 10^{-7} \ln \frac{D_m}{D_s}$$

$$= 4\pi f \times 10^{-7} \ln \frac{D_m}{D_s} \qquad \Omega/\text{m} \qquad (3.54)$$

or

$$X_L = 2.022 \times 10^{-3} f \ln \frac{D_m}{D_s} \qquad \Omega/\text{mi} \qquad (3.55)$$

where D_m is the distance between conductors. Both D_m and D_s must be in the same units, usually either meters or feet. The GMR found in tables is an equivalent D_s which accounts for skin effect where it is appreciable enough to affect inductance. Of course skin effect is greater at higher frequencies for a conductor of a given diameter. Values of D_s listed in Table A.1 are for a frequency of 60 Hz.

Some tables give values of inductive reactance in addition to GMR. One method is to expand the logarithmic term of Eq. (3.55) as follows:

$$X_L = 2.022 \times 10^{-3} f \ln \frac{1}{D_s} + 2.022 \times 10^{-3} f \ln D_m \qquad \Omega/\text{mi} \qquad (3.56)$$

If both D_s and D_m are in feet, the first term in Eq. (3.56) is the inductive reactance of one conductor of a two-conductor line having a distance of 1 ft between conductors, as may be seen by comparing Eq. (3.56) with Eq. (3.55). Therefore, the first term of Eq. (3.56) is called the *inductive reactance at 1 ft spacing* X_a. It depends upon the GMR of the conductor and the frequency. The second term of Eq. (3.56) is called the *inductive reactance spacing factor* X_d. This second term is independent of the type of conductor and depends on frequency and spacing only. The spacing factor is equal to zero when D_m is 1 ft. If D_m is less than 1 ft, the spacing factor is negative. The procedure for computing inductive reactance is to look up the inductive reactance at 1-ft spacing for the conductor under consideration and to add to this value the value of the inductive reactance spacing factor, both at the desired line frequency. Table A.1 includes values of inductive reactance at 1-ft spacing, and Table A.2 lists values of the inductive reactance spacing factor.

Example 3.3 Find the inductive reactance per mile of a single-phase line operating at 60 Hz. The conductor is *Partridge*, and spacing is 20 ft between centers.

SOLUTION For this conductor, Table A.1 lists $D_s = 0.0217$ ft. From Eq. (3.55), for one conductor,

$$X_L = 2.022 \times 10^{-3} \times 60 \ln \frac{20}{0.0217}$$

$$= 0.828 \ \Omega/\text{mi}$$

The above calculation is used if only D_s is known. Table A.1, however, lists inductive reactance at 1-ft spacing $X_a = 0.465 \ \Omega/\text{mi}$. From Table A.2 the inductive reactance spacing factor is $X_d = 0.3635 \ \Omega/\text{mi}$, and so the inductive reactance of one conductor is

$$0.465 + 0.3635 = 0.8285 \ \Omega/\text{mi}$$

Since the conductors composing the two sides of the line are identical, the inductive reactance of the line is

$$X_L = 2 \times 0.8285 = 1.657 \ \Omega/\text{mi}$$

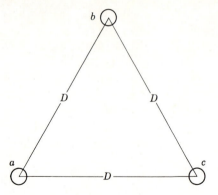

Figure 3.10 Cross-sectional view of the equilaterally spaced conductors of a three-phase line.

3.11 INDUCTANCE OF THREE-PHASE LINES WITH EQUILATERAL SPACING

So far in our discussion we have considered only single-phase lines. The equations we have developed are quite easily adapted, however, to the calculation of the inductance of three-phase lines. Figure 3.10 shows the conductors of a three-phase line spaced at the corners of an equilateral triangle. If we assume that there is no neutral wire, or if we assume balanced three-phase phasor currents, $I_a + I_b + I_c = 0$. Equation (3.45) determines the flux linkages of conductor a:

$$\psi_a = 2 \times 10^{-7}\left(I_a \ln \frac{1}{D_s} + I_b \ln \frac{1}{D} + I_c \ln \frac{1}{D}\right) \qquad \text{Wbt/m} \qquad (3.57)$$

Since $I_a = -(I_b + I_c)$, Eq. (3.57) becomes

$$\psi_a = 2 \times 10^{-7}\left(I_a \ln \frac{1}{D_s} - I_a \ln \frac{1}{D}\right) = 2 \times 10^{-7} I_a \ln \frac{D}{D_s} \qquad \text{Wbt/m} \quad (3.58)$$

and

$$L_a = 2 \times 10^{-7} \ln \frac{D}{D_s} \qquad \text{H/m} \qquad (3.59)$$

Equation (3.59) is the same in form as Eq. (3.34) for a single-phase line except that D_s replaces r'. Because of symmetry, the inductances of conductors b and c are the same as the inductance of conductor a. Since each phase consists of only one conductor, Eq. (3.59) gives the inductance per phase of the three-phase line.

3.12 INDUCTANCE OF THREE-PHASE LINES WITH UNSYMMETRICAL SPACING

When the conductors of a three-phase line are not spaced equilaterally, the problem of finding the inductance becomes more difficult. Then the flux linkages and inductance of each phase are not the same. A different inductance in each

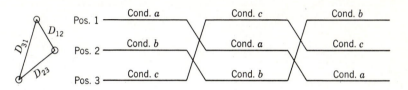

Figure 3.11 Transposition cycle.

phase results in an unbalanced circuit. Balance of the three phases can be restored by exchanging the positions of the conductors at regular intervals along the line so that each conductor occupies the original position of every other conductor over an equal distance. Such an exchange of conductor positions is called *transposition*. A complete transposition cycle is shown in Fig. 3.11. The phase conductors are designated *a*, *b*, and *c*, and the positions occupied are numbered 1, 2, and 3. Transposition results in each conductor having the same average inductance over the whole cycle.

Modern power lines are usually not transposed at regular intervals, although an interchange in the positions of the conductors may be made at switching stations in order to balance the inductance of the phases more closely. Fortunately, the dissymmetry between the phases of an untransposed line is small and is neglected in most calculations of inductance. If the dissymmetry is neglected, the inductance of the untransposed line is taken as equal to the average value of the inductive reactance of one phase of the same line correctly transposed. The derivations to follow are for transposed lines.

To find the average inductance of one conductor of a transposed line, the flux linkages of a conductor are found for each position it occupies in the transposition cycle, and the average flux linkages are determined. Let us apply Eq. (3.45) to conductor *a* of Fig. 3.11 to find the phasor expression for the flux linkages of *a* in position 1 when *b* is in position 2 and *c* is in position 3, as follows:

$$\psi_{a1} = 2 \times 10^{-7}\left(I_a \ln \frac{1}{D_s} + I_b \ln \frac{1}{D_{12}} + I_c \ln \frac{1}{D_{31}}\right) \qquad \text{Wbt/m} \qquad (3.60)$$

With *a* in position 2, *b* in position 3, and *c* in position 1,

$$\psi_{a2} = 2 \times 10^{-7}\left(I_a \ln \frac{1}{D_s} + I_b \ln \frac{1}{D_{23}} + I_c \ln \frac{1}{D_{12}}\right) \qquad \text{Wbt/m} \qquad (3.61)$$

and, with *a* in position 3, *b* in position 1, and *c* in position 2,

$$\psi_{a3} = 2 \times 10^{-7}\left(I_a \ln \frac{1}{D_s} + I_b \ln \frac{1}{D_{31}} + I_c \ln \frac{1}{D_{23}}\right) \qquad \text{Wbt/m} \qquad (3.62)$$

The average value of the flux linkages of a is

$$\psi_a = \frac{\psi_{a1} + \psi_{a2} + \psi_{a3}}{3}$$

$$= \frac{2 \times 10^{-7}}{3} \left(3I_a \ln \frac{1}{D_s} + I_b \ln \frac{1}{D_{12}D_{23}D_{31}} + I_c \ln \frac{1}{D_{12}D_{23}D_{31}} \right) \quad (3.63)$$

With the restriction that $I_a = -(I_b + I_c)$,

$$\psi_a = \frac{2 \times 10^{-7}}{3} \left(3I_a \ln \frac{1}{D_s} - I_a \ln \frac{1}{D_{12}D_{23}D_{31}} \right)$$

$$= 2 \times 10^{-7} I_a \ln \frac{\sqrt[3]{D_{12}D_{23}D_{31}}}{D_s} \quad \text{Wbt/m} \quad (3.64)$$

and the *average* inductance per phase is

$$L_a = 2 \times 10^{-7} \ln \frac{D_{eq}}{D_s} \quad \text{H/m} \quad (3.65)$$

where

$$D_{eq} = \sqrt[3]{D_{12}D_{23}D_{31}} \quad (3.66)$$

and D_s is the GMR of the conductor. D_{eq}, the geometric mean of the three distances of the unsymmetrical line, is the equivalent equilateral spacing, as may be seen by a comparison of Eq. (3.65) with Eq. (3.59). We should note the similarity between all the equations for the inductance of a conductor. If the inductance is in henrys per meter, the factor 2×10^{-7} appears in all the equations, and the denominator of the logarithmic term is always the GMR of the conductor. The numerator is the distance between wires of a two-wire line, the mutual GMD between sides of a composite-conductor single-phase line, the distance between conductors of an equilaterally spaced line, or the equivalent equilateral spacing of an unsymmetrical line.

Example 3.4 A single-circuit three-phase line operated at 60 Hz is arranged as shown in Fig. 3.12. The conductors are ACSR *Drake*. Find the inductive reactance per mile per phase.

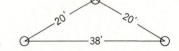

Figure 3.12 Arrangement of conductors for Example 3.4

SOLUTION From Table A.1,

$$D_s = 0.0373 \text{ ft} \qquad D_{eq} = \sqrt[3]{20 \times 20 \times 38} = 24.8 \text{ ft}$$

$$L = 2 \times 10^{-7} \ln \frac{24.8}{0.0373} = 13.00 \times 10^{-7} \text{ H/m}$$

$$X_L = 2\pi 60 \times 1609 \times 13.00 \times 10^{-7} = 0.788 \text{ } \Omega/\text{mi per phase}$$

Equation (3..55) may be used also, or, from Tables A.1 and A.2,

$$X_a = 0.399$$

and for 24.8 ft,

$$X_d = 0.389$$

$$X_L = 0.399 + 0.389 = 0.788 \text{ } \Omega/\text{mi per phase}$$

3.13 BUNDLED CONDUCTORS

At extra-high voltages (EHV), that is, voltages above 230 kV, corona with its resultant power loss and particularly its interference with communications is excessive if the circuit has only one conductor per phase. The high-voltage gradient at the conductor in the EHV range is reduced considerably by having two or more conductors per phase in close proximity compared with the spacing between phases. Such a line is said to be composed of *bundled* conductors. The bundle consists of two, three, or four conductors. The three-conductor bundle usually has the conductors at the vertices of an equilateral triangle, and the four-conductor bundle usually has its conductors at the corners of a square. Figure 3.13 shows these arrangements. The current will not divide exactly between the conductors of the bundle unless there is a transposition of the conductors within the bundle, but the difference is of no practical importance, and the GMD method is accurate for calculations.

Reduced reactance is the other equally important advantage of bundling. Increasing the number of conductors in a bundle reduces the effects of corona and reduces the reactance. The reduction of reactance results from the increased GMR of the bundle. The calculation of GMR is, of course, exactly the same as that of a stranded conductor. Each conductor of a two-conductor bundle, for instance, is treated as one strand of a two-strand conductor. If we let D_s^b indicate the GMR of a bundled conductor and D_s the GMR of the individual conductors composing the bundle, we find, referring to Fig. 3.13,

Figure 3.13 Bundle arrangements.

For a two-strand bundle

$$D_s^b = \sqrt[4]{(D_s \times d)^2} = \sqrt{D_s \times d} \qquad (3.67)$$

For a three-strand bundle

$$D_s^b = \sqrt[9]{(D_s \times d \times d)^3} = \sqrt[3]{D_s \times d^2} \qquad (3.68)$$

For a four-strand bundle

$$D_s^b = \sqrt[16]{(D_s \times d \times d \times d \times 2^{1/2})^4} = 1.09\sqrt[4]{D_s \times d^3} \qquad (3.69)$$

In computing inductance using Eq. (3.65), D_s^b of the bundle replaces D_s of a single conductor. To compute D_{eq}, the distance from the center of one bundle to the center of another bundle is sufficiently accurate for D_{ab}, D_{bc}, and D_{ca}. Obtaining the actual GMD between conductors of one bundle and those of another would be almost indistinguishable from the center-to-center distances for the usual spacing.

Example 3.5 Each conductor of the bundled-conductor line shown in Fig. 3.14 is ACSR, 1,272,000-cmil *Pheasant*. Find the inductive reactance in ohms per km (and per mile) per phase for $d = 45$ cm. Also find the per-unit series reactance of the line if its length is 160 km and base is 100 MVA, 345 kV.

SOLUTION From Table A.1 $D_s = 0.0466$ ft, and we multiply feet by 0.3048 to convert to meters.

$$D_s^b = \sqrt{0.0466 \times 0.3048 \times 0.45} = 0.080 \text{ m}$$

$$D_{eq} = \sqrt[3]{8 \times 8 \times 16} = 10.08 \text{ m}$$

$$X_L = 2\pi 60 \times 2 \times 10^{-7} \times 10^3 \ln \frac{10.08}{0.08}$$

$$= 0.365 \ \Omega/\text{km per phase}$$

$$(0.365 \times 1.609 = 0.587 \ \Omega/\text{mi per phase})$$

$$\text{Base } Z = \frac{(345)^2}{100} = 1190 \ \Omega$$

$$X = \frac{0.365 \times 160}{1190} = 0.049 \text{ per unit}$$

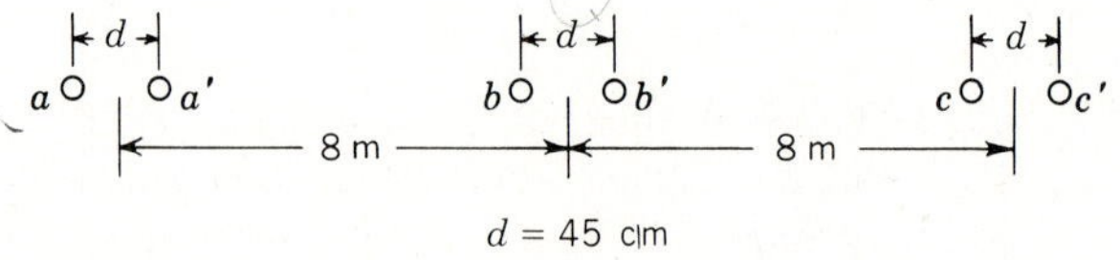

Figure 3.14 Spacing of conductors of a bundled-conductor line.

$$a \quad a' \qquad b \quad b' \qquad c \quad c'$$

$$|\leftarrow d \rightarrow| \qquad |\leftarrow d \rightarrow| \qquad |\leftarrow d \rightarrow|$$

$$|\leftarrow \quad 8 \text{ m} \quad \rightarrow| \leftarrow \quad 8 \text{ m} \quad \rightarrow|$$

$$d = 45 \text{ cm}$$

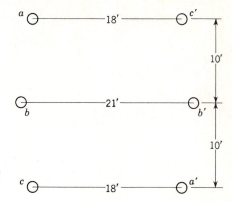

Figure 3.15 Typical arrangement of conductors of a parallel-circuit three-phase line.

3.14 PARALLEL-CIRCUIT THREE-PHASE LINES

Two three-phase circuits that are identical in construction and electrically in parallel have the same inductive reactance. The inductive reactance of the single equivalent circuit, however, is half that of each of the individual circuits considered alone only if they are so widely separated that there is negligible mutual inductance between them. If the two circuits are on the same tower, the method of GMD can be used to find the inductance per phase by considering all the conductors of any particular phase to be strands of one composite conductor.

Figure 3.15 shows a typical arrangement of a parallel-circuit three-phase line. Although the line will probably not be transposed, we obtain a practical value for inductance and the calculations are simplified if transposition is assumed. Conductors a and a' are in parallel to compose phase a. Phases b and c are similar. We assume that a and a' take the positions of b and b' and then of c and c' as those conductors are rotated similarly in the transposition cycle.

To calculate D_{eq} the GMD method requires that we use D_{ab}^p, D_{bc}^p, and D_{ca}^p, where the superscript indicates that these quantities are themselves GMD values and where D_{ab}^p means the GMD between the conductors of phase a and those of phase b.

The D_s of Eq. (3.65) is replaced by D_s^p, which is the geometric mean of the GMR values of the two conductors occupying first the positions of a and a', then the positions of b and b', and finally the positions of c and c'. Following each step of Example 3.6 is possibly the best means of understanding the procedure.

Example 3.6 A three-phase double-circuit line is composed of 300,000-cmil 26/7 ACSR *Ostrich* conductors arranged as shown in Fig. 3.15. Find the 60-Hz inductive reactance in ohms per mile per phase.

SOLUTION From Table A.1 for *Ostrich*

$$D_s = 0.0229 \text{ ft}$$

Distance a to b: Original position $= \sqrt{10^2 + 1.5^2} = 10.1$ ft

Distance a to b':   Original position $= \sqrt{10^2 + 19.5^2} = 21.9$ ft

The GMDs between phases are

$$D_{ab}^p = D_{bc}^p = \sqrt[4]{(10.1 \times 21.9)^2} = 14.88 \text{ ft}$$

$$D_{ca}^p = \sqrt[4]{(20 \times 18)^2} = 18.97 \text{ ft}$$

$$D_{eq} = \sqrt[3]{14.88 \times 14.88 \times 18.97} = 16.1 \text{ ft}$$

The GMR for the parallel-circuit line is found after first obtaining the GMR values for the three positions. The actual distance from a to a' is $\sqrt{20^2 + 18^2} = 26.9$ ft. Then GMR of each phase is

In position a-a': $\sqrt{26.9 \times 0.0229} = 0.785$ ft

In position b-b': $\sqrt{21 \times 0.0229} = 0.693$ ft

In position c-c': $\sqrt{26.9 \times 0.0229} = 0.785$ ft

Therefore

$$D_s^p = \sqrt[3]{0.785 \times 0.693 \times 0.785} = 0.753 \text{ ft}$$

$$L = 2 \times 10^{-7} \ln \frac{16.1}{0.753} = 6.13 \times 10^{-7} \text{ H/m per phase}$$

$$X_L = 2\pi 60 \times 1609 \times 6.13 \times 10^{-7} = 0.372 \ \Omega/\text{mi per phase}$$

3.15 SUMMARY OF INDUCTANCE CALCULATIONS FOR THREE-PHASE LINES

Although computer programs are usually available or written rather easily for calculating inductance of all kinds of lines, some understanding of the development of the equations used is rewarding from the standpoint of appreciating the effect of variables in designing a line. However, tables like A.1 and A.2 make the calculations quite simple except for parallel-circuit lines. Table A.1 also lists resistance.

The important equation for inductance per phase of single-circuit three-phase lines is given here for convenience

$$L = 2 \times 10^{-7} \ln \frac{D_{eq}}{D_s} \qquad \text{H/m per phase} \qquad (3.70)$$

Inductive reactance in ohms per kilometer at 60 Hz is found by multiplying inductance in henrys per meter by $2\pi60 \times 1000$:

$$X_L = 0.0754 \times \ln \frac{D_{eq}}{D_s} \qquad \Omega/\text{km per phase} \tag{3.71}$$

or

$$X_L = 0.1213 \ln \frac{D_{eq}}{D_s} \qquad \Omega/\text{mi per phase} \tag{3.72}$$

Both D_{eq} and D_s must be in the same units, usually feet. If the line has one conductor per phase, D_s is found directly from tables. For bundled conductors D_s^b, as defined in Sec. 3.13, is substituted for D_s. For both single- and bundled-conductor lines

$$D_{eq} = \sqrt[3]{D_{ab} D_{bc} D_{ca}} \tag{3.73}$$

For bundled-conductor lines D_{ab}, D_{bc}, and D_{ca} are distances between the centers of the bundles of phases a, b, and c.

For lines with one conductor per phase it is convenient to determine X_L by adding X_a for the conductor as found in tables like A.1 to X_d as found in Table A.2 corresponding to D_{eq}.

Inductance and inductive reactance of parallel-circuit lines are calculated by following the procedure of Example 3.6.

PROBLEMS

3.1 The all-aluminum conductor identified by the code word *Bluebell* is composed of 37 strands of diameter 0.1672 in. Tables of characteristics of all-aluminum conductors list an area of 1,033,500 cmil for this conductor. Are these values consistent with each other? Find the area in square millimeters.

3.2 Determine the dc resistance in ohms per km of *Bluebell* at 20°C by Eq. (3.2) and the information in Prob. 3.1, and check the result against the value listed in tables of 0.01678 Ω per 1000 ft. Compute the dc resistance in ohms per kilometer at 50°C and compare the result with the ac 60-Hz resistance of 0.1024 Ω/mi listed in tables for this conductor at 50°C. Explain any difference in values.

3.3 An all-aluminum conductor is composed of 37 strands each having a diameter of 0.333 cm. Compute the dc resistance in ohms per kilometer at 75°C.

3.4 A single-phase 60-Hz power line is supported on a horizontal crossarm. Spacing between conductors is 2.5 m. A telephone line is supported on a horizontal crossarm 1.8 m directly below the power line with a spacing of 1.0 m between the centers of its conductors. Find the mutual inductance between the power and telephone circuits and the 60-Hz voltage per kilometer induced in the telephone line if the current in the power line is 150 A.

3.5 If the power and telephone lines described in Prob. 3.4 are in the same horizontal plane and the distance between the nearest conductors of the two lines is 18 m, find the mutual inductance between the circuits and the voltage per mile induced in the telephone line for 150 A in the power line.

3.6 The conductor of a single-phase 60-Hz line is a solid round aluminum wire having a diameter of 0.412 cm. The conductor spacing is 3 m. Determine the inductance of the line in millihenrys per mile. How much of the inductance is due to internal flux linkages? Assume skin effect is negligible.

3.7 Find the GMR of a three-strand conductor in terms of r of an individual strand.

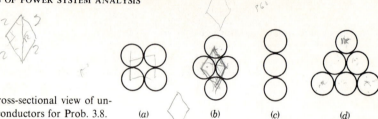

Figure 3.16 Cross-sectional view of unconventional conductors for Prob. 3.8.   (a)   (b)   (c)   (d)

3.8 Find the GMR of each of the unconventional conductors shown in Fig. 3.16 in terms of the radius r of an individual strand.

3.9 The distance between conductors of a single-phase line is 10 ft. Each conductor is composed of seven equal strands. The diameter of each strand is 0.1 in. Show that D_s for the conductor is 2.177 times the radius of each strand. Find the inductance of the line in millihenrys per mile.

3.10 Find the inductive reactance of ACSR *Rail* in ohms per kilometer at 1-m spacing.

3.11 Which conductor listed in Table A.1 has an inductive reactance at 7-ft spacing of 0.651 Ω/mi?

3.12 A three-phase line is designed with equilateral spacing of 16 ft. It is decided to build the line with horizontal spacing ($D_{13} = 2D_{12} = 2D_{23}$). The conductors are transposed. What should be the spacing between adjacent conductors in order to obtain the same inductance as in the original design?

3.13 A three-phase 60-Hz transmission line has its conductors arranged in a triangular formation so that two of the distances between conductors are 25 ft and the third distance is 42 ft. The conductors are ACSR *Osprey*. Determine the inductance and inductive reactance per phase per mile.

3.14 A three-phase 60-Hz line has flat horizontal spacing. The conductors have a GMR of 0.0133 m with 10 m between adjacent conductors. Determine the inductive reactance per phase in ohms per kilometer. What is the name of this conductor?

3.15 For short transmission lines if resistance is neglected the maximum power which can be transmitted per phase is equal to

$$\frac{|V_S| \cdot |V_R|}{|X|}$$

where V_S and V_R are the line-to-neutral voltages at the sending and receiving ends of the line and X is the inductive reactance of the line. This relationship will become apparent in the study of Chap. 5. If the magnitudes of V_S and V_R are held constant and if the cost of a conductor is proportional to its cross-sectional area, find the conductor in Table A.1 which has the maximum power-handling capacity per cost of conductor.

3.16 A three-phase underground distribution line is operated at 23 kV. The three conductors are insulated with 0.5 cm solid black polyethylene insulation laid flat and side by side directly in a dirt trench. The conductor is circular in cross section and has 33 strands of aluminum. The diameter of the conductor is 1.46 cm. The manufacturer gives the GMR as 0.561 cm and the cross section of the conductor as 1.267 cm². The thermal rating of the line buried in normal soil whose maximum temperature is 30°C is 350 A. Find the dc and ac resistance at 50°C and the inductive reactance in ohms per kilometer. To decide whether to consider skin effect in calculating resistance determine the percent skin effect at 50°C in the ACSR conductor of size nearest that of the underground conductor. Note that the series impedance of the distribution line is dominated by R rather than X_L because of the very low inductance due to the close spacing of the conductors.

3.17 The single-phase power line of Prob. 3.4 is replaced by a three-phase line on a horizontal crossarm in the same position as that of the original single-phase line. Spacing of the conductors of the power line is $D_{13} = 2D_{12} = 2D_{23}$, and equivalent equilateral spacing is 3 m. The telephone line remains in the position described in Prob. 3.4. If the current in the power line is 150 A, find the voltage per kilometer induced in the telephone line. Discuss the phase relation of the induced voltage with respect to the power-line current.

3.18 A 60-Hz three-phase line composed of one ACSR *Bluejay* conductor per phase has flat horizontal spacing of 11 m between adjacent conductors. Compare the inductive reactance in ohms per kilometer per phase of this line with that of a line using a two-conductor bundle of ACSR 26/7 conductors having the same total cross-cectional area of aluminum as the single-conductor line and 11 m spacing measured from the center of the bundles. The spacing between conductors in the bundle is 40 cm.

3.19 Calculate the inductive reactance in ohms per kilometer of a bundled 60-Hz three-phase line having three ACSR *Rail* conductors per bundle with 45 cm between conductors of the bundle. The spacing between bundle centers is 9, 9, and 18 m.

3.20 Six conductors of ACSR *Drake* constitute a 60-Hz double-circuit three-phase line arranged as shown in Fig. 3.15. The vertical spacing, however, is 14 ft; the longer horizontal distance is 32 ft; and the shorter horizontal distances are 25 ft. Find the inductance per phase per mile and the inductive reactance in ohms per mile.

CHAPTER
FOUR

CAPACITANCE OF TRANSMISSION LINES

As we discussed briefly at the beginning of Chap. 3, the shunt admittance of a transmission line consists of conductance and capacitive reactance. We have also mentioned that conductance is usually neglected because its contribution to shunt admittance is very small. For this reason this chapter has been given the title of capacitance rather than shunt admittance.

Another reason for neglecting conductance is that there is no good way of taking it into account because it is quite variable. Leakage at insulators, the principal source of conductance, changes appreciably with atmospheric conditions and with the conducting properties of dirt that collects on the insulators. Corona, which results in leakage between lines, is also quite variable with atmospheric conditions. It is fortunate that the effect of conductance is such a negligible component of shunt admittance.

Capacitance of a transmission line is the result of the potential difference between the conductors; it causes them to be charged in the same manner as the plates of a capacitor when there is a potential difference between them. The capacitance between conductors is the charge per unit of potential difference. Capacitance between parallel conductors is a constant depending on the size and spacing of the conductors. For power lines less than about 80 km (50 mi) long, the effect of capacitance is slight and is usually neglected. For longer lines of higher voltage, capacitance becomes increasingly important.

An alternating voltage impressed on a transmission line causes the charge on the conductors at any point to increase and decrease with the increase and decrease of the instantaneous value of the voltage between conductors at the point. The flow of charge is current, and the current caused by the alternate charging and discharging of a line due to an alternating voltage is called the *charging current* of the line. Charging current flows in a transmission line even when it is open-circuited. It affects the voltage drop along the line as well as the efficiency and power factor of the line and the stability of the system of which the line is a part.

4.1 ELECTRIC FIELD OF A
LONG STRAIGHT CONDUCTOR

Just as the magnetic field is important in considering inductance, so the electric field is important in studying capacitance. In the preceding chapter we discussed both the magnetic and electric fields of a two-wire line. Lines of electric flux originate on the positive charges of one conductor and terminate on the negative charges of the other conductor. The total electric flux emanating from a conductor is numerically equal to the number of coulombs of charge on the conductor. Electric flux density is the electric flux per square meter and is measured in coulombs per square meter.

If a long straight cylindrical conductor lies in a uniform medium such as air, has a uniform charge throughout its length, and is isolated from other charges so that the charge is uniformly distributed around its periphery, the flux is radial. All points equidistant from such a conductor are points of equipotential and have the same electric flux density. Figure 4.1 shows such an isolated conductor carrying a uniformly distributed charge. The electric flux density at x meters from the conductor can be computed by imagining a cylindrical surface concentric with the conductor and x meters in radius. Since all parts of the surface are equidistant from the conductor, which has a uniformly distributed charge, the cylindrical surface is a surface of equipotential and the electric flux density on

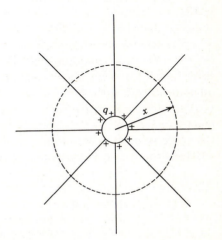

Figure 4.1 Lines of electric flux originating on the positive charges uniformly distributed over the surface of an isolated cylindrical conductor.

the surface is equal to the flux leaving the conductor per meter of length divided by the area of the surface in an axial length of 1 m. The electric flux density is

$$D = \frac{q}{2\pi x} \quad \text{C/m}^2 \tag{4.1}$$

where q is the charge on the conductor in coulombs per meter of length and x is the distance in meters from the conductor to the point where the electric flux density is computed. The electric field intensity, or the negative of the potential gradient, is equal to the electric flux density divided by the permittivity† of the medium. Therefore, the electric field intensity is

$$\mathscr{E} = \frac{q}{2\pi x k} \quad \text{V/m} \tag{4.2}$$

4.2 THE POTENTIAL DIFFERENCE BETWEEN TWO POINTS DUE TO A CHARGE

The potential difference between two points in volts is numerically equal to the work in joules per coulomb necessary to move a coulomb of charge between the two points. Electric field intensity is a measure of the force on a charge in the field. Electric field intensity in volts per meter is equal to the force in newtons per coulomb on a coulomb of charge at the point considered. Between two points the line integral of the force in newtons acting on a coulomb of positive charge is the work done in moving the charge from the point of lower potential to the point of higher potential and is numerically equal to the potential difference between the two points.

Consider a long straight wire carrying a positive charge of q C/m, as shown in Fig. 4.2. Points P_1 and P_2 are located at distances D_1 and D_2 meters from the center of the wire. The positive charge on the wire will exert a repelling force on a positive charge placed in the field. For this reason and because D_2 in this case is greater than D_1, work must be done on a positive charge to move it from P_2 to P_1, and P_1 is at a higher potential than P_2. The difference in potential is the amount of work done per coulomb of charge moved. On the other hand, if the coulomb moves from P_1 to P_2, it expends energy, and the amount of work, or energy, in newton-meters is the voltage *drop* from P_1 to P_2. The potential difference is independent of the path followed. The simplest way to compute the voltage drop between the two points is to compute the voltage between the

---

† In SI units the permittivity of free space k_0 is 8.85×10^{-12} F/m. Relative permittivity k_r is the ratio of the actual permittivity k of a material to the permittivity of free space. Thus, $k_r = k/k_0$. For dry air k_r is 1.00054 and is assumed equal to 1.0 in calculations for overhead lines.

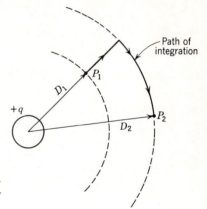

Figure 4.2 Path of integration between two points external to a cylindrical conductor having a uniformly distributed positive charge.

equipotential surfaces passing through P_1 and P_2 by integrating the field intensity over a radial path between the equipotential surfaces. Thus the instantaneous voltage drop between P_1 and P_2 is

$$v_{12} = \int_{D_1}^{D_2} \mathscr{E} \, dx = \int_{D_1}^{D_2} \frac{q}{2\pi kx} \, dx = \frac{q}{2\pi k} \ln \frac{D_2}{D_1} \quad \text{V} \tag{4.3}$$

where q is the instantaneous charge on the wire in coulombs per meter of length. Note that the voltage drop between two points, as given by Eq. (4.3), may be positive or negative depending on whether the charge causing the potential difference is positive or negative and on whether the voltage drop is computed from a point near the conductor to a point farther away, or vice versa. The sign of q may be either positive or negative, and the logarithmic term is either positive or negative depending on whether D_2 is greater or less than D_1.

4.3 CAPACITANCE OF A TWO-WIRE LINE

Capacitance between the two conductors of a two-wire line was defined as the charge on the conductors per unit of potential difference between them. In the form of an equation, capacitance per unit length of the line is

$$C = \frac{q}{v} \quad \text{F/m} \tag{4.4}$$

where q is the charge on the line in coulombs per meter and v is the potential difference between the conductors in volts. Hereafter, for convenience, we shall refer to capacitance per unit length as capacitance and indicate the correct dimensions for equations derived. The capacitance between two conductors can be found by substituting in Eq. (4.4) the expression for v in terms of q from

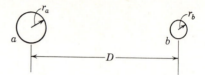

Figure 4.3 Cross section of a parallel-wire line.

Eq. (4.3). The voltage v_{ab} between the two conductors of the two-wire line shown in Fig. 4.3 can be found by determining the potential difference between the two conductors of the line, first by computing the voltage drop due to the charge q_a on conductor a and then by computing the voltage drop due to the charge q_b on conductor b. By the principle of superposition the voltage drop from conductor a to conductor b due to the charges on both conductors is the sum of the voltage drops caused by each charge alone.

Consider the charge q_a on conductor a, and assume that conductor b is uncharged and merely an equipotential surface in the electric field created by the charge on a. The equipotential surface of conductor b and the equipotential surfaces due to the charge on a are shown in Fig. 4.4. The distortion of the equipotential surfaces near conductor b is caused by the fact that conductor b is also an equipotential surface. Equation (4.3) was derived by assuming all the equipotential surfaces due to a uniform charge on a round conductor to be cylindrical and concentric with the conductor. Such is actually true for the case under discussion except in the region near conductor b. The potential of conductor b is that of the equipotential surface intersecting b. Therefore, in determining v_{ab} a path may be followed from conductor a through a region of undistorted equipotential surfaces to the equipotential surface intersecting conductor b. Then, moving along the equipotential surface to b gives no further change in voltage. This path of integration is indicated in Fig. 4.4 together with the direct path. Of course, the potential difference is the same regardless of the path over which the integration of the field intensity is taken. By following the path through the undistorted region, we see that the distances corresponding to D_2 and D_1 of Eq. (4.3) are D and r_a, respectively, in determining v_{ab} due to q_a. Similarly, in determining v_{ab} due to q_b, the distances corresponding to D_2 and D_1 of Eq. (4.3) are r_b and D, respectively. Converting to phasor notation (q_a and q_b become complex numbers), we obtain

$$V_{ab} = \underbrace{\frac{q_a}{2\pi k} \ln \frac{D}{r_a}}_{\text{due to } q_a} + \underbrace{\frac{q_b}{2\pi k} \ln \frac{r_b}{D}}_{\text{due to } q_b} \quad \text{V} \tag{4.5}$$

and, since $q_a = -q_b$ for a two-wire line,

$$V_{ab} = \frac{q_a}{2\pi k}\left(\ln \frac{D}{r_a} - \ln \frac{r_b}{D}\right) \quad \text{V} \tag{4.6}$$

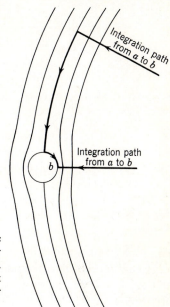

Figure 4.4 Equipotential surfaces of a portion of the electric field caused by a charged conductor a (not shown). Conductor b causes the equipotential surfaces to become distorted. Arrows indicate optional paths of integration between a point on the equipotential surface of conductor b and the conductor a, whose charge q_a creates the equipotential surfaces shown.

or, by combining the logarithmic terms,

$$V_{ab} = \frac{q_a}{2\pi k} \ln \frac{D^2}{r_a r_b} \quad \text{V} \tag{4.7}$$

The capacitance between conductors is

$$C_{ab} = \frac{q_a}{V_{ab}} = \frac{2\pi k}{\ln (D^2 / r_a r_b)} \quad \text{F/m} \tag{4.8}$$

If $r_a = r_b = r$,

$$C_{ab} = \frac{\pi k}{\ln (D/r)} \quad \text{F/m} \tag{4.9}$$

Equation (4.9) gives the capacitance between the conductors of a two-wire line. Sometimes it is desirable to know the capacitance between one of the conductors and a neutral point between them. For instance, if the line is supplied by a transformer having a grounded center tap, the potential difference between each conductor and the ground is half the potential difference between the two conductors and the *capacitance to ground*, or *capacitance to neutral*, is the charge on a conductor per unit of potential difference between the conductor and ground. Thus, the capacitance to neutral for the two-wire line is *twice the line-to-line capacitance* (capacitance between conductors). If the line-to-line capacitance is considered to be composed of two equal capacitances in series, the voltage across the line divides equally between them and their junction is at the ground potential. Thus, the capacitance to neutral is that of one of the two equal series

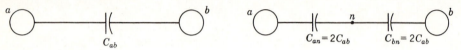

(*a*) Representation of line-to-line capacitance (*b*) Representation of line-to-neutral capacitance

Figure 4.5 Relationship between the concepts of line-to-line capacitance and line-to-neutral capacitance.

capacitances, or twice the line-to-line capacitance. Therefore,

$$C_n = C_{an} = C_{bn} = \frac{2\pi k}{\ln (D/r)} \qquad \text{F/m to neutral} \qquad (4.10)$$

The concept of capacitance to neutral is illustrated in Fig. 4.5.

Equation (4.10) corresponds to Eq. (3.34) for inductance. One difference between the equations for capacitance and inductance should be noted carefully. The radius in the equation for capacitance is the actual outside radius of the conductor and not the GMR of the conductor, as in the inductance formula.

Equation (4.3), from which Eqs. (4.5) to (4.10) were derived, is based on the assumption of uniform charge distribution over the surface of the conductor. When other charges are present, the distribution of charge on the surface of the conductor is not uniform and the equations derived from Eq. (4.3) are not strictly correct. The nonuniformity of charge distribution, however, can be neglected entirely in overhead lines since the error in Eq. (4.10) is only 0.01% even for such a close spacing as that where the ratio $D/r = 50$.

A question arises about the value to be used in the denominator of the argument of the logarithm in Eq. (4.10) when the conductor is a stranded cable, since the equation was derived for a solid round conductor. Since electric flux is perpendicular to the surface of a perfect conductor, the electric field at the surface of a stranded conductor is not the same as the field at the surface of a cylindrical conductor. Therefore, the capacitance calculated for a stranded conductor by substituting the outside radius of the conductor for r in Eq. (4.10) will be slightly in error because of the difference between the field in the neighborhood of such a conductor and the field near a solid conductor for which Eq. (4.10) was derived. The error is very small, however, since only the field very close to the surface of the conductor is affected. The outside radius of the stranded conductor is used in calculating the capacitance.

After the capacitance to neutral has been found, the capacitive reactance existing between one conductor and neutral for relative permittivity $k_r = 1$ is found by using the expression for C given in Eq. (4.10) to yield

$$X_c = \frac{1}{2\pi fC} = \frac{2.862}{f} \times 10^9 \ln \frac{D}{r} \qquad \Omega \cdot \text{m to neutral} \qquad (4.11)$$

Since C in Eq. (4.11) is in farads per meter, the proper units for X_C must be ohm-meters. We should also note that Eq. (4.11) expresses the reactance from line to neutral for 1 m of line. Since capacitive reactance is in parallel along the line, X_C in ohm-meters must be *divided* by the length of the line in meters to obtain the capacitive reactance in ohms to neutral for the entire length of the line.

When Eq. (4.11) is divided by 1609 to convert to ohm-miles, we obtain

$$X_C = \frac{1.779}{f} \times 10^6 \ln \frac{D}{r} \qquad \Omega\cdot\text{mi to neutral} \qquad (4.12)$$

Table A.1 lists the outside diameters of the most widely used sizes of ACSR. If D and r in Eq. (4.12) are in feet, *capacitive reactance at 1-ft spacing X'_a* is the first term and *capacitive reactance spacing factor X'_d* is the second term when the equation is expanded as follows:

$$X_C = \frac{1.779}{f} \times 10^6 \ln \frac{1}{r} + \frac{1.779}{f} \times 10^6 \ln D \qquad \Omega\cdot\text{mi to neutral} \quad (4.13)$$

Table A.1 includes values of X'_a for common sizes of ACSR, and similar tables are readily available for other types and sizes of conductors. Table A.3 lists values of X'_d.

Example 4.1 Find the capacitive susceptance per mile of a single-phase line operating at 60 Hz. The conductor is *Partridge*, and spacing is 20 ft between centers.

SOLUTION For this conductor Table A.1 lists an outside diameter of 0.642 in, and so

$$r = \frac{0.642}{2 \times 12} = 0.0268 \text{ ft}$$

and from Eq. (4.12)

$$X_C = \frac{1.779}{60} \times 10^6 \ln \frac{20}{0.0268} = 0.1961 \times 10^6 \ \Omega\cdot\text{mi to neutral}$$

$$B_C = \frac{1}{X_C} = 5.10 \times 10^{-6} \ \mho/\text{mi to neutral}$$

or in terms of capacitive reactance at 1-ft spacing and capacitive reactance spacing factor from Tables A.1 and A.3

$$X'_a = 0.1074 \text{ M}\Omega\cdot\text{mi}$$

$$X'_d = 0.0889 \text{ M}\Omega\cdot\text{mi}$$

$$X_C = 0.1074 + 0.0889 = 0.1963 \text{ M}\Omega\cdot\text{mi per conductor}$$

Line-to-line capacitive reactance and susceptance are

$$X_C = 2 \times 0.1963 \times 10^6 = 0.3926 \times 10^6 \ \Omega \cdot \text{mi}$$

$$B_C = \frac{1}{X_C} = 2.55 \times 10^{-6} \ \mho/\text{mi}$$

4.4 CAPACITANCE OF A THREE-PHASE LINE WITH EQUILATERAL SPACING

The three identical conductors of radius r of a three-phase line with equilateral spacing are shown in Fig. 4.6. Equation (4.5) expresses the voltage between two conductors due to the charges on each one if the charge distribution on the conductors can be assumed to be uniform. Thus the voltage V_{ab} of the three-phase line due only to the charges on conductors a and b is

$$V_{ab} = \frac{1}{2\pi k} \underbrace{\left(q_a \ln \frac{D}{r} + q_b \ln \frac{r}{D} \right)}_{\text{due to } q_a \text{ and } q_b.} \quad \text{V} \tag{4.14}$$

Equation (4.3) enables us to include the effect of q_c since uniform charge distribution over the surface of a conductor is equivalent to a concentrated charge at the center of the conductor. Therefore, due only to the charge q_c,

$$V_{ab} = \frac{q_c}{2\pi k} \ln \frac{D}{D} \quad \text{V}$$

which is zero since q_c is equidistant from a and b. However to show that we are considering all three charges we can write

$$V_{ab} = \frac{1}{2\pi k} \left(q_a \ln \frac{D}{r} + q_b \ln \frac{r}{D} + q_c \ln \frac{D}{D} \right) \quad \text{V} \tag{4.15}$$

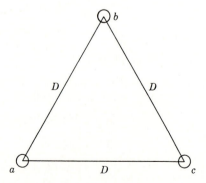

Figure 4.6 Cross section of a three-phase line with equilateral spacing.

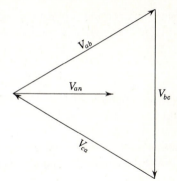

Figure 4.7 Phasor diagram of the balanced voltages of a three-phase line.

Similarly

$$V_{ac} = \frac{1}{2\pi k}\left(q_a \ln \frac{D}{r} + q_b \ln \frac{D}{D} + q_c \ln \frac{r}{D}\right) \quad \text{V} \tag{4.16}$$

Adding Eqs. (4.15) and (4.16) gives

$$V_{ab} + V_{ac} = \frac{1}{2\pi k}\left[2q_a \ln \frac{D}{r} + (q_b + q_c) \ln \frac{r}{D}\right] \quad \text{V} \tag{4.17}$$

In deriving these equations we have assumed that ground is far enough away to have negligible effect. Since the voltages are assumed to be sinusoidal and expressed as phasors, the charges are sinusoidal and expressed as phasors. If there are no other charges in the vicinity, the sum of the charges on the three conductors is zero and we can substitute $-q_a$ in Eq. (4.17) for $q_b + q_c$ and obtain

$$V_{ab} + V_{ac} = \frac{3q_a}{2\pi k} \ln \frac{D}{r} \quad \text{V} \tag{4.18}$$

Figure 4.7 is the phasor diagram of voltages. From this figure we obtain the following relations between the line voltages V_{ab} and V_{ac} and the voltage V_{an} from line a to the neutral of the three-phase circuit:

$$V_{ab} = \sqrt{3}V_{an}(0.866 + j0.5) \tag{4.19}$$

$$V_{ac} = -V_{ca} = \sqrt{3}V_{an}(0.866 - j0.5) \tag{4.20}$$

Adding Eqs. (4.19) and (4.20) gives

$$V_{ab} + V_{ac} = 3V_{an} \tag{4.21}$$

Substituting $3V_{an}$ for $V_{ab} + V_{ac}$ in Eq. (4.18), we obtain

$$V_{an} = \frac{q_a}{2\pi k} \ln \frac{D}{r} \quad \text{V} \tag{4.22}$$

Since capacitance to neutral is the ratio of the charge on a conductor to the voltage between that conductor and neutral,

$$C_n = \frac{q_a}{V_{an}} = \frac{2\pi k}{\ln (D/r)} \qquad \text{F/m to neutral} \qquad (4.23)$$

Comparison of Eqs. (4.23) and (4.10) shows that the two are identical. These equations express the capacitance to neutral for single-phase and equilaterally spaced three-phase lines, respectively. We saw in Chap. 3 that the equations for inductance per conductor were the same for single-phase and equilaterally spaced three-phase lines.

The term *charging current* is applied to the current associated with the capacitance of a line. For a *single-phase* circuit, the charging current is the product of the line-to-line voltage and the line-to-line susceptance, or, as a phasor,

$$I_{chg} = j\omega C_{ab} V_{ab} \qquad (4.24)$$

For a three-phase line, the charging current is found by multiplying the voltage to neutral by the capacitive susceptance to neutral. This gives the charging current per phase and is in accord with the calculation of balanced three-phase circuits on the basis of a single phase with neutral return. The phasor charging current in phase a is

$$I_{chg} = j\omega C_n V_{an} \qquad \text{A/mi} \qquad (4.25)$$

Since the rms voltage varies along the line, the charging current is not the same everywhere. Often the voltage used to obtain a value for charging current is the normal voltage for which the line is designed, such as 220 or 500 kV, which is probably not the actual voltage at either a generating station or a load.

4.5 CAPACITANCE OF A THREE-PHASE LINE WITH UNSYMMETRICAL SPACING

When the conductors of a three-phase line are not equilaterally spaced, the problem of calculating capacitance becomes more difficult. In the usual un-transposed line the capacitances of each phase to neutral are unequal. In a transposed line the average capacitance to neutral of any phase for the complete transposition cycle is the same as the average capacitance to neutral of any other phase, since each phase conductor occupies the same position as every other phase conductor over an equal distance along the transposition cycle. The dissymmetry of the untransposed line is slight for the usual configuration, and capacitance calculations are carried out as though all lines were transposed.

For the line shown in Fig. 4.8 three equations are found for V_{ab} for the three different parts of the transposition cycle. With phase a in position 1, b in position 2, and c in position 3,

$$V_{ab} = \frac{1}{2\pi k}\left(q_a \ln \frac{D_{12}}{r} + q_b \ln \frac{r}{D_{12}} + q_c \ln \frac{D_{23}}{D_{31}}\right) \qquad \text{V} \qquad (4.26)$$

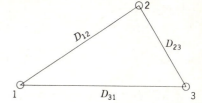

Figure 4.8 Cross section of a three-phase line with unsymmetrical spacing.

With a in position 2, b in position 3, and c in position 1,

$$V_{ab} = \frac{1}{2\pi k}\left(q_a \ln \frac{D_{23}}{r} + q_b \ln \frac{r}{D_{23}} + q_c \ln \frac{D_{31}}{D_{12}} \right) \qquad \text{V} \qquad (4.27)$$

and, with a in position 3, b in position 1, and c in position 2,

$$V_{ab} = \frac{1}{2\pi k}\left(q_a \ln \frac{D_{31}}{r} + q_b \ln \frac{r}{D_{31}} + q_c \ln \frac{D_{12}}{D_{23}} \right) \qquad \text{V} \qquad (4.28)$$

Equations (4.26) to (4.28) are similar to Eqs. (3.60) to (3.62) for the flux linkages of one conductor of a transposed line. However, in the equations for flux linkages we noted that the current in any phase was the same in every part of the transposition cycle. In Eqs. (4.26) to (4.28), if we disregard the voltage drop along the line, the voltage to neutral of a phase in one part of a transposition cycle is equal to the voltage to neutral of that phase in any part of the cycle. Hence, the voltage between any two conductors is the same in all parts of the transposition cycle. It follows that the charge on a conductor must be different when the position of the conductor changes with respect to other conductors. A treatment of Eqs. (4.26) to (4.28) analogous to that of Eqs. (3.60) to (3.62) is not rigorous.

The rigorous solution for capacitance is too involved to be practical except perhaps for flat spacing with equal distances between adjacent conductors. With the usual spacings and conductors, sufficient accuracy is obtained by assuming that the charge per unit length on a conductor is the same in every part of the transposition cycle. When the above assumption is made with regard to charge, the voltage between a pair of conductors is different for each part of the transposition cycle. Then an average value of voltage between the conductors can be found, and the capacitance calculated from the average voltage. We obtain the average voltage by adding Eqs. (4.26), (4.27), and (4.28) and dividing the result by 3. The average voltage between conductors a and b, based on the assumption of the same charge on a conductor regardless of its position in the transposition cycle, is

$$V_{ab} = \frac{1}{6\pi k}\left(q_a \ln \frac{D_{12}D_{23}D_{31}}{r^3} + q_b \ln \frac{r^3}{D_{12}D_{23}D_{31}} + q_c \ln \frac{D_{12}D_{23}D_{31}}{D_{12}D_{23}D_{31}} \right)$$

$$= \frac{1}{2\pi k}\left(q_a \ln \frac{D_{eq}}{r} + q_b \ln \frac{r}{D_{eq}} \right) \qquad \text{V} \qquad (4.29)$$

where

$$D_{eq} = \sqrt[3]{D_{12}D_{23}D_{31}} \qquad (4.30)$$

Similarly, the average voltage drop from conductor a to conductor c is

$$V_{ac} = \frac{1}{2\pi k}\left(q_a \ln \frac{D_{eq}}{r} + q_c \ln \frac{r}{D_{eq}}\right) \qquad \text{V} \qquad (4.31)$$

Applying Eq. (4.21) to find the voltage to neutral, we have

$$3V_{an} = V_{ab} + V_{ac} = \frac{1}{2\pi k}\left(2q_a \ln \frac{D_{eq}}{r} + q_b \ln \frac{r}{D_{eq}} + q_c \ln \frac{r}{D_{eq}}\right) \qquad \text{V} \quad (4.32)$$

Since $q_a + q_b + q_c = 0$ in a balanced three-phase circuit,

$$3V_{an} = \frac{3}{2\pi k} q_a \ln \frac{D_{eq}}{r} \qquad \text{V} \qquad (4.33)$$

and

$$C_n = \frac{q_a}{V_{an}} = \frac{2\pi k}{\ln (D_{eq}/r)} \qquad \text{F/m to neutral} \qquad (4.34)$$

Equation (4.34) for capacitance to neutral of a transposed three-phase line corresponds to Eq. (3.65) for the inductance per phase of a similar line. In finding capacitive reactance to neutral corresponding to C_n the reactance can be split into components of capacitive reactance to neutral at 1-ft spacing X'_a and capacitive reactance spacing factor X'_d as defined by Eq. (4.13).

Example 4.2 Find the capacitance and the capacitive reactance for 1 mi of the line described in Example 3.4. If the length of the line is 175 mi and the normal operating voltage is 220 kV, find capacitive reactance to neutral for the entire length of the line, the charging current per mile, and the total charging megavolt-amperes.

SOLUTION

$$r = \frac{1.108}{2 \times 12} = 0.0462 \text{ ft}$$

$$D_{eq} = 24.8 \text{ ft}$$

$$C_n = \frac{2\pi \times 8.85 \times 10^{-12}}{\ln (24.8/0.0462)} = 8.8466 \times 10^{-12} \text{ F/m}$$

$$X_C = \frac{10^{12}}{2\pi \times 60 \times 8.8466 \times 1609} = 0.1864 \times 10^6 \ \Omega \cdot \text{mi}$$

or from tables

$$X'_a = 0.0912 \times 10^6 \qquad X'_d = 0.0953 \times 10^6$$

$$X'_C = (0.0912 + 0.0953) \times 10^6 = 0.1865 \times 10^6 \ \Omega \cdot \text{mi to neutral}$$

For a length of 175 mi

$$\text{Capacitive reactance} = \frac{0.1865 \times 10^6}{175} = 1066 \ \Omega \ \text{to neutral}$$

$$I_{\text{chg}} = 2\pi 60 \ \frac{220{,}000}{\sqrt{3}} \times 8.8466 \times 10^{-12} \times 1609 = 0.681 \ \text{A/mi}$$

or $0.681 \times 175 = 119$ A for the line. Reactive power is $Q = \sqrt{3} \times 220 \times 119 \times 10^{-3} = 45.3$ Mvar. This amount of reactive power absorbed by the distributed capacitance is negative in keeping with the convention discussed in Chap. 2. In other words, positive reactive power is being *generated* by the distributed capacitance of the line.

4.6 EFFECT OF EARTH ON THE CAPACITANCE OF THREE-PHASE TRANSMISSION LINES

Earth affects the capacitance of a transmission line because its presence alters the electric field of the line. If we assume that the earth is a perfect conductor in the form of a horizontal plane of infinite extent, we realize that the electric field of charged conductors above the earth is not the same as it would be if the equipotential surface of the earth were not present. The electric field of the charged conductors is forced to conform to the presence of the earth's surface. The assumption of a flat, equipotential surface is, of course, limited by the irregularity of terrain and the type of surface of the earth. The assumption enables us, however, to understand the effect of a conducting earth on capacitance calculations.

Consider a circuit consisting of a single overhead conductor with a return path through the earth. In charging the conductor, charges come from the earth to reside on the conductor, and a potential difference exists between the conductor and earth. The earth has a charge equal in magnitude to that on the conductor but of opposite sign. Electric flux from the charges on the conductor to the charges on the earth is perpendicular to the earth's equipotential surface, since the surface is assumed to be a perfect conductor. Let us imagine a fictitious conductor of the same size and shape as the overhead conductor lying directly below the original conductor at a distance equal to twice the distance of the conductor above the plane of the ground. The fictitious conductor is below the surface of the earth by a distance equal to the distance of the overhead conductor above the earth. If the earth is removed and a charge equal and opposite to that on the overhead conductor is assumed on the fictitious conductor, the plane midway between the original conductor and the fictitious conductor is an equipotential surface and occupies the same position as the equipotential surface of the earth. The electric flux between the overhead conductor and this equipotential surface is the same as that which existed between the conductor and the earth. Thus, for purposes of calculation of capacitance, the earth may be replaced by a fictitious charged conductor below the surface of the earth by a distance equal to that of the overhead conductor above the earth. Such a conductor has a charge equal in magnitude and opposite in sign to that of the original conductor and is called the *image conductor*.

The method of calculating capacitance by replacing the earth by the image of an overhead conductor can be extended to more than one conductor. If we locate an image conductor for each overhead conductor, the flux between the original conductors and their images is perpendicular to the plane which replaces the earth, and that plane is an equipotential surface. The flux above the plane is the same as it is when the earth is present instead of the image conductors.

To apply the method of images to the calculation of capacitance for a three-phase line, refer to Fig. 4.9. We shall assume that the line is transposed and

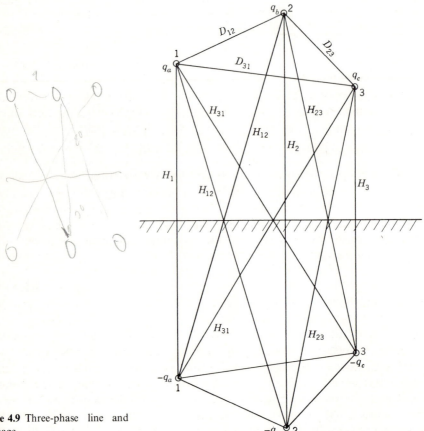

Figure 4.9 Three-phase line and its image.

that conductors a, b, and c carry the charges q_a, q_b, and q_c and occupy positions 1, 2, and 3, respectively, in the first part of the transposition cycle. The plane of the earth is shown, and below it are the conductors with the image charges $-q_a$, $-q_b$, and $-q_c$. Equations for the three parts of the transposition cycle can be written for the voltage drop from conductor a to conductor b as determined by the three charged conductors and their images. With conductor a in position 1, b in position 2, and c in position 3,

$$V_{ab} = \frac{1}{2\pi k}\left[q_a\left(\ln \frac{D_{12}}{r} - \ln \frac{H_{12}}{H_1}\right) + q_b\left(\ln \frac{r}{D_{12}} - \ln \frac{H_2}{H_{12}}\right)\right.$$

$$\left. + q_c\left(\ln \frac{D_{23}}{D_{31}} - \ln \frac{H_{23}}{H_{31}}\right)\right] \quad (4.35)$$

Similar equations for V_{ab} are written for the other parts of the transposition cycle. Accepting the approximately correct assumption of constant charge per unit length of each conductor throughout the transposition cycle allows us to obtain an average value of the phasor V_{ab}. The equation for the average value of the phasor V_{ac} is found in a similar manner, and $3V_{an}$ is obtained by adding the average values of V_{ab} and V_{ac}. Knowing that the sum of the charges is zero, we then find

$$C_n = \frac{2\pi k}{\ln (D_{eq}/r) - \ln (\sqrt[3]{H_{12}H_{23}H_{31}}/\sqrt[3]{H_1 H_2 H_3})} \quad \text{F/m to neutral} \quad (4.36)$$

Comparison of Eqs. (4.34) and (4.36) shows that the effect of the earth is to increase the capacitance of a line. To account for the earth the denominator of Eq. (4.34) must have subtracted from it the term

$$\log (\sqrt[3]{H_{12}H_{23}H_{31}}/\sqrt[3]{H_1 H_2 H_3}).$$

If the conductors are high above ground compared with the distances between them, the diagonal distances in the numerator of the correction term are nearly equal to the vertical distances in the denominator and the term is very small. This is the usual case, and the effect of ground is generally neglected for three-phase lines except for calculations by symmetrical components when the sum of the three line currents is not zero.

4.7 BUNDLED CONDUCTORS

Figure 4.10 shows a bundled-conductor line for which we can write an equation for the voltage from conductor a to conductor b as we did in deriving Eq. (4.26) except that now we must consider the charges on all six individual conductors. The conductors of any one bundle are in parallel, and we can assume the charge per bundle divides equally between the conductors of the bundle since the separation between bundles is usually more than 15 times the spacing between the conductors of the bundle. Also, since D_{12} is much greater than d, we can use D_{12}

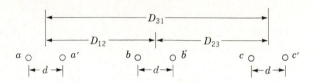

Figure 4.10 Cross section of a bundled-conductor three-phase line.

in place of the distances $D_{12} - d$ and $D_{12} + d$ and make other similar substitutions of bundle separation distances instead of using the more exact expressions that occur in finding V_{ab}. The difference due to this approximation cannot be detected in the final result for usual spacings even when the calculation is carried to five or six significant figures.

If the charge on phase a is q_a, conductors a and a' each have the charge $q_a/2$; similar division of charge is assumed for phases b and c. Then

$$V_{ab} = \frac{1}{2\pi k}\left[\frac{q_a}{2}\underbrace{\left(\ln\frac{D_{12}}{r}}_{a} + \underbrace{\ln\frac{D_{12}}{d}\right)}_{a'}\right.$$

$$\left. + \frac{q_b}{2}\left(\underbrace{\ln\frac{r}{D_{12}}}_{b} + \underbrace{\ln\frac{d}{D_{12}}}_{b'}\right) + \frac{q_c}{2}\left(\underbrace{\ln\frac{D_{23}}{D_{31}}}_{c} + \underbrace{\ln\frac{D_{23}}{D_{31}}}_{c'}\right)\right] \quad (4.37)$$

The letters under each logarithmic term indicate the conductor whose charge is accounted for by that therm. Combining terms gives

$$V_{ab} = \frac{1}{2\pi k}\left(q_a \ln\frac{D_{12}}{\sqrt{rd}} + q_b \ln\frac{\sqrt{rd}}{D_{12}} + q_c \ln\frac{D_{23}}{D_{31}}\right) \quad (4.38)$$

Equation (4.38) is the same as Eq. (4.26) except the $\sqrt{rd}$ has replaced r. It therefore follows that if we considered the line to be transposed, we would find

$$C_n = \frac{2\pi k}{\ln(D_{eq}/\sqrt{rd})} \quad \text{F/m to neutral} \quad (4.39)$$

The $\sqrt{rd}$ is the same as D_s^b for a two-conductor bundle except that r has replaced D_s. This leads us to the very important conclusion that a modified GMD method applies to the calculation of capacitance of a bundled-conductor three-phase line having two conductors per bundle. The modification is that we are using outside radius in place of the GMR of a single conductor.

It is logical to conclude that the modified GMD method applies to other bundling configurations. If we let D_{sC}^b stand for the modified GMR to be used in capacitance calculations to distinguish it from D_s^b used in inductance calculations, we have

$$C_n = \frac{2\pi k}{\ln(D_{eq}/D_{sC}^b)} \quad \text{F/m to neutral} \quad (4.40)$$

Then for a two-strand bundle

$$D_{sC}^b = \sqrt[4]{(r \times d)^2} = \sqrt{rd} \qquad (4.41)$$

for a three-strand bundle

$$D_{sC}^b = \sqrt[9]{(r \times d \times d)^3} = \sqrt[3]{rd^2} \qquad (4.42)$$

and for a four-strand bundle

$$D_{sC}^b = \sqrt[16]{(r \times d \times d \times d \times 2^{1/2})^4} = 1.09\sqrt[4]{rd^3} \qquad (4.43)$$

Example 4.3 Find the capacitive reactance to neutral of the line described in Example 3.5 in ohm-kilometers (and in ohm-miles) per phase.

SOLUTION Computed from the diameter given in Table A.1

$$r = \frac{1.382 \times 0.3048}{2 \times 12} = 0.01755 \text{ m}$$

$$D_{sC}^b = \sqrt{0.01755 \times 0.45} = 0.0889 \text{ m}$$

$$D_{eq} = \sqrt[3]{8 \times 8 \times 16} = 10.08 \text{ m}$$

$$C_m = \frac{2\pi \times 8.85 \times 10^{-12}}{\ln (10.08/0.0889)} = 11.754 \times 10^{-12} \text{ F/m}$$

$$X_C = \frac{10^{12} \times 10^{-3}}{2\pi 60 \times 11.754} = 0.2257 \times 10^6 \ \Omega \cdot \text{km per phase to neutral}$$

$$\left(X_C = \frac{0.2257 \times 10^6}{1.609} = 0.1403 \times 10^6 \ \Omega \cdot \text{mi per phase to neutral} \right)$$

4.8 PARALLEL-CIRCUIT THREE-PHASE LINES

Throughout our discussion of capacitance we have noted the similarity of the equations for inductance and capacitance. A modified GMD method has been found to apply in finding capacitance of bundled-conductor lines. We could show that this method is equally good for transposed three-phase lines with equilateral spacing (conductors at the vertices of a hexagon) and for flat vertical spacing (conductors of the three phases of each circuit lying in the same vertical plane). It is reasonable to assume that the modified GMD method can be used for arrangements intermediate between equilateral and flat-vertical spacing. Even though transpositions are not made, the method is generally used. An example should be sufficient to illustrate the method.

Example 4.4 Find the 60-Hz capacitive susceptance to neutral per mile per phase of the double-circuit line described in Example 3.6

SOLUTION From Example 3.6, $D_{eq} = 16.1$ ft.

The calculation of D_{sC}^{p} is the same as that of D_{s}^{p} in Example 3.6 except that the outside radius of the *Ostrich* conductor is used instead of its GMR. The outside diameter of 26/17, ACSR *Ostrich* is 0.680 in.

$$r = \frac{0.680}{2 \times 12} = 0.0283 \text{ ft}$$

$$D_{sC}^{p} = (\sqrt{26.9 \times 0.0283}\sqrt{21 \times 0.0283}\sqrt{26.9 \times 0.0283})^{1/3}$$
$$= \sqrt{0.0283}\,(26.9 \times 21 \times 26.9)^{1/6} = 0.837 \text{ ft}$$

$$C_n = \frac{2\pi \times 8.85 \times 10^{-12}}{\ln(16.1/0.837)} = 18.807 \times 10^{-12} \text{ F/m}$$

$$B_C = 2\pi \times 60 \times 18.807 \times 1609 = 11.41 \times 10^{-6} \text{ ℧/mi per phase to neutral}$$

4.9 SUMMARY

The similarity between inductance and capacitance calculations has been emphasized throughout our discussions. As in inductance calculations, computer programs are recommended if a large number of calculations of capacitance are required. Tables like A.1 and A.3 make the calculations quite simple, however, except for parallel-circuit lines.

The important equation for capacitance to neutral for a single-circuit, three-phase line is

$$C_n = \frac{2\pi k}{\ln D_{eq}/D_{sC}} \quad \text{F/m to neutral} \tag{4.44}$$

D_{sC} is the outside radius r of the conductor for a line consisting of one conductor per phase. For overhead lines k is 8.85×10^{-12} since k_r for air is 1.0. Capacitive reactance in ohm-meters is $1/2\pi fC$ where C is in farads/meter. So at 60 Hz

$$X_C = 4.77 \times 10^4 \ln \frac{D_{eq}}{D_{sC}} \quad \Omega \cdot \text{km to neutral} \tag{4.45}$$

or upon dividing by 1.609 km/mi

$$X_C = 2.965 \times 10^4 \ln \frac{D_{eq}}{D_{sC}} \quad \Omega \cdot \text{mi to neutral} \tag{4.46}$$

Values for capacitive susceptance in mhos/kilometer (siemens/kilometer) and mhos/mile (siemens/mile) are the reciprocals of Eqs. (4.45) and (4.46) respectively.

Both D_{eq} and D_{sC} must be in the same units, usually feet. For bundled conductors D_{sC}^{b}, as defined in Sec. 4.7, is substituted for D_{sC}. For both single- and bundled-conductor lines

$$D_{eq} = \sqrt[3]{D_{ab}D_{bc}D_{ca}}$$

For bundled-conductor lines D_{ab}, D_{bc}, and D_{ca} are distances between the centers of the bundles of phases a, b, and c.

For lines with one conductor per phase it is convenient to determine X_C by adding X'_a for the conductor as found in tables like A.1 to X'_d as found in Table A.3 corresponding to D_{eq}.

Capacitance and capacitive reactance of parallel-circuit lines are found by following the procedure of Example 4.4.

PROBLEMS

4.1 A three-phase transmission line has flat horizontal spacing with 2 m between adjacent conductors. At a certain instant the charge on one of the outside conductors is 60 μC/km, and the charge on the center conductor and on the other outside conductor is -30 μC/km. The radius of each conductor is 0.8 cm. Neglect the effect of ground, and find the voltage drop between the two identically charged conductors at the instant specified.

4.2 The 60-Hz capacitive reactance to neutral of a solid conductor, which is one conductor of a three-phase line with an equivalent equilateral spacing of 5 ft, is 196.1 kΩ-mi. What value of reactance would be specified in a table listing the capacitive reactance in ohm-miles to neutral of the conductor at 1-ft spacing for 25 Hz? What is the cross-sectional area of the conductor in circular mils?

4.3 Derive an equation for the capacitance to neutral in farads per meter of a single-phase line, taking into account the effect of ground. Use the same nomenclature as in the equation derived for the capacitance of a three-phase line where the effect of ground is represented by image charges.

4.4 Calculate the capacitance to neutral in farads per meter of a single-phase line composed of two single-strand conductors each having a diameter of 0.229 in. The conductors are 10 ft apart and 25 ft above ground. Compare the values obtained by Eq. (4.10) and by the equation derived in Prob. 4.3.

4.5 A three-phase 60-Hz transmission line has its conductors arranged in a triangular formation so that two of the distances between conductors are 25 ft and the third is 42 ft. The conductors are ACSR *Osprey*. Determine the capacitance to neutral in microfarads per mile and the capacitive reactance to neutral in ohm-miles. If the line is 150 mi long, find the capacitance to neutral and capacitive reactance of the line.

4.6 A three-phase 60-Hz line has flat horizontal spacing. The conductors have an outside diameter of 3.28 cm with 12 m between conductors. Determine the capacitive reactance to neutral in ohm-meters and the capacitive reactance of the line in ohms if its length is 125 mi.

4.7 A 60-Hz three-phase line composed of one ACSR *Bluejay* conductor per phase has flat horizontal spacing of 11 m between adjacent conductors. Compare the capacitive reactance in ohm-kilometers per phase of this line with that of a line using a two-conductor bundle of ACSR 26/7 conductors having the same total cross-sectional area of aluminum as the single-conductor line and the 11-m spacing measured between bundles. The spacing between conductors in the bundle is 40 cm.

4.8 Calculate the capacitive reactance in ohm-kilometers of a bundled 60-Hz three-phase line having three ACSR *Rail* conductors per bundle with 45 cm between conductors of the bundle. The spacing between bundle centers is 9, 9, and 18 m.

4.9 Six conductors of ACSR *Drake* constitute a 60-Hz double-circuit three-phase line arranged as shown in Fig. 3.15. The vertical spacing, however, is 14 ft; the longer horizontal distance is 32 ft; and the shorter horizontal distances are 25 ft. Find the capacitive reactance to neutral in ohm-miles and the charging current per mile per phase and per conductor at 138 kV.

FIVE

CURRENT AND VOLTAGE RELATIONS ON A TRANSMISSION LINE

We have examined the parameters of a transmission line and are ready to consider the line as an element of a power system. Figure 5.1 shows a 500-kV line having bundled conductors. In overhead lines the conductors are suspended from the tower and insulated from it and from each other by insulators, the number of which is determined by the voltage of the line. Each insulator string in Fig. 5.1 has 22 insulators. The two shorter arms above the phase conductors support wires usually made of steel. These wires being of much smaller diameter than the phase conductors are not visible in the picture, but they are electrically connected to the tower and are therefore at ground potential. These wires are referred to as ground wires and shield the phase conductors from lightning strokes.

A very important problem in the design and operation of a power system is the maintenance of the voltage within specified limits at various points in the system. In this chapter we shall develop formulas by which we can calculate the voltage, current, and power at any point on a transmission line provided we know these values at one point, usually at one end of the line. The chapter also provides an introduction to the study of transients on lossless lines in order to indicate how problems arise due to surges caused by lightning and switching.

The purpose of this chapter, however, is not merely to develop the pertinent equations; it also provides an opportunity to understand the effects of the parameters of the line on bus voltages and the flow of power. In this way we can see the importance of the design of the line and better understand the discussions to come in later chapters.

Figure 5.1 A 500-kV transmission line. Conductors are 76/19 ACSR with aluminum cross section of 2,515,000 cmil. Spacing between phases is 30 ft 3 in and the two conductors per bundle are 18 in apart. (*Courtesy Carolina Power and Light Company.*)

In the modern power system data from all over the system are being fed continuously into on-line computers for control purposes and for information. Load-flow studies performed by a computer readily supply answers to questions concerning the effect of switching lines into and out of the system or of changes in line parameters. Equations derived in this chapter remain important, however, in developing an overall understanding of what is occurring on a system and in calculating efficiency of transmission, losses, and limits of power flow over a line for both steady-state and transient conditions.

5.1 REPRESENTATION OF LINES

The general equations relating voltage and current on a transmission line recognize the fact that all four of the parameters of a transmission line discussed in the two preceding chapters are uniformly distributed along the line. We shall derive these general equations later but first we shall use lumped parameters which give good accuracy for short lines and for lines of medium length. If an overhead line is classified as short, shunt capacitance is so small that it can be omitted entirely with little loss of accuracy, and we need to consider only the series resistance R and the series inductance L for the total length of the line. Figure 5.2 shows a Y-connected generator supplying a balanced-Y load through a short transmission line. R and L are shown as concentrated, or lumped, parameters. It makes no difference, as far as measurements at the ends of the line are concerned, whether the parameters are lumped or uniformly distributed if the shunt admittance is neglected since the current is the same throughout the line in that case. The generator is represented by an impedance connected in series with the generated emf of each phase.

A medium-length line can be represented sufficiently well by R and L as lumped parameters, as shown in Fig. 5.3, with half the capacitance to neutral of the line lumped at each end of the equivalent circuit. Shunt conductance G, as mentioned previously, is usually neglected in overhead power transmission lines when calculating voltage and current.

Insofar as the handling of capacitance is concerned, open-wire 60-Hz lines

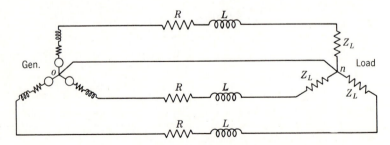

Figure 5.2 Generator supplying a balanced-Y load through a transmission line where the resistance R and inductance L are values for the entire length of the line. Line capacitance is omitted.

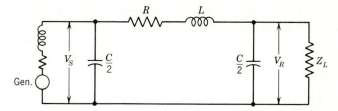

Figure 5.3 Single-phase equivalent of the circuit of Fig. 5.2 with the addition of capacitance to neutral for the entire length of the line divided between the two ends of the line.

less than about 80 km (50 mi) long are short lines. Medium-length lines are roughly between 80 km (50 mi) and 240 km (150 mi) long. Lines longer than 240 km (150 mi) require calculations in terms of distributed constants if a high degree of accuracy is required, although for some purposes a lumped-parameter representation can be used for lines up to 320 km (200 mi) long.

Normally, transmission lines are operated with balanced three-phase loads. Although the lines are not spaced equilaterally and not transposed, the resulting dissymmetry is slight and the phases are considered to be balanced.

In order to distinguish between the total series impedance of a line and the series impedance per unit length, the following nomenclature is adopted:

$$z = \text{series impedance per unit length per phase}$$
$$y = \text{shunt admittance per unit length per phase to neutral}$$
$$l = \text{length of line}$$
$$Z = zl = \text{total series impedance per phase}$$
$$Y = yl = \text{total shunt admittance per phase to neutral}$$

5.2 THE SHORT TRANSMISSION LINE

The equivalent circuit of a short transmission line is shown in Fig. 5.4, where I_S and I_R are the sending- and receiving-end currents and V_S and V_R are the sending- and receiving-end line-to-neutral voltages.

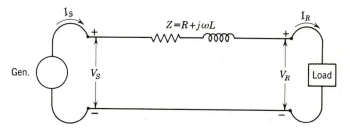

Figure 5.4 Equivalent circuit of a short transmission line where the resistance R and inductance L are values for the entire length of the line.

The circuit is solved as a simple series ac circuit. Since there are no shunt arms, the current is the same at the sending and receiving ends of the line and

$$I_S = I_R \tag{5.1}$$

The voltage at the sending end is

$$V_S = V_R + I_R Z \tag{5.2}$$

where Z is zl, the total series impedance of the line.

The effect of the variation of the power factor of the load on the *voltage regulation* of a line is most easily understood for the short line and therefore will be considered at this time. Voltage regulation of a transmission line is the rise in voltage at the receiving end, expressed in percent of full-load voltage when full load at a specified power factor is removed while the sending-end voltage is held constant. In the form of an equation

$$\text{Percent regulation} = \frac{|V_{R,\,NL}| - |V_{R,\,FL}|}{|V_{R,\,FL}|} \times 100 \tag{5.3}$$

where $|V_{R,\,NL}|$ is the magnitude of receiving-end voltage at no load and $|V_{R,\,FL}|$ is the magnitude of receiving-end voltage at full load with $|V_S|$ constant. After the load on a short transmission line, represented by the circuit of Fig. 5.4, is removed, the voltage at the receiving end is equal to the voltage at the sending end. In Fig. 5.4, with the load connected, the receiving-end voltage is designated by V_R, and $|V_R| = |V_{R,\,FL}|$. The sending-end voltage is V_S, and $|V_S| = |V_{R,\,NL}|$. The phasor diagrams of Fig. 5.5 are drawn for the same magnitudes of receiving-end voltage and current and show that a larger value of sending-end voltage is required to maintain a given receiving-end voltage when the receiving-end current is lagging the voltage than when the same current and voltage are in phase. A still smaller sending-end voltage is required to maintain the given receiving-end voltage when the receiving-end current leads the voltage. The voltage drop is the same in the series impedance of the line in all cases, but because of the different power factors the voltage drop is added to the receiving-end voltage at a different angle in each case. The regulation is greatest for lagging power factors and least, or even negative, for leading power factors. The inductive reactance of a transmission line is larger than the resistance, and the principle of regulation illustrated in Fig. 5.5 is true for any load supplied by a predominantly inductive

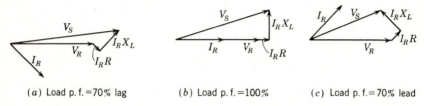

(*a*) Load p. f. = 70% lag   (*b*) Load p. f. = 100%   (*c*) Load p. f. = 70% lead

Figure 5.5 Phasor diagrams of a short transmission line. All diagrams are drawn for the same magnitudes of V_R and I_R.

circuit. The magnitudes of the voltage drops $I_R R$ and $I_R X_L$ for a short line have been exaggerated with respect to V_R in drawing the phasor diagrams in order to illustrate the point more clearly. The relation between power factor and regulation for longer lines is similar to that for short lines but is not visualized so easily.

5.3 THE MEDIUM-LENGTH LINE

The shunt admittance, usually pure capacitance, is included in the calculations for a line of medium length. If the total shunt admittance of the line is divided into two equal parts placed at the sending and receiving ends of the line, the circuit is called a nominal π. We shall refer to Fig. 5.6 to derive equations. To obtain an expression for V_S we note that the current in the capacitance at the receiving end is $V_R Y/2$ and the current in the series arm is $I_R + V_R Y/2$. Then

$$V_S = \left(V_R \frac{Y}{2} + I_R\right) Z + V_R \tag{5.4}$$

$$V_S = \left(\frac{ZY}{2} + 1\right) V_R + Z I_R \tag{5.5}$$

To derive I_S we note that the current in the shunt capacitance at the sending end is $V_S Y/2$, which added to the current in the series arm gives

$$I_S = V_S \frac{Y}{2} + V_R \frac{Y}{2} + I_R \tag{5.6}$$

Substituting V_S, as given by Eq. (5.5), in Eq. (5.6) gives

$$I_S = V_R Y\left(1 + \frac{ZY}{4}\right) + \left(\frac{ZY}{2} + 1\right) I_R \tag{5.7}$$

Corresponding equations can be derived for the nominal T which has all of the shunt admittance of the line lumped in the shunt arm of the T and the series impedance divided equally between the two series arms.

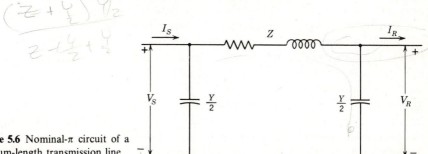

Figure 5.6 Nominal-π circuit of a medium-length transmission line.

Equations (5.5) and (5.7) may be expressed in the general form

$$V_S = AV_R + BI_R \tag{5.8}$$

$$I_S = CV_R + DI_R \tag{5.9}$$

where

$$A = D = \frac{ZY}{2} + 1$$

$$B = Z \quad C = Y\left(1 + \frac{ZY}{4}\right) \tag{5.10}$$

These *ABCD* constants are sometimes called the *generalized circuit constants* of the transmission line. In general they are complex numbers. *A* and *D* are dimensionless and equal each other if the line is the same when viewed from either end. The dimensions of *B* and *C* are ohms and mhos, respectively. The constants apply to linear, passive, and bilateral four-terminal networks having two pairs of terminals.

A physical meaning is easily assigned to the constants. By letting I_R be zero in Eq. (5.8) we see that *A* is the ratio V_S/V_R at no load. Similarly, *B* is the ratio V_S/I_R when the receiving end is short-circuited. The constant *A* is useful in computing regulation. If $V_{R,FL}$ is the receiving-end voltage at full load for a sending-end voltage of V_S, Eq. (5.3) becomes

$$\text{Percent regulation} = \frac{|V_s|/|A| - |V_{R,FL}|}{|V_{R,FL}|} \times 100 \tag{5.11}$$

ABCD constants are not widely used. They are introduced here because they simplify working with the equations. Appendix Table A.6 lists *ABCD* constants for various networks and combinations of networks.

5.4 THE LONG TRANSMISSION LINE: SOLUTION OF THE DIFFERENTIAL EQUATIONS

The exact solution of any transmission line and the one required for a high degree of accuracy in calculating 60-Hz lines more than approximately 150 mi long must consider the fact that the parameters of the lines are not lumped but are distributed uniformly throughout the length of the line.

Figure 5.7 shows one phase and the neutral connection of a three-phase line. Lumped parameters are not shown because we are ready to consider the solution of the line with the impedance and admittance uniformly distributed. The same diagram also represents a single-phase line if the series impedance of the line is the loop series impedance of the single-phase line instead of the series impedance per phase of the three-phase line and if the shunt admittance is the line-to-line shunt admittance of the single-phase line instead of the shunt admittance to neutral of the three-phase line.

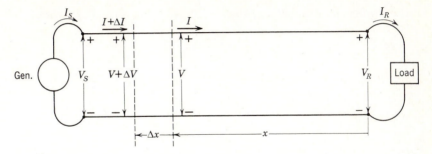

Figure 5.7 Schematic diagram of a transmission line showing one phase and the neutral return. Nomenclature for the line and the elemental length are indicated.

Let us consider a very small element in the line and calculate the difference in voltage and the difference in current between the ends of the element. We shall let x be the distance measured from the *receiving end* of the line to the small element of line, and we shall let the length of the element be Δx. Then $z\,\Delta x$ is the series impedance of the elemental length of the line, and $y\,\Delta x$ is its shunt admittance. The voltage to neutral at the end of the element toward the load is V, and V is the complex expression of the rms voltage, whose magnitude and phase vary with distance along the line. The voltage at the end of the element toward the generator is $V + \Delta V$. The rise in voltage over the elemental length of line in the direction of increasing x is ΔV, which is the voltage at the end toward the generator minus the voltage at the end toward the load. The rise in voltage in the direction of increasing x is also the product of the current in the element flowing opposite to the direction of increasing x and the impedance of the element, or $Iz\,\Delta x$. Thus

$$\Delta V = Iz\,\Delta x \qquad\qquad (5.12)$$

or

$$\frac{\Delta V}{\Delta x} = Iz \qquad\qquad (5.13)$$

and as $\Delta x \to 0$ the limit of the above ratio becomes

$$\frac{dV}{dx} = Iz \qquad\qquad (5.14)$$

Similarly, the current flowing out of the element toward the load is I. The magnitude and phase of the current I vary with distance along the line because of the distributed shunt admittance along the line. The current flowing into the element from the generator is $I + \Delta I$. The current entering the element from the generator end is higher than the current flowing away from the element in the direction of the load by the amount ΔI. This difference in current is the current $Vy\,\Delta x$ flowing in the shunt admittance of the element. Thus

$$\Delta I = Vy\,\Delta x$$

and pursuing steps similar to those of Eqs. (5.12) and (5.13) we obtain

$$\frac{dI}{dx} = Vy \qquad (5.15)$$

Let us differentiate Eqs. (5.14) and (5.15) with respect to x, and obtain

$$\frac{d^2V}{dx^2} = z\frac{dI}{dx} \qquad (5.16)$$

and

$$\frac{d^2I}{dx^2} = y\frac{dV}{dx} \qquad (5.17)$$

If we substitute the values of dI/dx and dV/dx from Eqs. (5.15) and (5.14) in Eqs. (5.16) and (5.17), respectively, we obtain

$$\frac{d^2V}{dx^2} = yzV \qquad (5.18)$$

and

$$\frac{d^2I}{dx^2} = yzI \qquad (5.19)$$

Now we have an equation (5.18) in which the only variables are V and x, and another equation (5.19) in which the only variables are I and x. The solutions of Eqs. (5.18) and (5.19) for V and I, respectively, must be expressions which when differentiated twice with respect to x yield the original expression times the constant yz. For instance, the solution for V when differentiated twice with respect to x must yield yzV. This suggests an exponential form of solution. Assume that the solution of Eq. (5.18) is†

$$V = A_1 \exp\left(\sqrt{yz}\ x\right) + A_2 \exp\left(-\sqrt{yz}\ x\right) \qquad (5.20)$$

Taking the second derivative of V with respect to x in Eq. (5.20) yields

$$\frac{d^2V}{dx^2} = yz[A_1 \exp\left(\sqrt{yz}\ x\right) + A_2 \exp\left(-\sqrt{yz}\ x\right)] \qquad (5.21)$$

which is yz times the assumed solution for V. Therefore, Eq. (5.20) is the solution of Eq. (5.18). When we substitute in Eq. (5.14) the value for V given by Eq. (5.20), we obtain,

$$I = \frac{1}{\sqrt{z/y}} A_1 \exp\left(\sqrt{yz}\ x\right) - \frac{1}{\sqrt{z/y}} A_2 \exp\left(-\sqrt{yz}\ x\right) \qquad (5.22)$$

The constants A_1 and A_2 can be evaluated by using the conditions at the

---

† The term $\exp\left(\sqrt{yz}\ x\right)$ in Eq. (5.20) and similar equatons is equivalent to ε raised to the power $\sqrt{yz}\ x$.

receiving end of the line, namely when $x = 0$, $V = V_R$ and $I = I_R$. Substitution of these values in Eqs. (5.20) and (5.22) yields

$$V_R = A_1 + A_2 \quad \text{and} \quad I_R = \frac{1}{\sqrt{z/y}}(A_1 - A_2)$$

Substituting $Z_c = \sqrt{z/y}$ and solving for A_1 give

$$A_1 = \frac{V_R + I_R Z_c}{2} \quad \text{and} \quad A_2 = \frac{V_R - I_R Z_c}{2}$$

Then, substituting the values found for A_1 and A_2 in Eqs. (5.20) and (5.22) and letting $\gamma = \sqrt{yz}$, we obtain

$$V = \frac{V_R + I_R Z_c}{2}\varepsilon^{\gamma x} + \frac{V_R - I_R Z_c}{2}\varepsilon^{-\gamma x} \tag{5.23}$$

$$I = \frac{V_R/Z_c + I_R}{2}\varepsilon^{\gamma x} - \frac{V_R/Z_c - I_R}{2}\varepsilon^{-\gamma x} \tag{5.24}$$

where $Z_c = \sqrt{z/y}$ and is called the *characteristic impedance* of the line, and $\gamma = \sqrt{yz}$ and is called the *propagation constant*.

Equations (5.23) and (5.24) give the rms values of V and I and their phase angles at any specified point along the line in terms of the distance x from the receiving end to the specified point, provided V_R, I_R, and the parameters of the line are known.

5.5 THE LONG TRANSMISSION LINE: INTERPRETATION OF THE EQUATIONS

Both γ and Z_c are complex quantities. The real part of the propagation constant γ is called the *attenuation constant* α and is measured in nepers per unit length. The quadrature part of γ is called the *phase constant* β and is measured in radians per unit length. Thus

$$\gamma = \alpha + j\beta \tag{5.25}$$

and Eqs. (5.23) and (5.24) become

$$V = \frac{V_R + I_R Z_c}{2}\varepsilon^{\alpha x}\varepsilon^{j\beta x} + \frac{V_R - I_R Z_c}{2}\varepsilon^{-\alpha x}\varepsilon^{-j\beta x} \tag{5.26}$$

and

$$I = \frac{V_R/Z_c + I_R}{2}\varepsilon^{\alpha x}\varepsilon^{j\beta x} - \frac{V_R/Z_c - I_R}{2}\varepsilon^{-\alpha x}\varepsilon^{-j\beta x} \tag{5.27}$$

The properties of $\varepsilon^{\alpha x}$ and $\varepsilon^{j\beta x}$ help to explain the variation of the phasor values of voltage and current as a function of distance along the line. The term $\varepsilon^{\alpha x}$ changes in magnitude as x changes, but $\varepsilon^{j\beta x}$, which is identical to $\cos \beta x + $

$j \sin \beta x$, always has a magnitude of 1 and causes a shift in phase of β rad per unit length of line.

The first term in Eq. (5.26), $[(V_R + I_R Z_c)/2]\varepsilon^{\alpha x}\varepsilon^{j\beta x}$, increases in magnitude and advances in phase as distance from the receiving end increases. Conversely, as progress along the line from the sending end toward the receiving end is considered, the term diminishes in magnitude and is retarded in phase. This is the characteristic of a traveling wave and is similar to the behavior of a wave in water, which varies in magnitude with time at any point, whereas its phase is retarded and its maximum value diminishes with distance from the origin. The variation in instantaneous value is not expressed in the term but is understood since V_R and I_R are phasors. The first term in Eq. (5.26) is called the *incident voltage*.

The second term in Eq. (5.26), $[(V_R - I_R Z_c)/2]\varepsilon^{-\alpha x}\varepsilon^{-j\beta x}$, diminishes in magnitude and is retarded in phase from the receiving end toward the sending end. It is called the *reflected voltage*. At any point along the line the voltage is the sum of the component incident and reflected voltages at that point.

Since the equation for current is similar to the equation for voltage, the current may be considered to be composed of incident and reflected currents.

If a line is terminated in its characteristic impedance Z_c, the receiving-end voltage V_R is equal to $I_R Z_c$ and there is no reflected wave of either voltage or current, as may be seen by substituting $I_R Z_c$ for V_R in Eqs. (5.26) and (5.27). A line terminated in its characteristic impedance is called a *flat line* or an *infinite line*. The latter term arises from the fact that a line of infinite length cannot have a reflected wave. Usually power lines are not terminated in their characteristic impedance, but communication lines are frequently so terminated in order to eliminate the reflected wave. A typical value of Z_c is 400 Ω for a single-circuit overhead line and 200 Ω for two circuits in parallel. The phase angle of Z_c is usually between 0 and $-15°$. Bundled-conductor lines have lower values of Z_c since such lines have lower L and higher C than lines with a single conductor per phase.

In power-system work, characteristic impedance is sometimes called *surge impedance*. The term surge impedance, however, is usually reserved for the special case of a lossless line. If a line is lossless, its resistance and conductance are zero and the characteristic impedance reduces to $\sqrt{L/C}$, a pure resistance. When dealing with high frequencies or with surges due to lightning, losses are often neglected and the surge impedance becomes important. Surge-impedance loading (SIL) of a line is the power delivered by a line to a purely resistive load equal to its surge impedance. When so loaded, the line supplies a current of

$$|I_L| = \frac{|V_L|}{\sqrt{3} \times \sqrt{L/C}} \qquad \text{A}$$

where $|V_L|$ is the line-to-line voltage at the load. Since the load is pure resistance,

$$\text{SIL} = \sqrt{3}|V_L| \frac{|V_L|}{\sqrt{3} \times \sqrt{L/C}} \qquad \text{W}$$

or, with $|V_L|$ in kilovolts,

$$\text{SIL} = \frac{|V_L|^2}{\sqrt{L/C}} \quad \text{MW} \tag{5.28}$$

Power-system engineers sometimes find it convenient to express the power transmitted by a line in terms of per unit of SIL, that is, as the ratio of the power transmitted to the surge-impedance loading. For instance, the permissible loading of a transmission line may be expressed as a fraction of its SIL, and SIL provides a comparison of load-carrying capabilities of lines.†

A *wavelength* λ is the distance along a line between two points of a wave which differ in phase by 360°, or 2π rad. If β is the phase shift in radians per mile, the wavelength in miles is

$$\lambda = \frac{2\pi}{\beta} \tag{5.29}$$

At a frequency of 60 Hz, a wavelength is approximately 3000 mi. The velocity of propagation of a wave in miles per second is the product of the wavelength in miles and the frequency in hertz, or

$$\text{Velocity} = f\lambda \tag{5.30}$$

If there is no load on a line, I_R is equal to zero, and, as determined by Eqs. (5.26) and (5.27), the incident and reflected voltages are equal in magnitude and in phase at the receiving end. In this case the incident and reflected currents are equal in magnitude but 180° out of phase at the receiving end. Thus, the incident and reflected currents cancel each other at the receiving end of an open line but not at any other point unless the line is entirely lossless so that the attenuation α is zero.

5.6 THE LONG TRANSMISSION LINE: HYPERBOLIC FORM OF THE EQUATIONS

The incident and reflected waves of voltage are seldom found when calculating the voltage of a power line. The reason for discussing the voltage and current of a line in terms of the incident and reflected components is that such an analysis is helpful in obtaining a fuller understanding of some of the phenomena of transmission lines. A more convenient form of the equations for computing current and voltage of a power line is found by introducing hyperbolic functions. Hyperbolic functions are defined in exponential form as follows:

$$\sinh \theta = \frac{\varepsilon^\theta - \varepsilon^{-\theta}}{2} \tag{5.31}$$

$$\cosh \theta = \frac{\varepsilon^\theta + \varepsilon^{-\theta}}{2} \tag{5.32}$$

---

† See R. D. Dunlop, R. Gutman, and P. P. Marchenko, "Analytical Development of Loadability Characteristics for EHV and UHV Transmission Lines," *IEEE Trans. PAS*, vol. 98, no. 2, 1979, pp. 606–617.

By rearranging Eqs. (5.23) and (5.24) and substituting hyperbolic functions for the exponential terms, a new set of equations is found. The new equations, giving voltage and current anywhere along the line, are

$$V = V_R \cosh \gamma x + I_R Z_c \sinh \gamma x \tag{5.33}$$

and

$$I = I_R \cosh \gamma x + \frac{V_R}{Z_c} \sinh \gamma x \tag{5.34}$$

Letting $x = l$ to obtain the voltage and current at the sending end, we have

$$V_S = V_R \cosh \gamma l + I_R Z_c \sinh \gamma l \tag{5.35}$$

and

$$I_S = I_R \cosh \gamma l + \frac{V_R}{Z_c} \sinh \gamma l \tag{5.36}$$

From examination of these equations we see that the generalized circuit constants for a long line are

$$A = \cosh \gamma l \qquad C = \frac{\sinh \gamma l}{Z_c} \tag{5.37}$$

$$B = Z_c \sinh \gamma l \qquad D = \cosh \gamma l$$

Equations (5.35) and (5.36) can be solved for V_R and I_R in terms of V_S and I_S to give

$$V_R = V_S \cosh \gamma l - I_S Z_c \sinh \gamma l \tag{5.38}$$

and

$$I_R = I_S \cosh \gamma l - \frac{V_S}{Z_c} \sinh \gamma l \tag{5.39}$$

For balanced three-phase lines the current in the above equations is the line current, and the voltage is the line-to-neutral voltage, that is, the line voltage divided by $\sqrt{3}$. In order to solve the equations, the hyperbolic functions must be evaluated. Since γl is usually complex, the hyperbolic functions are also complex and cannot be found directly from ordinary tables or electronic calculators. Before the widespread use of the digital computer, various charts, some of them especially adapted to the values usually encountered in transmission-line calculations, were frequently used to evaluate hyperbolic functions of complex arguments. Now the digital computer provides the usual means of incorporating such functions into our calculations.

For solving an occasional problem without resorting to a computer or charts there are several choices. The following equations give the expansions of hyperbolic sines and cosines of complex arguments in terms of circular and hyperbolic functions of real arguments:

$$\cosh (\alpha l + j\beta l) = \cosh \alpha l \cos \beta l + j \sinh \alpha l \sin \beta l \qquad (5.40)$$

$$\sinh (\alpha l + j\beta l) = \sinh \alpha l \cos \beta l + j \cosh \alpha l \sin \beta l \qquad (5.41)$$

Equations (5.40) and (5.41) make possible the computation of hyperbolic functions of complex arguments. The correct mathematical unit for βl is the radian, and the radian is the unit found for βl by computing the quadrature component of γl. Equations (5.40) and (5.41) can be verified by substituting in them the exponential forms of the hyperbolic functions and the similar exponential forms of the circular functions.

Another convenient method of evaluating a hyperbolic function is to expand it in a power series. Expansion by Maclaurin's series yields

$$\cosh \theta = 1 + \frac{\theta^2}{2!} + \frac{\theta^4}{4!} + \frac{\theta^6}{6!} + \cdots \qquad (5.42)$$

and

$$\sinh \theta = \theta + \frac{\theta^3}{3!} + \frac{\theta^5}{5!} + \frac{\theta^7}{7!} + \cdots \qquad (5.43)$$

The series converge rapidly for the values of γl usually found for power lines, and sufficient accuracy may be found by evaluating only the first few terms.

A third method of evaluating complex hyperbolic functions is suggested by Eqs. (5.31) and (5.32). Substituting $\alpha + j\beta$ for θ, we obtain

$$\cosh (\alpha + j\beta) = \frac{\varepsilon^\alpha \varepsilon^{j\beta} + \varepsilon^{-\alpha}\varepsilon^{-j\beta}}{2} = \tfrac{1}{2}(\varepsilon^\alpha \ \underline{/\beta} + \varepsilon^{-\alpha} \ \underline{/-\beta} \qquad (5.44)$$

and

$$\sinh (\alpha + j\beta) = \frac{\varepsilon^\alpha \varepsilon^{j\beta} - \varepsilon^{-\alpha} \varepsilon^{-j\beta}}{2} = \tfrac{1}{2}(\varepsilon^\alpha \ \underline{/\beta} - \varepsilon^{-\alpha} \ \underline{/-\beta)} \qquad (5.45)$$

Example 5.1 A single-circuit 60-Hz transmission line is 370 km (230 mi) long. The conductors are *Rook* with flat horizontal spacing and 7.25 m (23.8 ft) between conductors. The load on the line is 125 MW at 215 kV with 100% power factor. Find the voltage current, and power at the sending end and the voltage regulation of the line. Also determine the wavelength and velocity of propagation of the line.

SOLUTION Feet and miles rather than meters and kilometers are chosen for the calculations in order to use Tables A.1–A.3.

$$D_{eq} = \sqrt[3]{23.8 \times 23.8 \times 47.6} \cong 30.0 \text{ ft}$$

and from the tables for *Rook*

$$z = 0.1603 + j(0.415 + 0.4127) = 0.8431 \underline{/79.04°} \ \Omega/\text{mi}$$

$$y = j[1/(0.0950 + 0.1009)] \times 10^{-6} = 5.105 \times 10^{-6} \underline{/90°} \ \text{U}/\text{mi}$$

$$\gamma l = \sqrt{yz} \ l = 230 \sqrt{0.8431 \times 5.105 \times 10^{-6}} \ \underline{/\dfrac{79.04° + 90°}{2}}$$

$$= 0.4772 \underline{/84.52°} = 0.0456 + j0.4750$$

$$Z_c = \sqrt{\dfrac{z}{y}} = \sqrt{\dfrac{0.8431}{5.105 \times 10^{-6}}} \ \underline{/\dfrac{79.04° - 90°}{2}} = 406.4 \underline{/-5.48°} \ \Omega$$

$$V_R = \dfrac{215,000}{\sqrt{3}} = 124,130 \underline{/0°} \ \text{V to neutral}$$

$$I_R = \dfrac{125,000,000}{\sqrt{3} \times 215,000} = 335.7 \underline{/0°} \ \text{A}$$

From Eqs. (5.40) and (5.41)

$$\cosh \gamma l = \cosh 0.0456 \cos 0.475 + j \sinh 0.0456 \sin 0.475†$$

$$= 1.0010 \times 0.8893 + j0.0456 \times 0.4573$$

$$= 0.8902 + j0.0209 = 0.8904 \underline{/1.34°}$$

$$\sinh \gamma l = \sinh 0.0456 \cos 0.475 + j \cosh 0.0456 \sin 0.475$$

$$= 0.0456 \times 0.8893 + j1.0010 \times 0.4573$$

$$= 0.0405 + j0.4578 = 0.4596 \underline{/84.94°}$$

Then from Eq. (5.35)

$$V_S = 124,130 \times 0.8904 \underline{/1.34°} + 335.7 \times 406.4 \ \underline{/-5.48°} \times 0.4596 \underline{/84.94°}$$

$$= 110,495 + j2,585 + 11,480 + j61,642$$

$$= 137,851 \underline{/27.77} \ \text{V}$$

and from Eq. (5.36)

$$I_S = 335.7 \times 0.8904 \underline{/1.34°} + \dfrac{124,130}{406.4 \ \underline{/-5.48°}} \times 0.4596 \underline{/84.94°}$$

$$= 298.83 + j6.99 - 1.03 + j140.33$$

$$= 332.27 \underline{/26.33°}$$

† 0.475 rad = 27.2°

At the sending end

$$\text{Line voltage} = \sqrt{3} \times 137.85 = 238.8 \text{ kV}$$

$$\text{Line current} = 332.3 \text{ A}$$

$$\text{Power factor} = \cos(27.78° - 26.33°) = 0.9997 \cong 1.0$$

$$\text{Power} = \sqrt{3} \times 238.8 \times 332.3 \times 1.0 = 137,440 \text{ kW}$$

From Eq. (5.35) we see that at no load $(I_R = 0)$

$$V_R = \frac{V_S}{\cosh \gamma l}$$

So the voltage regulation is

$$\frac{137.85/0.8904 - 124.13}{124.13} \times 100 = 24.7\%$$

The wavelength and velocity of propagation are computed as follows:

$$\beta = \frac{0.4750}{230} = 0.002065 \text{ rad/mi}$$

$$\lambda = \frac{2\pi}{\beta} = \frac{2\pi}{0.002065} = 3043 \text{ mi}$$

$$\text{Velocity} = f\lambda = 60 \times 3043 = 182,580 \text{ mi/s}$$

We note particularly in this example that in the equations for V_S and I_S the value of voltage must be expressed in volts and must be the line-to-neutral voltage.

Example 5.2 Solve for the sending-end voltage and current found in Example 5.1 using per-unit calculations.

SOLUTION We choose a base of 125 MVA, 215 kV to achieve the simplest per-unit values and compute base impedance and base current as follows:

$$\text{Base impedance} = \frac{(215)^2}{125} = 370 \text{ } \Omega$$

$$\text{Base current} = \frac{125,000}{\sqrt{3} \times 215} = 335.7 \text{ A}$$

So

$$Z_c = \frac{406.4 \text{ } \underline{/-5.48°}}{370} = 1.098 \text{ } \underline{/-5.48°} \text{ per unit}$$

$$V_R = \frac{215}{215} = \frac{215/\sqrt{3}}{215/\sqrt{3}} = 1.0 \text{ per unit}$$

For use in Eq. (5.35) we chose V_R as the reference voltage. So

$$V_R = 1.0\underline{/0°} \text{ per unit (as a line-to-neutral voltage)}$$

and since the load is at unity power factor

$$I_R = \frac{337.5\underline{/0°}}{337.5} = 1.0\underline{/0°}$$

If the power factor had been less than 100%, I_R would have been greater than 1.0 and would have been at an angle determined by the power factor.

By Eq. (5.35)

$$V_S = 1.0 \times 0.8904 + 1.0 \times 1.098 \underline{/-5.48°} \times 0.4596\underline{/84.94°}$$

$$= 0.8902 + j0.0208 + 0.0923 + j0.4961$$

$$= 1.1102\underline{/27.75°} \text{ per unit}$$

and by Eq. (5.36)

$$I_S = 1.0 \times 0.8904\underline{/1.34°} + \frac{1.0\underline{/0°}}{1.098\underline{/-5.48°}} \times 0.4596\underline{/84.94°}$$

$$= 0.8902 + j0.0208 - 0.0031 + j0.4186$$

$$= 0.990\underline{/26.35°} \text{ per unit}$$

At the sending end

$$\text{Line voltage} = 1.1102 \times 215 = 238.7 \text{ V}$$

$$\text{Line current} = 0.990 \times 335.7 = 332.3 \text{ A}$$

Note that we multiply line-to-line voltage base by the per-unit magnitude of the voltage to find the line-to-line voltage magnitude. We could have multiplied line-to-neutral voltage base by the per-unit voltage to find line-to-neutral voltage magnitude. The factor $\sqrt{3}$ does not enter the calculations after we have expressed all quantities in per unit.

5.7 THE EQUIVALENT CIRCUIT OF A LONG LINE

The nominal-T and nominal-π circuits do not represent a transmission line exactly because they do not account for the parameters of the line being uniformly distributed. The discrepancy between the nominal T and π and the actual line becomes larger as the length of line increases. It is possible, however, to find the equivalent circuit of a long transmission line and to represent the line accurately, insofar as measurements at the ends of the line are concerned, by a network of lumped parameters. Let us assume that a π circuit similar to that of Fig. 5.6 is the equivalent circuit of a long line, but let us call the series arm of our equivalent-π circuit Z' and the shunt arms $Y'/2$ to distinguish them from the arms of the nominal-π circuit. Equation (5.5) gives the sending-end voltage of a symmetrical-π circuit in terms of its series and shunt arms and the voltage and current at the receiving end. By substituting Z' and $Y'/2$ for Z and $Y/2$ in Eq. (5.5), we obtain the sending-end voltage of our equivalent circuit in terms of its series and shunt arms and the voltage and current at the receiving end:

$$V_S = \left(\frac{Z'Y'}{2} + 1\right)V_R + Z'I_R \tag{5.46}$$

For our circuit to be equivalent to the long transmission line, the coefficients of V_R and I_R in Eq. (5.46) must be identical, respectively, to the coefficients of V_R and I_R in Eq. (5.35). Equating the coefficients of I_R in the two equations yields

$$Z' = Z_c \sinh \gamma l \tag{5.47}$$

$$Z' = \sqrt{\frac{z}{y}} \sinh \gamma l = zl \frac{\sinh \gamma l}{\sqrt{zy}\, l}$$

$$Z' = Z \frac{\sinh \gamma l}{\gamma l} \tag{5.48}$$

where Z is equal to zl, the total series impedance of the line. The term $(\sinh \gamma l)/\gamma l$ is the factor by which the series impedance of the nominal π must be multiplied to convert the nominal π to the equivalent π. For small values of γl, $\sinh \gamma l$ and γl are almost identical, and this fact shows that the nominal π represents the medium-length transmission line quite accurately, insofar as the series arm is concerned.

To investigate the shunt arms of the equivalent-π circuit, we equate the coefficients of V_R in Eqs. (5.35) and (5.46) and obtain

$$\frac{Z'Y'}{2} + 1 = \cosh \gamma l \tag{5.49}$$

Substituting $Z_c \sinh \gamma l$ for Z' gives

$$\frac{Y'Z_c \sinh \gamma l}{2} + 1 = \cosh \gamma l \tag{5.50}$$

and

$$\frac{Y'}{2} = \frac{1}{Z_c} \frac{\cosh \gamma l - 1}{\sinh \gamma l} \tag{5.51}$$

Another form of the expression for the shunt admittance of the equivalent circuit can be found by substituting in Eq. (5.51) the identity

$$\tanh \frac{\gamma l}{2} = \frac{\cosh \gamma l - 1}{\sinh \gamma l} \tag{5.52}$$

The identity can be verified by substituting the exponential forms of Eqs. (5.31)

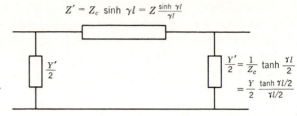

$$Z' = Z_c \sinh \gamma l = Z \frac{\sinh \gamma l}{\gamma l}$$

$$\frac{Y'}{2}$$

$$\frac{Y'}{2} = \frac{1}{Z_c} \tanh \frac{\gamma l}{2}$$

$$= \frac{Y}{2} \frac{\tanh \gamma l/2}{\gamma l/2}$$

Figure 5.8 Equivalent-π circuit of a transmission line.

and (5.32) for the hyperbolic functions and by recalling that $\tanh \theta = \sinh \theta / \cosh \theta$. Now

$$\frac{Y'}{2} = \frac{1}{Z_c} \tanh \frac{\gamma l}{2} \tag{5.53}$$

$$\frac{Y'}{2} = \frac{Y}{2} \frac{\tanh (\gamma l/2)}{\gamma l/2} \tag{5.54}$$

where Y is equal to yl, the total shunt admittance of the line. Equation (5.54) shows the correction factor used to convert the admittance of the shunt arms of the nominal π to that of the equivalent π. Since $\tanh (\gamma l/2)$ and $\gamma l/2$ are very nearly equal for small values of γl, the nominal π represents the medium-length transmission line quite accurately, for we have seen previously that the correction factor for the series arm is negligible for medium-length lines. The equivalent-π circuit is shown in Fig. 5.8. An equivalent-T circuit can also be found for a transmission line.

Example 5.3 Find the equivalent-π circuit for the line described in Example 5.1 and compare it with the nominal π.

SOLUTION Since sinh γl and cosh γl are already known from Example 5.1, Eqs. (5.47) and (5.51) will be used.

$$Z' = 406.4 \underline{/-5.48°} \times 0.4596 \underline{/84.94°} = 186.78 \underline{/79.46°} \; \Omega \text{ in series arm}$$

$$\frac{Y'}{2} = \frac{0.8902 + j0.0208 - 1}{186.78 \underline{/79.46°}} = \frac{0.1118 \underline{/169.27°}}{186.78 \underline{/79.46°}}$$

$$= 0.000599 \underline{/89.81°} \; \mho \text{ in each shunt arm}$$

The nominal-π circuit has a series impedance of

$$Z = 230 \times 0.8431 \underline{/79.04°} = 193.9 \underline{/79.04°}$$

and equal shunt arms of

$$\frac{Y}{2} = \frac{5.105 \times 10^{-6} \underline{/90°}}{2} \times 230 = 0.000587 \underline{/90°} \; \Omega$$

For this line the impedance of the series arm of the nominal π exceeds that of the equivalent π by 3.8%. The conductance of the shunt arms of the nominal π is 2.0% less than that of the equivalent π. So we conclude that the nominal π may represent long lines sufficiently well if a high degree of accuracy is not required.

5.8 POWER FLOW THROUGH A TRANSMISSION LINE

Although power flow at any point along a transmission line can always be found if the voltage, current, and power factor are known or can be calculated, very interesting equations for power can be derived in terms of $ABCD$ constants. Of course, the equations apply to any two-terminal-pair network. Repeating Eq. (5.8) and solving for the receiving-end current I_R yields

$$V_S = AV_R + BI_R \tag{5.55}$$

$$I_R = \frac{V_S - AV_R}{B} \tag{5.56}$$

Letting

$$A = |A|\underline{/\alpha} \qquad B = |B|\underline{/\beta}$$
$$V_R = |V_R|\underline{/0°} \qquad V_S = |V_S|\underline{/\delta}$$

we obtain

$$I_R = \frac{|V_S|}{|B|}\underline{/\delta - \beta} - \frac{|A|\cdot|V_R|}{|B|}\underline{/\alpha - \beta} \tag{5.57}$$

Then the complex power $V_R I_R^*$ at the receiving end is

$$P_R + jQ_R = \frac{|V_S|\cdot|V_R|}{|B|}\underline{/\beta - \delta} - \frac{|A|\cdot|V_R|^2}{|B|}\underline{/\beta - \alpha} \tag{5.58}$$

and real and reactive power at the receiving end are

$$P_R = \frac{|V_S|\cdot|V_R|}{|B|}\cos(\beta - \delta) - \frac{|A|\cdot|V_R|^2}{|B|}\cos(\beta - \alpha) \tag{5.59}$$

$$Q_R = \frac{|V_S|\cdot|V_R|}{|B|}\sin(\beta - \delta) - \frac{|A|\cdot|V_R|^2}{|B|}\sin(\beta - \alpha) \tag{5.60}$$

Noting that the expression for complex power $P_R + jQ_R$ is shown by Eq. (5.58) to be the resultant of combining two phasors expressed in polar form, we can plot these phasors in the complex plane whose horizontal and vertical coordinates are in power units (watts and vars). Figure 5.9 shows the two complex quantities and their difference as expressed by Eq. (5.58). Figure 5.10 shows the same phasors with the origin of the coordinate axes shifted. This figure is a power diagram with the resultant whose magnitude is $|P_R + jQ_R|$, or $|V_R|\cdot|I_R|$,

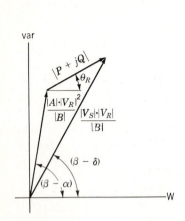

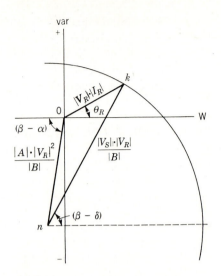

Figure 5.9 Phasors of Eq. 5.58 plotted in the complex plane, with magnitudes and angles as indicated.

Figure 5.10 Power diagram obtained by shifting the origin of the coordinate axes of Fig. 5.9.

at an angle θ_R with the horizontal axis. As expected, the real and imaginary components of $|P_R + jQ_R|$ are

$$P_R = |V_R| \cdot |I_R| \cos \theta_R \qquad (5.61)$$

and

$$Q_R = |V_R| \cdot |I_R| \sin \theta_R \qquad (5.62)$$

where θ_R is the phase angle by which V_R leads I_R, as discussed in Chap. 2. The sign of Q is consistent with the convention which assigns positive values to Q when current is lagging the voltage.

Now let us determine some points on the power diagram of Fig. 5.10 for various loads with fixed values of $|V_S|$ and $|V_R|$. First, we notice that the position of point n is not dependent on the current I_R and will not change so long as $|V_R|$ is constant. We note further that the distance from point n to point k is constant for fixed values of $|V_S|$ and $|V_R|$. Therefore, as the distance 0 to k changes with changing load, the point k, since it must remain at a constant distance from the fixed point n, is constrained to move in a circle whose center is at n. Any change in P_R will require a change in Q_R to keep k on the circle. If a different value of $|V_S|$ is held constant for the same value of $|V_R|$, the location of point n is unchanged but a new circle of radius nk is found.

In Eqs. (5.50) to (5.62) $|V_S|$ and $|V_R|$ are line-to-neutral voltages and coordinates in Fig. 5.10 are watts and vars per phase. However, if $|V_S|$ and $|V_R|$ are line-to-line voltages each distance in Fig. 5.10 is increased by a factor of 3 and the coordinates on the diagram are total three-phase watts and vars. If the voltages are kilovolts the coordinates are megawatts and megavars.

If the receiving-end voltage is held constant and receiving-end circles are drawn for different values of sending-end voltage, the resulting circles are concentric because the location of the center of the receiving-end power circles is independent of the sending-end voltage. A family of receiving-end circles is shown in Fig. 5.11 for a constant receiving-end voltage. The load line marked on Fig. 5.11 is convenient if the load changes in magnitude while its power factor remains constant. The angle between the load line through the origin and the horizontal axis is the angle whose cosine is the power factor of the load. The load line of Fig. 5.11 is drawn for lagging loads since all the points on the line are in the first quadrant and have positive vars.

Since the advent of digital computers circle diagrams have been of little practical use. They have been introduced here to illustrate some concepts of transmission-line operation. For instance, examination of Fig. 5.10 shows that there is a limit to the power that can be transmitted to the receiving end of the line for specified magnitudes of sending- and receiving-end voltages. An increase in power delivered means that the point k will move along the circle until the angle $\beta - \delta$ is zero; that is, more power will be delivered until δ equals β. Further increases in δ result in less power received. The maximum power is

$$P_{R,\,\text{max}} = \frac{|V_S| \cdot |V_R|}{|B|} - \frac{|A| \cdot |V_R|^2}{|B|} \cos (\beta - \alpha) \tag{5.63}$$

The load must draw a large leading current to achieve the condition of maximum power received.

In Fig. 5.11 the length of the vertical line from point a at the intersection of the load line and the $|V_{S4}|$ circle to point b on the $|V_{S3}|$ circle is the amount of

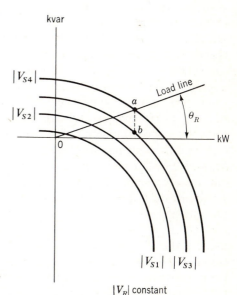

Figure 5.11 Receiving-end power circles for various values of $|V_S|$ and a constant $|V_R|$.

negative reactive power that must be drawn by capacitors added in parallel with the load to maintain constant $|V_R|$ when the sending-end voltage is reduced from $|V_{S4}|$ to $|V_{S3}|$. Addition of slightly more capacitive kilovars will result in a combined load of unity power factor and a further reduction in $|V_S|$ for the same $|V_R|$. Of course, this analysis means that a constant $|V_S|$ would result in higher $|V_R|$ capacitors are added in parallel with the lagging load.

One of many results made apparent by studying the circle diagram of Fig. 5.11 is the variation of sending-end voltage to maintain constant receiving-end voltage for different values of real and reactive power received. For instance, for a constant θ_R at the load the coordinates of the intersection of the load line with a circle of constant sending-end voltage are the P and Q of the load for that value of $|V_S|$ and the $|V_R|$ for which the diagram is drawn.

5.9 REACTIVE COMPENSATION OF TRANSMISSION LINES

The performance of transmission lines, especially those of medium length and longer, can be improved by reactive compensation of a series or parallel type. Series compensation consists of a capacitor bank placed in series with each phase conductor of the line. Shunt compensation refers to the placement of inductors from each line to neutral to reduce partially or completely the shunt susceptance of a high-voltage line, which is particularly important at light loads when the voltage at the receiving end may otherwise become very high.

Series compensation reduces the series impedance of the line, which is the principal cause of voltage drop and the most important factor in determining the maximum power which the line can transmit. In order to understand the effect of series impedance Z on maximum power transmission we examine Eq. (5.63) and see that maximum power transmitted is dependent upon the reciprocal of the generalized circuit constant B which for the nominal π equals Z and for the equivalent π equals $Z (\sinh \gamma l)/\gamma l$. Because the A, C, and D constants are functions of Z they will also change in value, but these changes will be small in comparison to the change in B.

The desired reactance of the capacitor bank can be determined by compensating for a specific amount of the total inductive reactance of the line. This leads to the term "compensation factor" which is defined by X_C/X_L where X_C is the capacitive reactance of the series capacitor bank per phase and X_L is the total inductive reactance of the line per phase.

When the nominal-π circuit is used to represent the line and capacitor bank, the physical location of the capacitor bank along the line is not taken into account. If only the sending- and receiving-end conditions of the line are of interest, this will not create any significant error. However, when the operating conditions along the line are of interest, the physical location of the capacitor bank must be taken into account. This can be accomplished most easily by determining the *ABCD* constants of the portions of line on each side of the capacitor bank and representing the capacitor bank by its *ABCD* constants. The equivalent constants of the series combination of line-capacitor-line can then be determined by applying the equations found in Table A.6 in the Appendix.

In the southwestern part of the United States series compensation is especially important because large generating plants are located hundreds of miles from load centers and large amounts of power must be transmitted over long distances. The lower voltage drop in the line with series compensation is an additional advantage. Series capacitors are also useful in balancing the voltage drop of two parallel lines.

Example 5.4 In order to show the relative changes in the B constant with respect to the change of the A, C, and D constants of a line when series compensation is applied, find the constants for the line of Example 5.1 uncompensated and for a series compensation factor of 70%.

SOLUTION The equivalent-π circuit and quantities found in Examples 5.1 and 5.3 can be used with Eqs. (5.37) to find, for the uncompensated line

$$A = D = \cosh \gamma l = 0.8904 \,\underline{/1.34°}$$

$$B = Z' = 186.78 \,\underline{/79.46°} \; \Omega$$

$$C = \frac{\sinh \gamma l}{Z_c} = \frac{0.4596 \,\underline{/84.94°}}{406.4 \,\underline{/-5.48°}}$$

$$= 0.001131 \,\underline{/90.42°} \; \mho$$

The series compensation alters only the series arm of the equivalent-π circuit. The new series arm impedance is also the generalized constant B. So

$$B = 186.78 \,\underline{/79.46°} - j0.7 \times 230(0.415 + 0.4127)$$

$$= 34.17 + j50.38 = 60.88 \,\underline{/55.85°} \; \Omega$$

and by Eqs. (5.10)

$$A = 60.88 \,\underline{/55.85°} \times 0.000599 \,\underline{/89.81°} + 1 = 0.970 \,\underline{/1.24°}$$

$$C = 2 \times 0.000599 \,\underline{/89.81°} + 60.88 \,\underline{/55.85°} (0.000599 \,\underline{/89.81})^2$$

$$= 0.001180 \,\underline{/90.41°} \; \mho$$

The example shows that compensation has reduced the constant B to about one-third of its value for the uncompensated line without affecting the A and C constants appreciably. Thus, maximum power which can be transmitted is increased by about 300%.

When a transmission line, with or without series compensation, has the desired load transmission capability, attention is turned to operation under light loads or at no load. Charging current is an important factor to be considered and should not be allowed to exceed the rated full-load current of the line.

Equation (4.25) shows us that the charging current is usually defined as $B_C|V|$ if B_C is the total capacitive susceptance of the line and $|V|$ is the rated voltage to neutral. As noted following Eq. (4.25) this calculation is not an exact determination of charging current because of the variation of $|V|$ along the line. If we connect inductors from line to neutral at various points along the line so that the total inductive susceptance is B_L, the charging current becomes

$$I_{chg} = (B_C - B_L)|V|$$

$$= B_C|V|\left(1 - \frac{B_L}{B_C}\right) \qquad (5.64)$$

We recognize that the charging current is reduced by the term in parentheses. The shunt compensation factor is B_L/B_C.

The other benefit of shunt compensation is the reduction of the receiving-end voltage of the line which on long high-voltage lines tends to become too high at no load. In the discussion preceding Eq. (5.11) we noted that $|V_S|/|A|$ equals $|V_{R, NL}|$. We also have seen that when shunt capacitance is neglected A equals 1.0. In the medium-length and longer lines, however, the presence of capacitance reduces A. Thus, the reduction of the shunt susceptance to a value of $(B_C - B_L)$ can limit the rise of the no-load voltage at the receiving-end of the line if shunt inductors are introduced as load is removed.

By applying both series and shunt compensation to long transmission lines it is possible to transmit large amounts of power efficiently and within the desired voltage constraints. Ideally the series and shunt elements should be placed at intervals along the line. Series capacitors can be bypassed and shunt inductors switched off when desirable. As with series compensation $ABCD$ constants provide a straightforward method of analysis of shunt compensation.

Example 5.5 Find the voltage regulation of the line of Example 5.1 when a shunt inductor is connected at the receiving end of the line during no-load conditions if the reactor compensates for 70% of the total shunt admittance of the line.

SOLUTION From Example 5.1 the shunt admittance of the line is

$$y = j5.105 \times 10^{-6} \ \text{℧/mi}$$

and for the entire line

$$B_C = j5.105 \times 10^{-6} \times 230 = 0.001174 \ \text{℧}$$

For 70% compensation

$$B_L = -j0.7 \times 0.001174 = 0.000822$$

We know the $ABCD$ constants of the line from Example 5.4. Table A.6 in the

Appendix tells us that the inductor alone is represented by the generalized constants

$$A = D = 1 \qquad B = 0 \qquad C = -j0.000822 \; \mho$$

The equation in Table A.6 for combining two networks in series tells us that for the line and inductor

$$A_{eq} = 0.8904 \underline{/1.34°} + 186.78 \underline{/79.46°}(0.000822 \underline{/-90°})$$

$$= 1.0411 \underline{/-0.40°}$$

The voltage regulation with the shunt reactor connected at no load becomes

$$\frac{137.85/1.0411 - 124.13}{124.13} = 6.67\%$$

which is a considerable reduction from the value of 24.7% for the regulation of the uncompensated line.

5.10 TRANSMISSION LINE TRANSIENTS

The transient overvoltages which occur on a power system are either of external origin (for example, a lightning discharge) or are generated internally by switching operations. In general the transients on transmission systems are caused by any sudden change in the operating condition or configuration of the systems. Lightning is always a potential hazard to power system equipment, but switching operations can also cause equipment damage. At voltages up to about 230 kV the insulation level of the lines and equipment is dictated by the need to protect against lightning. On systems where voltages are above 230 kV but less than 700 kV switching operations as well as lightning are potentially damaging to insulation. At voltages above 700 kV switching surges are the main determinant of the level of insulation.

Underground cables are, of course, immune to direct lightning strokes and can be protected against transients originating on overhead lines. However, for economic and technical reasons overhead lines at transmission voltage levels prevail except under unusual circumstances and for short distances such as under a river.

Overhead lines can be protected from direct strokes of lightning in most cases by one or more wires at ground potential strung above the power-line conductors as mentioned in the description of Fig. 5.1. These protecting wires, called ground wires, are connected to ground through the transmission towers supporting the line. The zone of protection is usually considered to be 30° on each side of vertical beneath a ground wire; that is, the power lines must come within this 60° sector. The ground wires, rather than the power line, receive the lightning strokes in most cases.

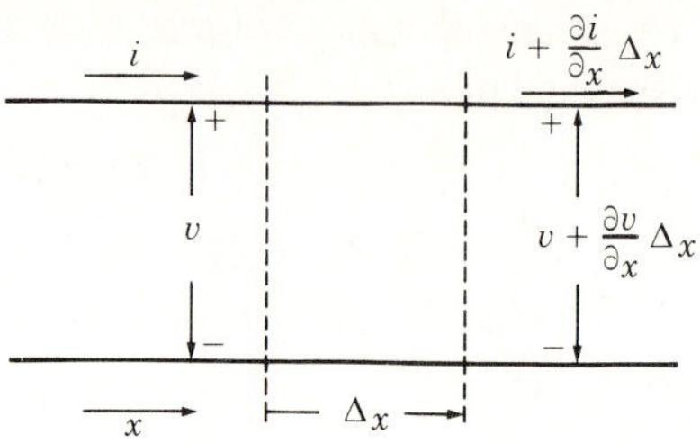

Figure 5.12 Schematic diagram of an elemental section of a transmission line showing one phase and neutral return. Voltage v and current i are functions of both x and t. The distance x is measured from the sending end of the line.

Lightning strokes hitting ground wires or power conductors cause an injection of current which divides with half the current flowing in one direction and half in the other. The crest value of current along the struck conductor varies widely because of the wide variation in the intensity of the strokes. Values of 10,000 A and upward are typical. In the case where a power line receives a direct stroke the damage to equipment at line terminals is caused by the voltages between line and ground resulting from the injected charges which travel along the line as current. These voltages are typically above a million volts. Strokes to the ground wires can also cause high-voltage surges on the power lines by electromagnetic induction.

5.11 TRANSIENT ANALYSIS: TRAVELING WAVES

The study of surges on transmission lines regardless of their origin is very complex and we can consider here only the case of a lossless line.†

A lossless line is a good representation for lines of high frequency where ωL and ωC become very large compared to R and G. For lightning surges on a power transmission line the study of a lossless line is a simplification that enables us to understand some of the phenomena without becoming too involved in complicated theory.

Our approach to the problem is similar to that used earlier for deriving the steady-state voltage and current relations for the long line with distributed constants. We shall now measure the distance x along the line from the sending end (rather than from the receiving end) to the differential element of length Δx shown in Fig. 5.12. The voltage v and the current i are functions of both x and t so that we need to use partial derivatives. The series voltage drop along the elemental length of line is

$$i(R\Delta x) + (L\Delta x)\frac{\partial i}{\partial t}$$

and we can write

$$\frac{\partial v}{\partial x}\Delta x = -\left(Ri + L\frac{\partial i}{\partial t}\right)\Delta x \tag{5.65}$$

† For further study see A. Greenwood, *Electrical Transients in Power Systems*, Wiley-Interscience, New York, 1971.

The negative sign is necessary because $v + (\partial v/\partial x)\,\Delta x$ must be less than v for positive values of i and $\partial i/\partial t$. Similarly

$$\frac{\partial i}{\partial x}\,\Delta x = -\left(Gv + C\frac{\partial v}{\partial t}\right)\Delta x \qquad (5.66)$$

We can divide through both Eqs. (5.65) and (5.66) by Δx, and since we are considering only a lossless line, R and G will equal zero to give

$$\frac{\partial v}{\partial x} = -L\frac{\partial i}{\partial t} \qquad (5.67)$$

and

$$\frac{\partial i}{\partial x} = -C\frac{\partial v}{\partial t} \qquad (5.68)$$

Now we can eliminate i by taking the partial derivative of both terms in Eq. (5.67) with respect to x and the partial derivative of both terms in Eq. (5.68) with respect to t. This procedure yields $\partial^2 i/\partial x\partial t$ in both resulting equations, and eliminating this second partial derivative of i between the two equations yields

$$\frac{1}{LC}\frac{\partial^2 v}{\partial x^2} = \frac{\partial^2 v}{\partial t^2} \qquad (5.69)$$

Equation (5.69) is the so-called traveling-wave equation of a lossless transmission line. A solution of the equation is a function of $(x - vt)$, and the voltage is expressed by

$$v = f(x - vt) \qquad (5.70)$$

The function is undefined but must be single valued. The constant v must have the dimensions of meters per second if x is in meters and t is in seconds. We can verify this solution by substituting this expression for v into Eq. (5.69) to determine v. First we make the change in variable

$$u = x - vt \qquad (5.71)$$

and write

$$v(x, t) = f(u) \qquad (5.72)$$

Then

$$\frac{\partial v}{\partial t} = \frac{\partial f(u)}{\partial u}\frac{\partial u}{\partial t}$$

$$= -v\frac{\partial f(u)}{\partial u} \qquad (5.73)$$

and

$$\frac{\partial^2 v}{\partial t^2} = v^2\frac{\partial^2 f(u)}{\partial u^2} \qquad (5.74)$$

Similarly we obtain

$$\frac{\partial^2 v}{\partial x^2} = \frac{\partial^2 f(u)}{\partial u^2} \qquad (5.75)$$

Substituting these second partial derivatives of v into Eq. (5.69) yields

$$\frac{1}{LC}\frac{\partial^2 f(u)}{\partial u^2} = v^2 \frac{\partial^2 f(u)}{\partial u^2} \qquad (5.76)$$

and we see that Eq. (5.70) is a solution of Eq. (5.69) if

$$v = \frac{1}{\sqrt{LC}} \qquad (5.77)$$

The voltage as expressed by Eq. (5.70) is a wave traveling in the positive x direction. Figure 5.13 shows a function of $(x - vt)$ which is similar to the shape of a wave of voltage traveling along a line which has been struck by lightning. The function is shown for two values of time t_1 and t_2 where $t_2 > t_1$. An observer traveling with the wave and staying at the same point on the wave sees no change in voltage at that point. To the observer

$$x - vt = \text{a constant}$$

from which it follows that

$$\frac{dx}{dt} = v = \frac{1}{\sqrt{LC}} \quad \text{m/s} \qquad (5.78)$$

for L and C in H/m and F/m, respectively. Thus the voltage wave travels in the positive x direction with the velocity v.

A function of $(x + vt)$ can also be shown to be a solution of Eq. (5.69) and, by similar reasoning, can be properly interpreted as a wave traveling in the negative x direction. The general solution of Eq. (5.69) is thus

$$v = f_1(x - vt) + f_2(x + vt) \qquad (5.79)$$

which is a solution for simultaneous occurrence of forward and backward components on the line. Initial conditions and boundary (terminal) conditions determine the particular values for each component.

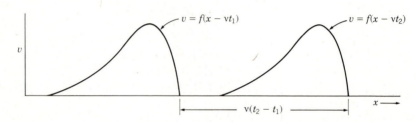

Figure 5.13 A voltage wave which is a function of $(x - vt)$ is shown for values of t equal to t_1 and t_2.

If we express a forward traveling wave, also called an incident wave, as

$$v^+ = f_1(x - vt) \tag{5.80}$$

a wave of current will result from the moving charges and will be expressed by

$$i^+ = \frac{1}{\sqrt{L/C}} f_1(x - vt) \tag{5.81}$$

which can be verified by substitution of these values of voltage and current in Eq. (5.67) and recalling that v is equal to $1/\sqrt{LC}$.

Similarly for a backward moving wave of voltage where

$$v^- = f_2(x + vt) \tag{5.82}$$

the corresponding current is

$$i^- = -\frac{1}{\sqrt{L/C}} f_2(x + vt) \tag{5.83}$$

From Eqs. (5.80) and (5.81) we note that

$$\frac{v^+}{i^+} = \sqrt{\frac{L}{C}} \tag{5.84}$$

and from Eqs. (5.82) and (5.83) that

$$\frac{v^-}{i^-} = -\sqrt{\frac{L}{C}} \tag{5.85}$$

If we had decided to assume the positive direction of current for i^- to be in the direction of travel of the backward traveling wave the minus signs would change to plus signs in Eqs. (5.83) and (5.85). We choose however to keep the positive x direction as the direction for positive current for both forward and backward traveling waves.

The ratio of v^+ to i^+ is called the characteristic impedance Z_c of the line. We have encountered characteristic impedance previously in the steady-state solution for the long line where Z_c was defined as $\sqrt{z/y}$ which equals $\sqrt{L/C}$ when R and G are zero.

5.12 TRANSIENT ANALYSIS: REFLECTIONS

We shall now consider what happens when a voltage is first applied to the sending end of a transmission line which is terminated in an impedance Z_R. For our very simple treatment we will consider Z_R to be a pure resistance. If the termination is other than a pure resistance we would resort to Laplace transforms. The transforms of voltage, current, and impedance would be functions of the Laplace transform variable s.

When the switch is closed applying a voltage to a line a wave of voltage v^+

accompanied by a wave of current i^+ starts to travel along the line. The ratio of the voltage v_R at the end of the line at any time to the current i_R at the end of the line must equal the terminating resistance Z_R. Therefore the arrival of v^+ and i^+ at the receiving end where their values are v_R^+ and i_R^+ must result in backward traveling or reflected waves v^- and i^- having values v_R^- and i_R^- at the receiving end such that

$$\frac{v_R}{i_R} = \frac{v_R^+ + v_R^-}{i_R^+ + i_R^-} = Z_R \tag{5.86}$$

where v_R^- and i_R^- are the reflected waves v^- and i^- measured at the receiving end.

If we let $Z_c = \sqrt{L/C}$ we find from Eqs. (5.84) and (5.85)

$$i_R^+ = \frac{v_R^+}{Z_c} \tag{5.87}$$

and

$$i_R^- = -\frac{v_R^-}{Z_c} \tag{5.88}$$

Then substituting these values of i_R^+ and i_R^- in Eq. (5.86) yields

$$v_R^- = \frac{Z_R - Z_c}{Z_R + Z_c} v_R^+ \tag{5.89}$$

The voltage v_R^- at the receiving end is evidently the same function of t as v_R^+ (but with diminished magnitude unless Z_R is zero or infinity). The reflection coefficient ρ_R for voltage at the receiving end of the line is defined as v_R^-/v_R^+ so, *for voltage,*

$$\rho_R = \frac{Z_R - Z_c}{Z_R + Z_c} \tag{5.90}$$

We note from Eqs. (5.87) and (5.88) that

$$\frac{i_R^+}{i_R^-} = -\frac{v_R^+}{v_R^-} \tag{5.91}$$

and that therefore the reflection coefficient for current is always the negative of the reflection coefficient for voltage.

If the line is terminated in its characteristic impedance Z_c, we see that the reflection coefficient for both voltage and current will be zero. There will be no reflected waves, and the line will behave as though it is infinitely long. Only when a reflected wave returns to the sending end does the source sense that the line is not either infinitely long or terminated in Z_c.

Termination in a short circuit results in a ρ_R for voltage of -1. If the termination is an open circuit, Z_R is infinite and ρ_R is found by dividing the numerator and denominator in Eq. (5.90) by Z_R and allowing Z_R to approach infinity to yield $\rho_R = 1$ in the limit for voltage.

We should note at this point that waves traveling back toward the sending end will cause new reflections as determined by the reflection coefficient at the sending end ρ_s. For impedance at the sending end equal to Z_s Eq. (5.90) becomes

$$\rho_s = \frac{Z_s - Z_c}{Z_s + Z_c} \tag{5.92}$$

With sending-end impedances of Z_s the value of the initial voltage impressed across the line will be the source voltage multiplied by $Z_c/(Z_s + Z_c)$. Equation (5.84) shows that the incident wave of voltage experiences a line impedance of Z_c, and at the instant when the source is connected to the line Z_c and Z_s in series act as a voltage divider.

Example 5.6 A dc source of 120 V with negligible resistance is connected through a switch S to a lossless transmission line having $Z_c = 30\ \Omega$. If the line is terminated in a resistance of 90 Ω, plot v_R versus time until $t = 5T$ where T is the time for a voltage wave to travel the length of the line. The circuit is shown in Fig. 5.14a.

SOLUTION When S is closed the incident wave of voltage starts to travel along the line and is expressed by the equation

$$v = 120U(vt - x)$$

where $U(vt - x)$ is the unit step function which equals zero when $(vt - x)$ is negative and equals unity when $(vt - x)$ is positive. There can be no reflected wave until the incident wave reaches the end of the line. Impedance to the incident wave is $Z_c = 30\ \Omega$. Since resistance of the source is zero, $v^+ = 120$ V

$$\rho_R = \frac{90 - 30}{90 + 30} = \frac{1}{2}$$

When v^+ reaches the end of the line a reflected wave originates of value

$$v^- = (\tfrac{1}{2})120 = 60 \text{ V}$$

and so

$$v_R = 120 + 60 = 180 \text{ V}$$

When $t = 2T$ the reflected wave arrives at the sending end where the sending end reflection coefficient ρ_s is calculated by Eq. (5.92). The line termination for the reflected wave is Z_s, the impedance in series with the source, or zero in this case. So

$$\rho_s = \frac{0 - 30}{0 + 30} = -1$$

and a reflected wave of -60 V starts toward the receiving end to keep the sending-end voltage equal to 120 V. This new wave reaches the receiving end at $t = 3T$ and reflects toward the sending end a wave of

$$\tfrac{1}{2}(-60) = -30 \text{ V}$$

and the receiving-end voltage becomes

$$180 - 60 - 30 = 90 \text{ V}$$

An excellent method of keeping track of the various reflections as they occur is the lattice diagram shown in Fig. 5.14b. Here time is measured along the vertical axis in intervals of T. On the slant lines are recorded the values of the incident and reflected waves. In the space between the slant lines is shown the sum of all the waves above and is the current or voltage for a point in that area of the chart. For instance, at x equal to three-fourths of the line length and $t = 4.25 \ T$ the intersection of the dashed lines through these points is within the area which indicates the voltage is 90 V.

Figure 5.14c shows the receiving-end voltage plotted against time. The voltage is approaching its steady-state value of 120 V.

Lattice diagrams for current may also be drawn. We must remember, however, that the reflection coefficient for current is always the negative of the reflection coefficient for voltage.

If the resistance at the end of the line of Example 5.6 is reduced to 10 Ω as shown in the circuit of Fig. 5.15a, the lattice diagram and plot of voltage are as shown in Figs. 5.15b and c. The resistance of 10 Ω gives us a negative value for the reflection coefficient for voltage, which will always occur for resistance where Z_R is less than Z_c. As we see by comparing Figs. 5.14 and 5.15 the negative ρ_R causes the receiving-end voltage to build up gradually to 120 V while a positive ρ_R causes an initial jump in voltage to a value greater than that of the voltage originally applied at the sending end.

Reflections do not necessarily arise only at the end of a line. If one line is joined to a second line of different characteristic impedance, as in the case of an overhead line connected to an underground cable, a wave incident to the junction will behave as though the first line is terminated in the Z_c of the second line. However, the part of the incident wave which is not reflected will travel (as a refracted wave) along the second line at whose termination a reflected wave will occur. Bifurcations of a line will also cause reflected and refracted waves.

It should now be obvious that a thorough study of transmission-line transients in general is a complicated problem. We realize, however, that a voltage surge such as that shown in Fig. 5.13 encountering an impedance at the end of a lossless line (for instance, at a transformer bus) will cause a voltage wave of the same shape to travel back toward the source of the surge. The reflected wave will be reduced in magnitude if the terminal impedance is other than a short or open circuit, but if Z_R is greater than Z_c our study has shown that the peak terminal voltage will be higher than, often close to, double the peak voltage of the surge.

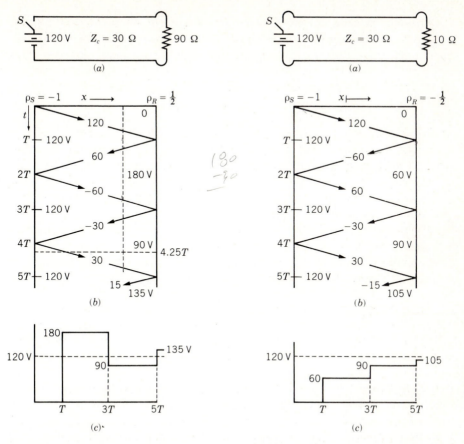

Figure 5.14 Circuit diagram, lattice diagram, and plot of voltage versus time for Example 5.6 where the receiving-end resistance is 90 Ω.

Figure 5.15 Circuit diagram, lattice diagram, and plot of voltage versus time when the receiving-end resistance for Example 5.6 is changed to 10 Ω.

Terminal equipment is protected by surge arresters which are also called lightning arresters. An ideal arrester connected from the line to a grounded neutral would (*a*) become conducting at a design voltage above the arrester rating, (*b*) limit the voltage across its terminals to this design value, and (*c*) becomes nonconducting again when the line-to-neutral voltage drops below the design value.

Originally, an arrester was simply an air gap. In this application when the surge voltage reaches a value for which the gap is designed an arc occurs to cause an ionized path to ground, essentially a short circuit. However, when the surge ends the 60-Hz current from the generators still flows through the arc to ground. The arc has to be extinguished by the opening of circuit breakers.

Arresters capable of extinguishing 60-Hz current after conducting surge current to ground were developed later. These arresters are made using non-linear resistors in series with air gaps to which an arc quenching capability has been added. The nonlinear resistance decreases rapidly as the voltage across it rises. Typical resistors made of silicon carbide conduct current proportional to approximately the fourth power of the voltage across the resistor. When the gaps arc over as a result of a voltage surge, a low-resistance current path to ground is provided through the nonlinear resistors. After the surge ends and the voltage across the arrester returns to the normal line-to-neutral level the resistance is sufficient to limit the arc current to a value which can be quenched by the series gaps. Quenching is usually accomplished by cooling and deionizing the arc by elongating it magnetically between insulating plates.

The most recent development in surge arresters is the use of zinc oxide in place of silicon carbide. The voltage across the zinc oxide resistor is extremely constant over a very high range of current, which means that its resistance at normal line voltage is so high that a series air gap is not necessary to limit the drain of 60-Hz current at normal voltage.†

5.13 DIRECT-CURRENT TRANSMISSION

The transmission of energy by direct current becomes economical when compared to ac transmission only when the extra cost of the terminal equipment required for dc lines is offset by the lower cost of building the lines. Converters at the two ends of the dc lines operate both as rectifiers to change the generated alternating to direct current and as inverters for converting direct to alternating current so that power can flow in either direction.

The year 1954 is generally recognized as the starting date for modern high-voltage dc transmission when a dc line began service at 100 kV from Vastervik on the mainland of Sweden to Visby on the island of Gotland, a distance of 100 km (62.5 mi) across the Baltic Sea. Static conversion equipment was in operation much earlier to transfer energy between systems of 25 and 60 Hz, essentially a dc transmission line of zero length. In the United States a dc line operating at 800 kV transfers power generated in the Pacific Northwest to the southern part of California. As the cost of conversion equipment decreases with respect to the cost of line construction the economical minimum length of dc lines also decreases and at this time is about 600 km (375 mi).

Operation of a dc line began in 1977 to transmit power from a mine-mouth generating plant burning lignite at Center, North Dakota to near Duluth,

---

† See E. C. Sakshaug, J. S. Kresge, and S. A. Miske, Jr., "A New Concept in Station Arrester Design," *IEEE Trans. Power Appar. Syst.*, vol. PAS-96, no. 2, March/April 1977, pp. 647–656.

Minnesota, a distance of 740 km (460 mi). Preliminary studies showed that the dc line including terminal facilities would cost about 30% less than the comparable ac line and auxiliary equipment. This line operates at ± 250 kV (500 kV line-to-line) and transmits 500 MW.

Direct-current lines usually have one conductor which is at a positive potential with respect to ground and a second conductor operating at an equal negative potential. Such a line is said to be bipolar. The line could be operated with one energized conductor with the return path through the earth, which has a much lower resistance to direct than to alternating current. In this case, or with a grounded return conductor, the line is said to be monopolar.

In addition to the lower cost of dc transmission over long distances there are other advantages. Voltage regulation is less of a problem since at zero frequency ωL is no longer a factor, whereas it is the chief contributor to voltage drop in an ac line. Another advantage of direct current is the possibility of monopolar operation in an emergency when one side of a bipolar line becomes grounded.

The fact that underground ac transmission is limited to about 50 km because of excessive charging current at longer distances, direct current was chosen to transfer power under the English Channel between Great Britain and France. The use of direct current for this installation also avoided the difficulty of synchronizing the ac systems of the two countries.

No network of dc lines is possible at this time because no circuit breaker is available for direct current comparable to the highly developed ac breakers. The ac breaker can extinguish the arc which is formed when the breaker opens because zero current occurs twice in each cycle. The direction and amount of power in the dc line is controlled by the converters in which grid-controlled mercury-arc devices are being displaced by the semiconductor rectifier (SCR). A rectifier unit will contain perhaps 200 SCRs.

Still another advantage of direct current is the smaller amount of right of way required. The distance between the two conductors of the North Dakota–Duluth 500-kV line is 25 feet. The 500-kV ac line shown in Fig. 5.1 has 60.5 feet between the outside conductors. Another consideration is the peak voltage of the ac line which is $\sqrt{2} \times 500 = 707$ kV. So the line requires more insulation between the tower and conductors as well as greater clearance above the earth.

We conclude that dc transmission has many advantages over alternating current, but dc transmission remains very limited in usage except for long lines because there is no dc device which can provide the excellent switching operations and protection of the ac circuit breaker. There is also no simple device to change the voltage level, which the transformer accomplishes for ac systems.

5.14 SUMMARY

The long-line equations given by Eqs. (5.35) and (5.36) are, of course, valid for a line of any length. The approximations for the short and medium-length lines make analysis easier in the absence of a computer.

Circle diagrams were introduced because of their instructional value in showing the maximum power which can be transmitted by a line and also in showing the effect of power factor of the load or the addition of capacitors.

ABCD constants provide a straightforward means of writing equations in a more concise form and are very convenient in problems involving network reduction. Their usefulness is apparent in the discussion of series and shunt reactive compensation.

The simple discussion of transients although confined to lossless lines and dc sources should give some idea of the complexity of the study of transients which arise from lightning and switching in power systems.

PROBLEMS

5.1 An 18-km 60-Hz single circuit three-phase line is composed of *Partridge* conductors equilaterally spaced with 1.6 m between centers. The line delivers 2500 kW at 11 kV to a balanced load. What must be the sending-end voltage when the power factor is (a) 80% lagging, (b) unity, and (c) 90% leading? Assume a wire temperature of 50°C.

5.2 A 100-mi, three-phase transmission line delivers 55 MVA at 0.8 power factor lagging to the load at 132 kV. The line is composed of *Drake* conductors with flat horizontal spacing of 11.9 ft between adjacent conductors. Determine the sending-end voltage, current, and power. Assume a wire temperature of 50°C.

5.3 Find the *ABCD* constants of a π circuit having a 600-Ω resistor for the shunt branch at the sending end, a 1-kΩ resistor for the shunt branch at the receiving end, and an 80-Ω resistor for the series branch.

5.4 The *ABCD* constants of a three-phase transmission line are

$$A = D = 0.936 + j0.016 = 0.936\underline{/0.98°}$$
$$B = 33.5 + j138 = 142.0\underline{/76.4°}\,\Omega$$
$$C = (-5.18 + j914) \times 10^{-6}\,\mho$$

The load at the receiving end is 50 MW at 220 kV with a power factor of 0.9 lagging. Find the magnitude of the sending-end voltage and the voltage regulation. Assume the magnitude of the sending-end voltage remains constant.

5.5 Use per-unit values on a base of 230 kV, 100 MVA to find voltage, current, power, and power factor at the sending end of a transmission line delivering a load of 60 MW at 230 kV with 0.8 power-factor lagging. The three-phase line is arranged in flat, horizontal spacing with 15 ft between adjacent *Ostrich* conductors. Line length is 70 mi. Assume a wire temperature of 50°C. Note that base admittance must be the reciprocal of base impedance.

5.6 Evaluate cosh θ and sinh θ for $\theta = 0.5\underline{/82°}$.

5.7 Justify Eq. (5.52) by substituting for the hyperbolic functions the equivalent exponential expressions.

5.8 A 60-Hz three-phase transmission is 175 mi long. It has a total series impedance of $35 + j140\,\Omega$ and a shunt admittance of $930 \times 10^{-6}\underline{/90°}\,\mho$. It delivers 40 MW at 220 kV, with 90% power factor lagging. Find the voltage at the sending end by (a) the short-line approximation, (b) the nominal-π approximation, (c) the long-line equation.

5.9 Determine the equivalent-π circuit for the line of Prob. 5.8.

5.10 Determine the voltage regulation for the line described in Prob. 5.8. Assume that the sending-end voltage remains constant.

5.11 A three-phase 60-Hz transmission line is 250 mi long. The voltage at the sending end is 220 kV. The parameters of the line are $R = 0.2\,\Omega/\text{mi}$, $X = 0.8\,\Omega/\text{mi}$, and $Y = 5.3\,\mu\mho/\text{mi}$. Find the sending-end current when there is no load on the line.

5.12 If the load on the line described in Prob. 5.11 is 80 MW at 220 kV, with unity power factor, calculate the current, voltage, and power at the sending end. Assume that the sending-end voltage is held constant, and calculate the voltage regulation of the line for the load specified above.

5.13 Rights of way for transmission circuits are becoming difficult to obtain in urban areas and existing lines are often upgraded by reconductoring the line with larger conductors or by reinsulating the line for operation at higher voltage. Thermal considerations and maximum power which the line can transmit are the important considerations. A 138-kV line is 50 km long and is composed of *Partridge* conductors with flat horizontal spacing of 5 m between adjacent conductors. Neglect resistance and find the percent increase in power which can be transmitted for constant $|V_S|$ and $|V_R|$ while δ is limited to 45° (a) if the *Partridge* conductor is replaced by *Osprey* which has more than twice the area of aluminum in square millimeters, (b) if a second *Partridge* conductor is placed in a two-conductor bundle 40 cm from the original conductor and a center-to-center distance between bundles of 5 m, and (c) if the voltage of the original line is raised to 230 kV with increased conductor spacing of 8 m.

5.14 Construct a receiving-end power-circle diagram similar to Fig. 5.10 for the line of Prob. 5.8. Locate the point corresponding to the load of Prob. 5.8, and locate the center of circles for various values of $|V_S|$ if $|V_R| = 220$ kV. Draw the circle passing through the load point. From the measured radius of the latter circle determine $|V_S|$, and compare this value with the values calculated for Prob. 5.8.

5.15 Use the diagram constructed for Prob. 5.14 to determine the sending-end voltage for various values of kilovars supplied by synchronous condensers in parallel with the designated load at the receiving end. Include values of added kilovars to give unity power factor and a power factor of 0.9 leading at the receiving end. Assume that the sending-end voltage is adjusted to maintain 220 kV at the load.

5.16 A receiving-end power-circle diagram is drawn for a constant receiving-end voltage. For a certain load at this receiving-end voltage the sending-end voltage is 115 kV. The receiving-end circle for $|V_S| = 115$ kV has a radius of 5 in. The horizontal and vertical coordinates of the receiving-end circles are -0.25 and -4.5 in, respectively. Find the voltage regulation for the load.

5.17 A three-phase transmission line is 300 mi long and serves a load of 400 MVA, 0.8 lagging-power factor at 345 kV. The *ABCD* constants of the line are

$$A = D = 0.8180\underline{/1.3°}$$
$$B = 172.2\underline{/84.2°} \ \Omega$$
$$C = 0.001933\underline{/90.4°} \ \mho$$

(a) Determine the sending-end line-to-neutral voltage, the sending-end current and the percent voltage drop at full load.

(b) Determine the receiving-end line-to-neutral voltage at no load, the sending-end current at no load, and the voltage regulation.

5.18 A series capacitor bank is to be installed at the midpoint of the 300-mi line of Prob. 5.17. The *ABCD* constants for 150 mi of line are

$$A = D = 0.9534\underline{/0.3°}$$
$$B = 90.33\underline{/84.1°} \ \Omega$$
$$C = 0.001014\underline{/90.1°} \ \mho$$

The *ABCD* constants of the series capacitor bank are

$$A = D = 1\underline{/0°}$$
$$B = 146.6\underline{/-90°} \ \Omega$$
$$C = 0$$

(a) Determine the equivalent *ABCD* constants of the series combination of the line-capacitor-line. (See Table A.6 in the Appendix.)

(b) Rework Prob. 5.17 using these equivalent *ABCD* constants.

5.19 The shunt admittance of a 300-mi transmission line is

$$y_c = 0 + j6.87 \times 10^{-6} \text{ } \mho/\text{mi}$$

Determine the *ABCD* constants of a shunt reactor that will compensate for 60% of the total shunt admittance.

5.20 A 250-Mvar, 345-kV shunt reactor whose admittance is 0.0021 $\underline{/-90°}$ $\mho$ is connected to the receiving end of the 300-mi line of Prob. 5.17 at no load.

     (*a*) Determine the equivalent *ABCD* constants of the line in series with the shunt reactor. (See Table A.6 in the Appendix.)

     (*b*) Rework part (*b*) of Prob. 5.17 using these equivalent *ABCD* constants and the sending-end voltage found in Prob. 5.17.

5.21 Draw the lattice diagram for current and plot current versus time at the sending end of the line of Example 5.6 for the line terminated in (*a*) an open circuit and (*b*) a short circuit.

5.22 Plot voltage versus time for the line of Example 5.6 at a point distant from the sending end equal to one-fourth of the length of the line if the line is terminated in a resistance of 10 Ω.

5.23 Solve Example 5.6 if a resistance of 54 Ω is in series with the source.

5.24 Voltage from a dc source is applied to an overhead transmission line by closing a switch. The end of the overhead line is connected to an underground cable. Assume both the line and the cable are lossless and that the initial voltage along the line is v^+. If the characteristic impedances of the line and cable are 400 Ω and 50 Ω, respectively, and the end of the cable is open-circuited, find in terms of v^+ (*a*) the voltage at the junction of the line and cable immediately after the arrival of the incident wave and (*b*) the voltage at the open end of the cable immediately after arrival of the first voltage wave.

SYSTEM MODELING

At this point in our discussion of power systems we have completed development of the circuit model of a transmission line and have looked at some calculations of voltage, current, and power on the line. In this chapter we shall develop circuit models for the synchronous machine and for the power transformer.

The synchronous machine as an ac generator driven by a turbine is the device which converts mechanical energy to electrical energy. Conversely as a motor the machine converts electrical energy to mechanical energy. We are chiefly concerned with the synchronous generator, but we shall give some consideration to the synchronous motor. The induction motor is more widely used than a synchronous motor in industry, but its treatment is beyond the scope of this book. We cannot treat the synchronous machine fully, but there are many books on the subject of ac machinery which provide quite adequate analysis of generators, motors, and transformers.† Our treatment of the synchronous machine will enable us to have confidence in the equivalent circuit sufficient for understanding our further study of the role of the generator in power system analysis and will constitute a review for those who have devoted previous study to the subject.

---

† For a much more detailed discussion of synchronous machines and transformers, consult any of the texts on electric machinery such as A. E. Fitzgerald, C. Kingsley, Jr., and A. Kusko, *Electric Machinery*, 3d ed., McGraw-Hill Book Company, New York, 1971, or L. W. Matsch, *Electromagnetic and Electromechanical Machines*, 2d ed., IEP, New York, 1977.

The transformer provides the link between the generator and the transmission line and between the transmission line and the distribution system which delivers energy through still other transformers to the loads on the system. As with the synchronous machine our treatment of the transformer will not be comprehensive but will provide us with the circuit model suitable to our further study.

Then we shall see how a one-line diagram describes the combination of the component models to form a complete system, and we shall devote some further study to the application of per-unit quantities to calculations for the system.

6.1 CONSTRUCTION OF THE SYNCHRONOUS MACHINE

The two principal parts of a synchronous machine are ferromagnetic structures. The stationary part which is essentially a hollow cylinder is called the *stator* or *armature* and has longitudinal slots in which are coils of the armature winding. This winding carries the current supplied to an electrical load or system by a generator or the current received from an ac supply by a motor. The rotor is the part of the machine which is mounted on the shaft and rotates inside the hollow stator. The winding on the rotor is called the field winding and is supplied with dc current. The very high magnetomotive force (mmf) produced by this current in the field winding combines with the mmf produced by current in the armature winding. The resultant flux across the air gap between the stator and rotor generates voltage in the coils of the armature winding and provides the electro-magnetic torque between the stator and rotor. Figure 6.1 shows the threading of a four-pole cylindrical rotor into the stator of a 1525-MVA generator.

The dc current is supplied to the field winding by an *exciter* which may be a generator mounted on the shaft or a separate dc source connected to the field winding through brushes bearing on slip rings. Large ac generators usually have exciters consisting of an ac source with solid-state rectifiers.

If the machine is a generator the shaft is driven by a prime mover which is usually a steam or hydraulic turbine. The electro-magnetic torque developed in the generator when it delivers power opposes the torque of the prime mover. The difference between these two torques is due to losses in the iron core and friction. In a motor the electromagnetic torque developed in the machine except for core and friction losses is delivered to the shaft which drives the mechanical load.

Figure 6.2 shows a very elementary three-phase generator. The field winding is merely indicated by a coil. The generator in this figure is called a nonsalient pole machine because it has a cylindrical rotor. The rotor of Fig. 6.1 is the cylindrical type. In the actual machine the winding has a large number of turns distributed in slots around the circumference of the rotor. The strong magnetic field produced links the stator coils to induce voltage in the armature winding as the shaft is turned by the prime mover.

The stator is shown in cross section in Fig. 6.2. Opposite sides of a coil, which is almost rectangular, are in slots a and a' 180° apart. Similar coils are in

Figure 6.1 Photograph showing the threading of a four-pole cylindrical rotor into the stator of a 1525-MVA generator. (*Courtesy Utility Power Corporation, Wisconsin.*)

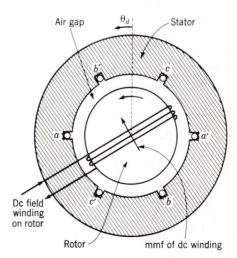

Figure 6.2 Elementary three-phase ac generator showing end view of the two-pole cylindrical rotor and cross section of the stator.

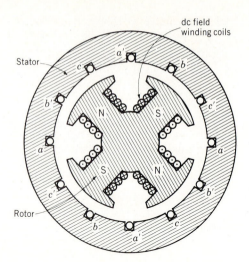

Figure 6.3 Cross section of an elementary stator and salient-pole rotor.

slots b and b' and slots c and c'. Coil sides in slots a, b, and c are 120° apart. The conductors shown in the slots indicate a coil of only one turn, but such a coil may have many turns and is usually in series with identical coils in adjacent slots to form a winding having ends designated a and a'. Windings designated b and b' and c and c' are the same as the a-a' winding except for their location around the armature.

Figure 6.3 shows a salient-pole machine which has four poles. Opposite sides of an armature coil are 90° apart. So there are two coils for each phase. Coil sides a, b, and c of adjacent coils are 60° apart. The two coils of each phase may be connected in series or in parallel.

Although not shown in Fig. 6.3 salient-pole machines usually have damper windings which consist of short-circuited copper bars through the pole face similar to part of a "squirrel cage" winding of an induction motor. The purpose of the damper winding is to reduce the mechanical oscillations of the rotor about synchronous speed which is determined, as we shall soon see, by the number of poles of the machine and the frequency of the system to which the machine is connected.

In the two-pole machine one cycle of voltage is generated for each revolution of the two-pole rotor. In the four-pole machine two cycles are generated in each coil per revolution. Since the number of cycles per revolution equals the number of pairs of poles the frequency of the generated voltage is

$$f = \frac{P}{2} \frac{N}{60} \qquad \text{Hz} \qquad (6.1)$$

where P is the number of poles and N is the rotor speed in revolutions per minute.

Since one cycle of voltage (360° of the voltage wave) is generated every time a pair of poles passes a coil we must distinguish between electrical degrees used to express voltage and current and mechanical degrees to express position of the rotor. In a two-pole machine electrical and mechanical degrees are equal. In a four-pole machine two cycles, or 720 electrical degrees, are produced per revolution of 360 mechanical degrees. The number of electrical degrees equals $P/2$ times the number of mechanical degrees in any machine.

By proper design of the rotor and by proper distribution of the stator windings around the armature very pure sinusoidal voltages will be generated. These voltages are called *no-load generated voltages* or simply *generated voltages*. For our analysis in later sections we shall consider the mmf produced by the rotor to be sinusoidally distributed.

If the ends of the windings are connected to each other and the junction is designated o, the generated voltages (labeled E_{ao}, E_{bo}, and E_{co} to agree with the notation adopted in Chap. 2) are displaced 120 electrical degrees from each other.

6.2 ARMATURE REACTION IN A SYNCHRONOUS MACHINE

If a balanced three-phase load is connected to a three-phase generator, balanced three-phase currents will flow in the phases of the armature winding. These armature currents will create additional mmfs which we need to investigate. We shall look at the mmf produced by each of the three windings which are spaced 120° apart around a two-pole machine as in Fig. 6.2. We shall also specify a cylindrical-rotor machine, again as in Fig. 6.2, so that all flux paths across the air gap of the machine will have essentially the same reluctance.

Since we may choose t to be zero when i_a has its maximum value the balanced three-phase currents in each phase may be expressed by the equations

$$i_a = I_m \cos \omega t \tag{6.2}$$

$$i_b = I_m \cos (\omega t - 120°) \tag{6.3}$$

$$i_c = I_m \cos (\omega t - 240°) \tag{6.4}$$

where ω is in electrical degrees per second. For our two-pole machine ω is also the angular velocity of the rotor in mechanical degrees per second. We shall assume that the positive direction chosen for current is toward the observer in conductor a in Fig. 6.2 and make similar assumptions for currents in conductors b and c.

Positions around the armature will be identified by the angular displacement θ_d measured from the axis of coil a as in Fig. 6.2. The subscript d is used to distinguish the displacement angle θ_d from the angle θ which we are accustomed to use to express the time phase between a voltage and a current. We recognize that the mmf around the armature produced by the armature current in any phase must be a function of both displacement θ_d and time t. We shall designate the mmf due to the current in phase a as $\mathscr{F}_a(\theta_d, t)$. Since this mmf is produced by i_a it must be a sinusoidal function of time in phase with i_a.

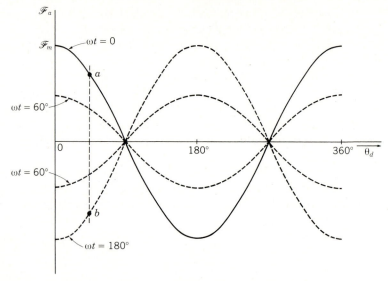

Figure 6.4 Distribution around the armature of mmf produced by the current in phase a of the generator of Fig. 6.2 for various values of ωt when i_a is expressed by Eq. (6.2).

The distribution of mmf around the periphery of the armature due to armature current is not sinusoidal. In a practical machine each coil occupies a number of slots and the distribution of mmf is nearly triangular, but in the analysis we are about to make it is customary to consider only the fundamental harmonic of the triangular wave, and this we shall do.

For the assumption of sinusoidal distribution of mmf the solid line in Fig. 6.4 shows the mmf of phase a around the armature as a function of θ_d at $t = 0$; that is when i_a has its maximum value. At this time the distribution of $\mathscr{F}_a$ around the armature is expressed by

$$\mathscr{F}_a(\theta_d, 0) = \mathscr{F}_m \cos \theta_d \qquad (6.5)$$

where $\mathscr{F}_m$ is the maximum value of $\mathscr{F}_a$.

For some values of t other than zero the dashed lines of Fig. 6.4 show the distribution of $\mathscr{F}_a$ around the armature. We note that the maximum value of $\mathscr{F}_a$ is $0.5\,\mathscr{F}_m$ when ωt is $60°$ because at that instant of time i_a equals $0.5\,I_m$ from Eq. (6.2). At the fixed position $\theta_d = 45°$, $\mathscr{F}_a$ is varying sinusoidally between values indicated by points a and b on Fig. 6.4. At $\theta_d = 90°$ $\mathscr{F}_a$ is always zero. So to account for time variation as well as spatial variation of $\mathscr{F}_a$ we have, since $\mathscr{F}_a$ is in time phase with i_a,

$$\mathscr{F}_a(\theta_d, t) = \mathscr{F}_m \times \underbrace{\cos \theta_d}_{\substack{\text{to account for} \\ \text{position on} \\ \text{the armature}}} \times \underbrace{\cos \omega t}_{\substack{\text{to account for} \\ \text{time variation}}} \qquad (6.6)$$

Similarly for phases b and c

$$\mathscr{F}_b(\theta_d, t) = \mathscr{F}_m \cos(\theta_d - 120°) \cos(\omega t - 120°) \qquad (6.7)$$

and

$$\mathscr{F}_c\,(\theta_d,\,t) = \mathscr{F}_m \cos\,(\theta_d - 240°) \cos\,(\omega t - 240°) \qquad (6.8)$$

The sum of these three mmfs is the resultant mmf $\mathscr{F}_{ar}$ which is called the mmf of *armature reaction.* By using the trigonometric identity

$$\cos\alpha\cos\beta = \tfrac{1}{2}\cos\,(\alpha - \beta) + \tfrac{1}{2}\cos\,(\alpha + \beta) \qquad (6.9)$$

and by recognizing that the sum of three sinusoidal terms of equal magnitude displaced in phase by 120° and 240° is zero we find

$$\mathscr{F}_{ar} = \mathscr{F}_a + \mathscr{F}_b + \mathscr{F}_c = \tfrac{3}{2}\mathscr{F}_m \cos\,(\theta_d - \omega t) \qquad (6.10)$$

Equation (6.10) describes a wave of mmf traveling around the armature in the direction of increasing θ_d. To an observer moving with any point on the wave the mmf is constant and

$$\theta_d - \omega t = \text{a constant} \qquad (6.11)$$

from which we obtain

$$\frac{d\theta_d}{dt} = \omega \qquad (6.12)$$

Equation (6.12) tells us that the mmf of armature reaction is rotating around the armature at the angular velocity ω equal to the angular velocity of the rotor. Therefore this mmf is stationary with respect to the mmf produced by the dc winding of the rotor. The net flux across the air gap between the stator and rotor is produced by the resultant of these two mmfs.

If we neglect saturation and recall that we are considering a cylindrical-rotor machine, we can consider separately the fluxes produced by each of these mmfs and speak of ϕ_f produced by the dc current in the rotor and ϕ_{ar} produced by armature reaction. When a flux having sinusoidal distribution is rotating around the armature the maximum flux linkages with a coil occur when the direction of the mmf causing the flux coincides with the axis of the coil, but the rate of change of flux linkages is then zero. Likewise when the rotor has turned through 90° the flux linkages become zero, but their rate of change is a maximum. Thus the voltage induced in the coil is 90° out of phase with the flux linkages. By applying Lenz's law and taking into account the assumed positive directions of current in the coil and of the mmf in Fig. 6.2 we could show that the induced voltage is lagging rather than leading the mmf.

6.3 THE CIRCUIT MODEL OF A SYNCHRONOUS MACHINE

To draw the phasor diagram for phase a we visualize the separate components of the flux at the axis of coil a; that is *where θ_d equals zero.* The rotor flux ϕ_f is the only one to be considered when the armature current is zero. This flux ϕ_f generates the no-load voltage E_{ao} which we shall designate here as E_f. The flux

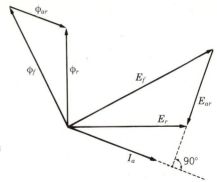

Figure 6.5 Phasor diagram showing time relation of components of flux at the axis of coil a $(\theta_d = 0)$ and the voltages and currents of phase a of the generator of Fig. 6.2. Similar diagrams may be drawn for phases b and c and apply to all cylindrical-rotor generators.

ϕ_{ar} due to the armature-reaction mmf $\mathscr{F}_{ar}$ is in phase with the current i_a (at $\theta_d = 0$) as may be seen by comparing Eqs. (6.2) and (6.10) and recognizing that $\cos \omega t = \cos (-\omega t)$. The sum of ϕ_f and ϕ_{ar}, saturation neglected, is ϕ_r, the resultant flux which generates the voltage E_r in the coil windings composing phase a. The phasor diagram for phase a is shown in Fig. 6.5. Voltages E_f and E_{ar} lag by 90° the fluxes ϕ_f and ϕ_{ar} which generate them. The resultant flux ϕ_r is the flux across the air gap of the machine and generates E_r in the stator. Similar diagrams can be drawn for each phase.

In Fig. 6.5 we note particularly that E_{ar} is lagging I_a by 90°. The magnitude of E_{ar} is determined by ϕ_{ar} which in turn is proportional to $|I_a|$ since it is the result of armature current. So we can specify an inductive reactance X_{ar} such that

$$E_{ar} = -jI_a X_{ar} \tag{6.13}$$

Equation (6.13) defines E_{ar} so that it has the proper phase angle with respect to I_a. Then the voltage generated in phase a by the air-gap flux is E_r where

$$E_r = E_f + E_{ar} = E_f - jI_a X_{ar} \tag{6.14}$$

The voltage generated in each phase by the resultant flux exceeds the terminal voltage V_t of the phase only by the voltage drop due to the armature current times the leakage reactance X_l of the winding if resistance is neglected. If the terminal voltage is V_t,

$$V_t = E_r - jI_a X_l \tag{6.15}$$

The product $I_a X_l$ accounts for the voltage drop caused by that portion of the flux (produced by armature current) which does not cross the air gap of the machine. So from Eqs. (6.14) and (6.15)

$$V_t = \underbrace{E_f}_{\substack{\text{generated} \\ \text{at no load}}} - \underbrace{jI_a X_{ar}}_{\substack{\text{due to armature} \\ \text{reaction}}} - \underbrace{jI_a X_l}_{\substack{\text{due to armature} \\ \text{leakage reactance}}} \tag{6.16}$$

or

$$V_t = E_f - jI_a X_s \tag{6.17}$$

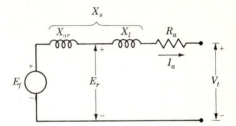

Figure 6.6 Equivalent circuit of an ac generator.

where X_s, called the synchronous reactance, is equal to $X_{ar} + X_l$. If the resistance of the armature R_a is to be considered, Eq. (6.17) becomes

$$V_t = E_f - I_a(R_a + jX_s) \qquad (6.18)$$

R_a is usually so much smaller than X_s that its omission is not of great consequence here, where we are most interested in a qualitative approach.

Now we have arrived at a relationship which allows us to represent the generator by the simple but very useful equivalent circuit shown in Fig. 6.6 which corresponds to Eq. (6.18).

Examination of Fig. 6.5 reveals a very important point about a synchronous generator. When the current I_a lags the no-load generated voltage E_f by 90°, ϕ_{ar} subtracts directly from ϕ_f, and ϕ_r is greatly reduced. Conversely, armature current leading the no-load voltage by 90° causes ϕ_{ar} to add directly to ϕ_f, and ϕ_r is greatly increased. The relationship between E_f, E_{ar}, and E_t for these two cases is shown in Fig. 6.7. If a highly inductive load is applied to a generator the terminal voltage will be considerably below the no-load terminal voltage. On the other hand a capacitive load will cause the terminal voltage to rise considerably above its no-load value. These results help to verify our phasor diagram and equivalent circuit.

In developing this theory we restricted ourselves to considering a two-pole machine. The theory applies equally well to multiple-pole machines but is a little more complicated to develop because of keeping in mind the differences between electrical and mechanical degrees.

The principles we have discussed could be extended to a synchronous motor.

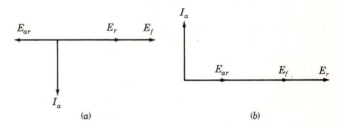

Figure 6.7 Phasor diagrams showing the relation between E_f and E_{ar} when current delivered by a generator is (a) lagging E_f by 90° and (b) leading E_f by 90°.

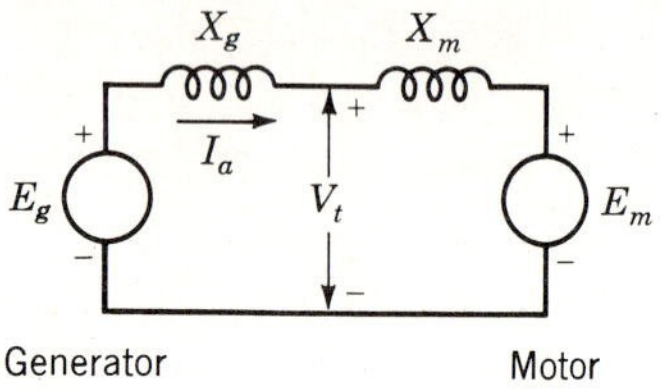

Figure 6.8 Circuit diagram for a generator and motor. I_a is the current delivered by the generator and received by the motor.

The equivalent circuit for the motor is identical with that of the generator with the direction shown for I_a reversed. The generated voltages of the generator and motor are often identified by single-subscript notation as E_g and E_m, respectively, instead of E_f, especially when they are in the same circuit as in Fig. 6.8, for which the equations are .

$$V_t = E_g - jI_a X_g \qquad (6.19)$$

and

$$V_t = E_m + jI_a X_m \qquad (6.20)$$

Synchronous reactances of the generator and motor are X_g and X_m, respectively, and armature resistance is neglected.

When we study faults on a synchronous machine in Chap. 10 we shall see that the current flowing immediately after the occurrence of a fault differs from the steady-state value. Instead of synchronous reactance X_s, we use *subtransient reactance X''* or *transient reactance X'* in modeling the synchronous machine for fault calculations. These reactances will be used in some problems before our further study of them in Chap. 10.

If we had considered salient-pole machines we would have had to account for the difference between the flux path directly into the pole face (called the direct axis) and the path between poles (called the quadrature axis). To do so the armature current is divided into two components. One component is 90° out of phase with the no-load generated voltage E_f, and the other is in phase with E_f. The first component produces the mmf whose flux causes a voltage drop accounted for by the product of the current and the so-called *direct-axis* synchronous reactance X_d. The other component is in phase with E_f and produces the mmf and flux which causes a voltage drop accounted for by the product of this component of current and the *quadrature-axis* synchronous reactance X_q.

Table A.4 in the Appendix lists per-unit values of various reactances for synchronous machines. As the table shows, values of X_d and X_q in cylindrical-rotor machines are essentially equal. For this reason we did not need to consider X_d and X_q separately in our discussion of armature reaction but simply called the synchronous reactance X_s. To simplify our work we shall continue to assume all synchronous machines to have cylindrical rotors. Two-reaction theory discussing direct- and quadrature-axis reactances can be found in most textbooks on ac machinery. Table A.4 also gives values of X'_d and X''_d.

6.4 THE EFFECT OF
SYNCHRONOUS-MACHINE EXCITATION

Changing the excitation of synchronous machines is an important factor in controlling the flow of reactive power. First we shall consider a generator connected at its terminals to a very large power system, a system so large that the voltage V_t at the terminals of the generator will not be altered by any changes in the excitation of the generator. The bus to which the generator is connected is sometimes called an *infinite bus*, which means that its voltage will remain constant and no frequency change will occur regardless of changes made in power input or field excitation of the synchronous machine connected to it. If we decide to maintain a certain power input from the generator to the system, $|V_t| \cdot |I_a| \cos \theta$ will remain constant as we vary the dc field excitation to vary $|E_g|$. Then, for a high and a low value of $|E_g|$ the phasor diagrams of the generator are given by Fig. 6.9. The angle δ is called the *torque angle* or *power angle* of the machine. Normal excitation is defined as the excitation when

$$|E_g| \cos \delta = V_t \tag{6.21}$$

For the condition of Fig. 6.9a the generator is overexcited and supplies lagging current to the system. The machine can also be considered to be drawing leading current from the system. Like a capacitor, it supplies reactive power to the system. Figure 6.9b is for an underexcited generator supplying leading current to the system, or it may be considered to be drawing lagging current from the system. The underexcited generator draws reactive power from the system. This action can be explained by the mmf of armature reaction. For instance, when the generator is overexcited, it must deliver lagging current since lagging current produces an opposing mmf to reduce the overexcitation.

We note that E_g leads V_t in Fig. 6.9, which is always true for a generator and is necessary to satisfy Eq. (6.19).

Figure 6.10 shows overexcited and underexcited synchronous motors drawing the same power at the same terminal voltage. The overexcited motor draws leading current and acts like a capacitive circuit when viewed from the network to which it supplies reactive power. The underexcited motor draws lagging current, absorbs reactive power, and is acting like an inductive circuit when

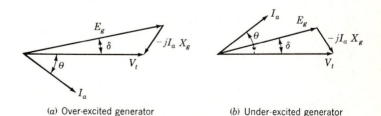

(a) Over-excited generator   (b) Under-excited generator

Figure 6.9 Phasor diagrams of (a) overexcited and (b) underexcited generator. I_a is current delivered by the generator.

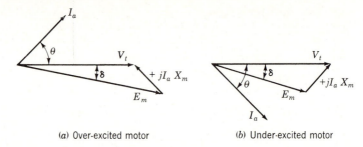

(a) Over-excited motor   (b) Under-excited motor

Figure 6.10 Phasor diagrams of (a) overexcited and (b) underexcited motor. I_a is current drawn by the motor.

viewed from the network. We see from Fig. 6.10 that E_m lags V_t in order to satisfy Eq. (6.20), and this is always true for a synchronous motor. Briefly, Figs. 6.9 and 6.10 show us that *overexcited* generators and motors *supply* reactive power to the system and *underexcited* generators and motors *absorb* reactive power from the system.

6.5 THE IDEAL TRANSFORMER

We now have models for transmission lines and synchronous machines and are ready to consider transformers which consist of two or more coils placed so that they are linked by the same flux. In a power transformer the coils are placed on an iron core in order to confine the flux so that almost all of the flux linking any one coil links all the others. Several coils may be connected in series or parallel to form one winding, the coils of which may be stacked on the core alternately with those of the other winding or windings.

Figure 6.11 is the photograph of a three-phase transformer which raises the voltage of a generator in a nuclear power station to the transmission-line voltage. The transformer is rated 750 MVA 525/22.8 kV.

Figure 6.12 shows how two windings may be placed on an iron core to form a single-phase transformer of the so-called shell type. The number of turns in a winding may be several hundred up to several thousand.

We shall begin our analysis by assuming that the flux varies sinusoidally in the core and that the transformer is *ideal*, which means that the permeability μ of the core is infinite and the resistance of the windings is zero. With infinite permeability of the core all of the flux is confined to the core and therefore links all of the turns of both windings. The voltage e induced in each winding by the changing flux is also the terminal voltage v of the winding since the winding resistance is zero.

We can see from the relationship of the windings shown in Fig. 6.12 that voltages e_1 and e_2 induced by the changing flux are in phase when they are defined by the $+$ and $-$ polarity marks indicated. Then by Faraday's law

$$v_1 = e_1 = N_1 \frac{d\phi}{dt} \tag{6.22}$$

Figure 6.11 Photograph of a three-phase transformer rated 750 MVA 525/22.8 kV. (*Courtesy Duke Power Company.*)

and

$$v_2 = e_2 = N_2 \frac{d\phi}{dt} \tag{6.23}$$

where ϕ is the instantaneous value of the flux and N_1 and N_2 are the number of turns on windings 1 and 2, as shown in Fig. 6.12. Since we have assumed sinusoidal variation of the flux we can convert to phasor form after dividing Eq. (6.22)

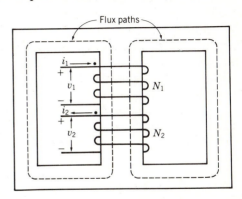

Figure 6.12 Two-winding transformer.

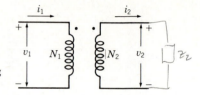

Figure 6.13 Schematic representation of a two-winding transformer.

by Eq. (6.23) to yield

$$\frac{V_1}{V_2} = \frac{E_1}{E_2} = \frac{N_1}{N_2} \tag{6.24}$$

Usually we do not know the direction in which the coils of a transformer are wound. One device to provide winding information is to place a dot at the end of each winding such that all dotted ends of windings are positive at the same time; that is, voltage drops from dotted to unmarked terminals of all windings are in phase. Dots are shown on the two-winding transformer in Fig. 6.12 according to this convention. We also note that the same result is achieved by placing the dots so that current flowing from the dotted terminal to the unmarked terminal of each winding produces a magnetomotive force acting in the same direction in the magnetic circuit. Figure 6.13 is a schematic representation of a transformer and provides the same information about the transformer as is provided by Fig. 6.12.

To find the relation between the currents i_1 and i_2 in the windings we apply Ampere's law which states that the magnetomotive force around a closed path is

$$\oint H \cdot ds = i \tag{6.25}$$

where i is the current enclosed by the line integral of the field intensity H around the path. In applying the law around each of the closed paths of flux shown by dotted lines in Fig. 6.12, i_1 is enclosed N_1 times and the current i_2 is enclosed N_2 times. However, $N_1 i_1$ and $N_2 i_2$ produce magnetomotive forces in opposite directions and

$$\oint H \cdot ds = N_1 i_1 - N_2 i_2 \tag{6.26}$$

The minus sign would change to plus if we had chosen the opposite direction for the current i_2. The integral of the field intensity H around the closed path is zero when permeability is infinite. So upon converting to phasor form we have

$$N_1 I_1 - N_2 I_2 = 0 \tag{6.27}$$

So

$$\frac{I_1}{I_2} = \frac{N_2}{N_1} \tag{6.28}$$

and I_1 and I_2 are in phase. Note then that I_1 and I_2 are in phase if we choose the current to be positive when entering the dotted terminal of one winding and leaving the dotted terminal of the other. If the direction chosen for either current is reversed they are 180° out of phase.

From Eq. (6.28)

$$I_1 = \frac{N_2}{N_1} I_2 \qquad (6.29)$$

and in the ideal transformer I_1 must be zero if I_2 is zero.

The winding across which an impedance or other load may be connected is called the *secondary* winding and any circuit elements connected to this winding are said to be on the secondary side of the transformer. Similarly the winding which is toward the source of energy is called the primary winding on the primary side. In the power system energy often will flow in either direction through a transformer and the designation of primary and secondary loses its meaning. The terms are in general use, however, and we shall use them wherever they do not cause confusion.

If an impedance Z_2 is connected across winding 2 of the circuit of Figs. 6.12 or 6.13

$$Z_2 = \frac{V_2}{I_2} \qquad (6.30)$$

but upon substituting for V_2 and I_2 values determined from Eqs. (6.24) and (6.28)

$$Z_2 = \frac{(N_2/N_1)V_1}{(N_1/N_2)I_1} \qquad (6.31)$$

and the impedance as measured across the primary winding is

$$Z_2' = \frac{V_1}{I_1} = \left(\frac{N_1}{N_2}\right)^2 Z_2 \qquad (6.32)$$

Thus the impedance connected to the secondary side is referred to the primary side by multiplying the impedance on the secondary side of the transformer by the square of the ratio of primary to secondary voltage.

We should note also that $V_1 I_1$ and $V_2 I_2$ are equal as shown by the following equation which again makes use of Eqs. (6.24) and (6.28)

$$V_1 I_1 = \frac{N_1}{N_2} V_2 \times \frac{N_2}{N_1} I_2 = V_2 I_2 \qquad (6.33)$$

and similarly

$$V_1 I_1^* = V_2 I_2^* \qquad (6.34)$$

so the voltamperes and complex power input to the primary winding equal the output of these same quantities from the secondary winding since we are considering an ideal transformer.

Example 6.1 If $N_1 = 2000$ and $N_2 = 500$ in the circuit of Fig. 6.13 and if $V_1 = 1200\underline{/0°}$ V and $I_1 = 5\underline{/-30°}$ A with an impedance Z_2 connected across winding 2, find V_2, I_2, Z_2 and the impedance Z_2' which is defined as the value of Z_2 referred to the primary side of the transformer.

SOLUTION

$$V_2 = \frac{500}{2000}(1200\underline{/0°}) = 300\underline{/0°} \text{ V}$$

$$I_2 = \frac{2000}{500}(5\underline{/-30°}) = 20\underline{/-30°} \text{ A}$$

$$Z_2 = \frac{300\underline{/0°}}{20\underline{/-30°}} = 15\underline{/30°} \ \Omega$$

$$Z_2' = (15\underline{/30°})\left(\frac{2000}{500}\right)^2 = 240\underline{/30°} \ \Omega$$

or

$$Z_2' = \frac{1200\underline{/0°}}{5\underline{/-30°}} = 240\underline{/30°} \ \Omega$$

6.6 THE EQUIVALENT CIRCUIT OF A PRACTICAL TRANSFORMER

The ideal transformer is a first step in studying a practical transformer where (1) permeability is not infinite, (2) winding resistance is present, (3) losses occur in the iron core due to the cyclic changing of direction of the flux, and (4) not all the flux linking any one winding links the other windings.

When a sinusoidal voltage is applied to a transformer winding on an iron core with the secondary winding open a small current will flow in the primary such that in a well-designed transformer the maximum flux density B_m occurs at the knee of the B-H, or saturation curve of the transformer. This current is called the magnetizing current. Losses in the iron occur due, first, to the fact that the cyclic changes of the direction of the flux in the iron require energy which is dissipated as heat and called *hysteresis loss*. The second loss is due to the fact that circulating currents are induced in the iron due to the changing flux, and these currents produce an $|I|^2R$ loss in the iron called *eddy-current loss*. Hysteresis loss is reduced by the use of certain high grades of alloy steel for the core. Eddy-current loss is reduced by building up the core with laminated sheets of steel. With the secondary open the transformer primary circuit is simply one of very high inductance due to the iron core. The current lags the applied voltage by slightly less than 90°, and the component of current in phase with the voltage accounts for the energy loss in the core. In the equivalent circuit-magnetizing

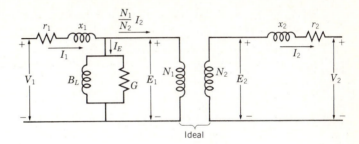

Figure 6.14 Transformer equivalent circuit using the ideal transformer concept.

current I_E is taken into account by an inductive susceptance B_L in parallel with a conductance G.

In the practical two-winding transformer some of the flux linking the primary winding does not link the secondary. This flux is proportional to the primary current and causes a voltage drop that is accounted for by an inductive reactance x_1, called *leakage reactance*, which is added in series with the primary winding of the ideal transformer. Similar leakage reactance x_2 must be added to the secondary winding to account for the voltage due to the flux linking the secondary but not the primary. When we also account for the resistances r_1 and r_2 of the windings we have the transformer model shown in Fig. 6.14. In this model the ideal transformer is the link between the circuit parameters r_1, x_1, G, and B_L added to the primary side of the transformer and r_2 and x_2 added to the secondary side.

The ideal transformer may be omitted in the equivalent circuit if we refer all quantities to either the high- or the low-voltage side of the transformer. For instance, if we refer all voltages, currents, and impedances of the circuit of Fig. 6.14 to the primary circuit of the transformer having N_1 turns, and for simplicity let $a = N_1/N_2$, we have the circuit of Fig. 6.15. Very often we neglect magnetizing current because it is so small compared to the usual load currents. To further simplify the circuit we let

$$R_1 = r_1 + a^2 r_2 \tag{6.35}$$

and

$$X_1 = x_1 + a^2 x_2 \tag{6.36}$$

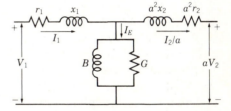

Figure 6.15 Transformer equivalent circuit with path for magnetizing current.

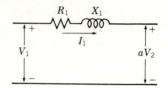

Figure 6.16 Transformer equivalent circuit with magnetizing current neglected.

to obtain the equivalent circuit of Fig. 6.16. All impedances and voltages in the part of the circuit connected to the secondary terminals must now be referred to the primary side.

Example 6.2 A single-phase transformer has 2000 turns on the primary winding and 500 turns on the secondary. Winding resistances are $r_1 = 2.0\ \Omega$ and $r_2 = 0.125\ \Omega$. Leakage reactances are $x_1 = 8.0\ \Omega$ and $x_2 = 0.50\ \Omega$. The resistance load Z_2 is 12 Ω. If applied voltage at the terminals of the primary winding is 1200 V, find V_2 and the voltage regulation. Neglect magnetizing current.

SOLUTION

$$a = \frac{N_1}{N_2} = \frac{2000}{500} = 4$$

$$R_1 = 2 + 0.125(4)^2 = 4.0\ \Omega$$

$$X_1 = 8 + 0.5(4)^2 = 16\ \Omega$$

$$Z_2' = 12 \times (4)^2 = 192\ \Omega$$

The equivalent circuit is shown in Fig. 6.17

$$I_1 = \frac{1200}{192 + 4 + j16} = 6.10\underline{/-4.67°}\ \text{A}$$

$$aV_2 = 6.10\underline{/-4.67°} \times 192 = 1171.6\underline{/-4.67°}\ \text{V}$$

$$V_2 = \frac{1171.6\underline{/-4.67°}}{4} = 292.9\underline{/-4.67°}\ \text{V}$$

$$\text{Voltage regulation} = \frac{1200/4 - 292.9}{292.9} = 0.0242 \quad \text{or} \quad 2.42\%$$

Although magnetizing current may be neglected as in Example 6.2 for most

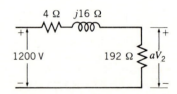

Figure 6.17 Circuit for Example 6.2.

power-system calculations, G and B_L can be calculated for the equivalent circuit by an open-circuit test. Rated voltage is applied to the primary winding with the other windings open. The impedance measured across the terminals of this winding whose resistance and leakage reactance are r_1 and x_1 is

$$Z = r_1 + jx_1 + \frac{1}{G + jB_L} \qquad (6.37)$$

and since r_1 and x_1 are very small compared to the measured impedance, G and B_L can be determined in this manner.

The parameters R and X of the two-winding transformer are determined by the short-circuit test where impedance is measured across the terminals of one winding when the other winding is short-circuited. Just enough voltage is applied to circulate rated current. Since only a small voltage is required the magnetizing current is insignificant and the measured impedance is essentially equivalent to $R + jX$.

6.7 THE AUTOTRANSFORMER

An autotransformer differs from the ordinary transformer in that the windings of the autotransformer are electrically connected as well as being coupled by a mutual flux. We shall examine the autotransformer by connecting electrically the windings of an ideal transformer. Figure 6.18a is a schematic diagram of an ideal transformer, and Fig. 6.18b shows how the windings are connected electrically to form an autotransformer. Here the windings are shown so their voltages are additive although they could have been connected to oppose each other. The great disadvantage of the autotransformer is that electrical isolation is lost, but the following example will demonstrate the increase in power rating obtained.

Example 6.3 A 30 kVA, single-phase transformer rated 240/120 V is connected as an autotransformer as shown in Fig. 6.18b. Rated voltage is applied to the low-tension winding of the transformer. Consider the transformer to be ideal and the load to be such that rated currents $|I_1|$ and $|I_2|$ flow in the windings. Determine $|V_2|$ and the kilovoltampere rating of the autotransformer.

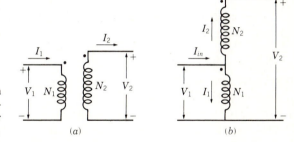

Figure 6.18 Schematic diagram of an ideal transformer connected (a) in the usual manner and (b) as an autotransformer.

SOLUTION

$$|I_1| = \frac{30,000}{120} = 250 \text{ A}$$

$$|I_2| = \frac{30,000}{240} = 125 \text{ A}$$

$$|V_2| = 240 + 120 = 360 \text{ V}$$

The directions chosen for positive current in defining I_1 and I_2 in relation to the dotted terminals show that these currents are in phase. So the input current is

$$|I_{in}|\underline{/\theta} = |I_1|\underline{/\theta} + |I_2|\underline{/\theta}$$

$$|I_{in}| = 250 + 125 = 375 \text{ A}$$

Input kVA is

$$375 \times 120 \times 10^{-3} = 45 \text{ kVA}$$

Output kVA is

$$125 \times 360 \times 10^{-3} = 45 \text{ kVA}$$

From this example we see that the autotransformer has given us a larger voltage ratio than the ordinary transformer and transmitted more kilovolt-amperes between the two sides of the transformer. So an autotransformer provides a higher rating for the same cost. It also operates more efficiently since the losses remain the same as for the ordinary connection. However, the loss of electrical isolation between the high- and low-voltage sides of the autotransformer is usually the decisive factor in favor of the ordinary connection in most applications. In power systems three-phase autotransformers are frequently used to make small adjustments of bus voltages.

6.8 PER-UNIT IMPEDANCES IN SINGLE-PHASE TRANSFORMER CIRCUITS

The ohmic values of resistance and leakage reactance of a transformer depend on whether they are measured on the high- or low-tension side of the transformer. If they are expressed in per unit, the base kilovoltamperes is understood to be the kilovoltampere rating of the transformer. The base voltage is understood to be the voltage rating of the low-tension winding if the ohmic values of resistance and leakage reactance are referred to the low-tension side of the transformer and to be the voltage rating of the high-tension winding if they are referred to the high-tension side of the transformer. The per-unit impedance of a transformer is the same regardless of whether it is determined from ohmic values referred to the high-tension or low-tension sides of the transformers, as shown by the following example.

Example 6.4 A single-phase transformer is rated 110/440 V, 2.5 kVA. Leakage reactance measured from the low-tension side is 0.06 Ω. Determine leakage reactance in per unit.

SOLUTION

$$\text{Low-tension base impedance} = \frac{0.110^2 \times 1000}{2.5} = 4.84 \ \Omega$$

In per unit

$$X = \frac{0.06}{4.84} = 0.0124 \text{ per unit}$$

If leakage reactance had been measured on the high-tension side, the value would be

$$X = 0.06 \left(\frac{440}{110} \right)^2 = 0.96 \ \Omega$$

$$\text{High-tension base impedance} = \frac{0.440^2 \times 1000}{2.5} = 77.5 \ \Omega$$

In per unit

$$X = \frac{0.96}{77.5} = 0.0124 \text{ per unit}$$

A great advantage in making per-unit computations is realized by the proper selection of different bases for circuits connected to each other through a transformer. To achieve the advantage in a single-phase system, the voltage bases for the circuit connected through the transformer must have the same ratio as the turns ratio of the transformer windings. With such a selection of voltage bases and the same kilovoltampere base, the per-unit value of an impedance will be the same when it is expressed on the base selected for its own side of the transformer as when it is referred to the other side of the transformer and expressed on the base of that side.

So the transformer is represented completely by its impedance $(R + jX)$ in per unit when magnetizing current is neglected. No per-unit voltage transformation occurs when this system is used, and the current will also have the same per-unit value on both sides of the transformer if magnetizing current is neglected.

Example 6.5 Three parts of a single-phase electric system are designated A, B, and C and are connected to each other through transformers, as shown in Fig. 6.19. The transformers are rated as follows:

A–B 10,000 kVA, 138/13.8 kV, leakage reactance 10%

B–C 10,000 kVA, 138/69 kV, leakage reactance 8%

If the base in circuit B is chosen as 10,000 kVA, 138 kV, find the per-unit impedance of the 300-Ω resistive load in circuit C referred to circuits C, B, and A. Draw the impedance diagram neglecting magnetizing current, transformer resistances, and line impedances. Determine the voltage regulation if the voltage at the load is 66 kV with the assumption that the voltage input to circuit A remains constant.

SOLUTION

$$\text{Base voltage for circuit } A = 0.1 \times 138 = 13.8 \text{ kV}$$

$$\text{Base voltage for circuit } C = 0.5 \times 138 = 69 \text{ kV}$$

$$\text{Base impedance of circuit } C = \frac{69^2 \times 1000}{10,000} = 476 \ \Omega$$

$$\text{Per-unit impedance of load in circuit } C = \frac{300}{476} = 0.63 \text{ per unit}$$

Because the selection of base in various parts of the system was determined by the turns ratio of the transformers, the per-unit impedance of the load referred to any part of the system will be the same. This is verified as follows:

$$\text{Base impedance of circuit } B = \frac{138^2 \times 1000}{10,000} = 1900 \ \Omega$$

$$\text{Impedance of load referred to circuit } B = 300 \times 2^2 = 1200 \ \Omega$$

$$\text{Pre-unit impedance of load referred to } B = \frac{1200}{1900} = 0.63 \text{ per unit}$$

$$\text{Base impedance of circuit } A = \frac{13.8^2 \times 1000}{10,000} = 19 \ \Omega$$

$$\text{Impedance of load referred to } A = 300 \times 2^2 \times 0.1^2 = 12 \ \Omega$$

$$\text{Per-unit impedance of load referred to } A = \frac{12}{19} = 0.63 \text{ per unit}$$

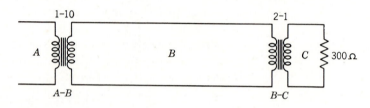

Figure 6.19 Circuit for Example 6.5.

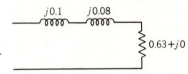

Figure 6.20 Impedance diagram for Example 6.5. Impedances are marked in per unit.

Figure 6.20 is the required impedance diagram with impedances marked in per unit.

The calculation of regulation proceeds as follows:

$$\text{Voltage at load} = \frac{66}{69} = 0.957 + j0 \text{ per unit}$$

$$\text{Load current} = \frac{0.957 + j0}{0.63 + j0} = 1.52 + j0 \text{ per unit}$$

$$\text{Voltage input} = (1.52 + j0)(j0.10 + j0.08) + 0.957$$
$$= 0.957 + j0.274 = 0.995 \text{ per unit}$$

$$\text{Voltage input} = \text{voltage at load with load removed}$$

Therefore

$$\text{Regulation} = \frac{0.995 - 0.957}{0.957} \times 100 = 3.97\%$$

Because of the advantage previously pointed out, the principle followed in the above example in selecting the base for various parts of the system is always followed in making computations by per unit or percent. The kilovoltampere base should be the same in all parts of the system, and the selection of the base kilovolts in one part of the system determines the base kilovolts to be assigned, according to the turns ratios of the transformers, to the other parts of the system. Following this principle of assigning base kilovolts allows us to combine on one impedance diagram the per-unit impedances determined in different parts of the system.

6.9 THREE-PHASE TRANSFORMERS

Three identical single-phase transformers may be connected so that the three windings of one voltage rating are Δ-connected and the three windings of the other voltage rating are Y-connected to form a three-phase transformer. Such a transformer is said to be connected Y-Δ or Δ-Y. The other possible connections are Y-Y and Δ-Δ. If the three single-phase transformers each have three windings (a primary, secondary, and tertiary), two sets might be connected in Y and one in Δ or two could be Δ-connected with one Y-connected. Rather than use three

identical single-phase transformers, a more usual unit is a three-phase transformer where all three phases are on one iron structure. The theory is the same for
a three-phase transformer as for a three-phase bank of single-phase transformers.

Let us consider a numerical example of a Y-Y transformer composed of
three single-phase transformers each rated 25 MVA, 38.1/3.81 kV. The rating as
a three-phase transformer is, therefore, 75 MVA, 66/6.6 kV ($\sqrt{3} \times 38.1 = 66$).
Figure 6.21 shows the transformer with a balanced resistive load of 0.6 Ω per
phase on the low-tension side. Windings of the primary and secondary drawn in
parallel directions are on the same single-phase transformer. Since the circuit is
balanced and we are assuming balanced three-phase voltages the neutral of the
load and the neutral of the low-voltage winding are at the same potential.
Therefore each 0.6-Ω resistor is considered as directly connected across a
3.81-kV winding whether the neutrals are connected or not. On the high-voltage
side the impedance measured from line to neutral is

$$0.6\left(\frac{38.1}{3.81}\right)^2 = 0.6\left(\frac{66}{6.6}\right)^2 = 60 \ \Omega$$

Figure 6.22a shows the same three transformers connected Y-Δ to the same
resistive load of 0.6 Ω per phase. So far as the voltage magnitude at the low-
tension terminals is concerned the Y-Δ transformer can be replaced by a Y-Y
transformer bank having a turns ratio for each individual transformer (or for
each pair of phase windings of a three-phase transformer) of 38.1/2.2 kV as
shown in Fig. 6.22b. The transformers of Fig. 6.22a and b are equivalent if we are
not concerned with phase shift. As we shall see in Chap. 11, there is a phase shift
of the voltages between sides of the Y-Δ transformer that need not be considered
here. Figure 6.22b shows us that, viewed from the high-tension side of the transformer, the resistance of each phase of the load is

$$0.6\left(\frac{38.1}{2.2}\right)^2 = 0.6\left(\frac{66}{3.81}\right)^2 = 180 \ \Omega$$

Here the multiplying factor is the square of the ratio of line-to-line voltages and

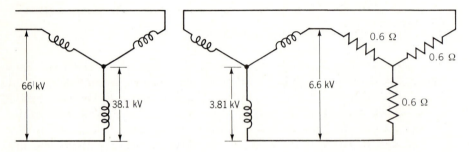

Figure 6.21 Y-Y transformer rated 66/6.6 kV.

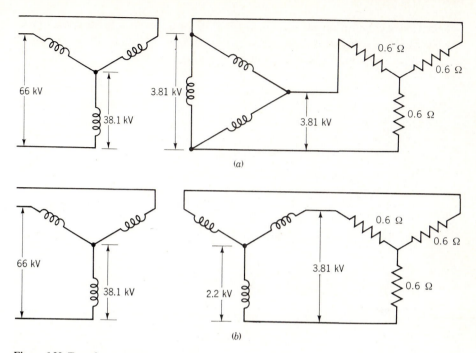

Figure 6.22 Transformer of Fig. 6.21 (*a*) connected Y-Δ and (*b*) replaced by a Y-Y transformer with line-to-line voltage ratio equal to that of the Y-Δ.

not the square of the turns ratios of the individual windings of the Y-Δ transformer.

This discussion leads to the conclusion that to transfer the ohmic value of impedance from the voltage level on one side of a three-phase transformer to the voltage level on the other, the multiplying factor is the square of the ratio of line-to-line voltages regardless of whether the transformer connection is Y-Y or Y-Δ. Therefore in per-unit calculations involving transformers in three-phase circuits we follow the same principle developed for single-phase circuits and require the base voltages on the two sides of the transformer to have the same ratio as the rated line-to-line voltages on the two sides of the transformer. The kVA base is the same on each side.

Example 6.6 The three transformers rated 25 MVA, 38.1/3.81 kV are connected Y-Δ as shown in Fig. 6.22a with the balanced load of three 0.6-Ω, Y-connected resistors. Choose a base of 75 MVA, 66 kV for the high-tension side of the transformer and specify the base for the low-tension side. Determine the per-unit resistance of the load on the base for the low-tension side. Then determine the load resistance R_L referred to the high-tension side and the per-unit value of this resistance on the chosen base.

SOLUTION The rating of the transformer as a three-phase bank is 75 MVA, 66Y/3.81Δ kV. So base for the low-tension side is 75 MVA, 3.81 kV.

Base impedance on the low-tension side is

$$\frac{(3.81)^2}{75} = 0.1935$$

and on the low-tension side

$$R_L = \frac{0.6}{0.1935} = 3.10 \text{ per unit}$$

Base impedance on the high-tension side is

$$\frac{(66)^2}{75} = 58.1 \ \Omega$$

and we have seen that the resistance per phase referred to the high-tension side is 180 Ω. So

$$R_L = \frac{180}{58.1} = 3.10 \text{ per unit}$$

The resistance R and leakage reactance X of a three-phase transformer are measured by the short-circuit test as discussed for single-phase transformers. In a three-phase equivalent circuit R and X are connected in each line to an ideal three-phase transformer. Since R and X will have the same per-unit value whether on the low-tension or the high-tension side of the transformer, the single-phase equivalent circuit will account for the transformer by the impedance $R + jX$ in per unit without the ideal transformer if all quantities in the circuit are in per unit with the proper selection of base.

Table A.5 in the Appendix lists typical values of transformer impedances which are essentially equal to the leakage reactance since the resistance is usually less than 0.01 per unit.

Example 6.7 A three-phase transformer is rated 400 MVA, 220Y/22Δ kV. The short-circuit impedance measured on the low-voltage side of the transformer is 0.121 Ω and because of the low resistance this value may be considered equal to the leakage reactance. Determine the per-unit reactance of the transformer and the value to be used to represent this transformer in a system whose base on the high-tension side of the transformer is 100 MVA, 230 kV.

SOLUTION On its own base the transformer reactance is

$$\frac{0.121}{(22)^2/400} = 0.10 \text{ per unit}$$

On the chosen base the reactance becomes

$$0.1\left(\frac{220}{230}\right)^2 \frac{100}{400} = 0.0228 \text{ per unit}$$

6.10 PER-UNIT IMPEDANCES OF
THREE-WINDING TRANSFORMERS

Both the primary and secondary windings of a two-winding transformer have the same kilovoltampere rating, but all three windings of a three-winding transformer may have different kilovoltampere ratings. The impedance of each winding of a three-winding transformer may be given in percent or per unit based on the rating of its own winding, or tests may be made to determine the impedances. In any case, however, all the per-unit impedances in the impedance diagram must be expressed on the same kilovoltamepre base.

Three impedances may be measured by the standard short-circuit test, as follows:

Z_{ps} = leakage impedance measured in primary with secondary short-circuited and tertiary open

Z_{pt} = leakage impedance measured in primary with tertiary short-circuited and secondary open

Z_{st} = leakage impedance measured in secondary with tertiary short-circuited and primary open

If the three impedances measured in ohms are referred to the voltage of one of the windings, the impedances of each separate winding referred to that same winding are related to the measured impedances so referred as follows:

$$Z_{ps} = Z_p + Z_s$$
$$Z_{pt} = Z_p + Z_t \qquad (6.38)$$
$$Z_{st} = Z_s + Z_t$$

where Z_p, Z_s, and Z_t are the impedances of the primary, secondary, and tertiary windings referred to the primary circuit if Z_{ps}, Z_{pt}, and Z_{st} are the measured impedances referred to the primary circuit. Solving Eqs. (6.38) simultaneously yields

$$Z_p = \tfrac{1}{2}(Z_{ps} + Z_{pt} - Z_{st})$$
$$Z_s = \tfrac{1}{2}(Z_{ps} + Z_{st} - Z_{pt}) \qquad (6.39)$$
$$Z_t = \tfrac{1}{2}(Z_{pt} + Z_{st} - Z_{ps})$$

The impedances of the three windings are connected in star to represent the single-phase equivalent circuit of the three-winding transformer with magnetizing current neglected, as shown in Fig. 6.23. The common point is fictitious and unrelated to the neutral of the system. The points p, s, and t are connected to the parts of the impedance diagrams representing the parts of the system connected to the primary, secondary, and tertiary windings of the transformer. Since the ohmic values of the impedances must be referred to the same voltage, it follows that conversion to per-unit impedance requires the same kilovoltampere base for all three circuits and requires voltage bases in the three circuits that are in the same ratio as the rated line-to-line voltages of the three circuits of the transformer.

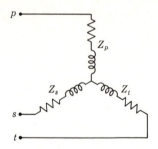

Figure 6.23 The equivalent circuit of a three-winding transformer. Points p, s, and t link the circuit of the transformer to the appropriate equivalent circuits representing parts of the system connected to the primary, secondary, and tertiary windings.

Example 6.8 The three-phase ratings of a three-winding transformer are:

Primary   Y-connected, 66 kV, 15 MVA
Secondary Y-connected, 13.2 kV, 10.0 MVA
Tertiary Δ-connected, 2.3 kV, 5 MVA

Neglecting resistance, the leakage impedances are

$$Z_{ps} = 7\% \text{ on 15-MVA 66-kV base}$$

$$Z_{pt} = 9\% \text{ on 15-MVA 66-kV base}$$

$$Z_{st} = 8\% \text{ on 10.0-MVA 13.2-kV base}$$

Find the per-unit impedances of the star-connected equivalent circuit for a base of 15 MVA, 66 kV in the primary circuit.

SOLUTION With a base of 15 MVA, 66 kV in the primary circuit, the proper bases for the per-unit impedances of the equivalent circuit are 15 MVA, 66 kV for primary-circuit quantities, 15 MVA, 13.2 kV for secondary-circuit quantities, and 15 MVA, 2.3 kV for tertiary-circuit quantities.

Z_{ps} and Z_{pt} were measured in the primary circuit and are therefore already expressed on the proper base for the equivalent circuit. No change of voltage base is required for Z_{st}. The required change in base kVA for Z_{st} is made as follows:

$$Z_{st} = 8\% \times 15/10 = 12\%$$

In per unit on the specified base

$$Z_p = \tfrac{1}{2}(j0.07 + j0.09 - j0.12) = j0.02 \text{ per unit}$$

$$Z_s = \tfrac{1}{2}(j0.07 + j0.12 - j0.09) = j0.05 \text{ per unit}$$

$$Z_t = \tfrac{1}{2}(j0.09 + j0.12 - j0.07) = j0.07 \text{ per unit}$$

Example 6.9 A constant-voltage source (infinite bus) supplies a purely resistive 5-MW 2.3-kV load and a 7.5-MVA 13.2-kV synchronous motor having a subtransient reactance of $X'' = 20\%$. The source is connected to the primary of the three-winding transformer described in Example 6.8. The motor and

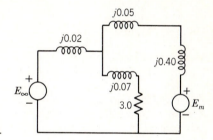

Figure 6.24 Impedance diagram for Example 6.9.

resistive load are connected to the secondary and tertiary of the transformer. Draw the impedance diagram of the system and mark the per-unit impedances for a base of 66 kV, 15 MVA in the primary.

SOLUTION The constant-voltage source can be represented by a generator having no internal impedance.

The resistance of the load is 1.0 per unit on a base of 5 MVA, 2.3 kV in the tertiary. Expressed on a 15-MVA 2.3-kV base the load resistance is

$$R = 1.0 \times \frac{15}{5} = 3.0 \text{ per unit}$$

Changing the reactance of the motor to a base of 15 MVA, 13.2 kV yields

$$X'' = 0.20 \frac{15}{7.5} = 0.40 \text{ per unit}$$

Figure 6.24 is the required impedance diagram.

6.11 THE ONE-LINE DIAGRAM

We now have the circuit models for transmission lines, synchronous machines, and transformers. We shall see next how to portray the assemblage of these components to model a complete system. Since a balanced three-phase system is always solved as a single-phase circuit composed of one of the three lines and a neutral return, it is seldom necessary to show more than one phase and the neutral return when drawing a diagram of the circuit. Often the diagram is simplified further by omitting the completed circuit through the neutral and by indicating the component parts by standard symbols rather than by their equivalent circuits. Circuit parameters are not shown, and a transmission line is represented by a single line between its two ends. Such a simplified diagram of an electric system is called a one-line diagram. It indicates by a single line and standard symbols the transmission lines and associated apparatus of an electric system.

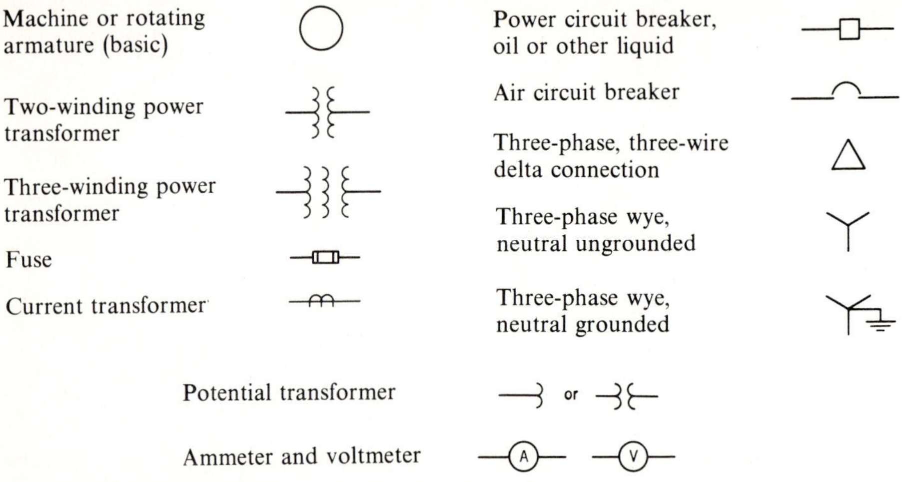

Machine or rotating armature (basic)	◯	Power circuit breaker, oil or other liquid	▭
Two-winding power transformer		Air circuit breaker	
Three-winding power transformer		Three-phase, three-wire delta connection	△
Fuse		Three-phase wye, neutral ungrounded	Y
Current transformer		Three-phase wye, neutral grounded	

Potential transformer   or

Ammeter and voltmeter   Ⓐ   Ⓥ

Figure 6.25 Apparatus symbols.

The purpose of the one-line diagram is to supply in concise form the significant information about the system. The importance of different features of a system varies with the problem under consideration, and the amount of information included on the diagram depends on the purpose for which the diagram is intended. For instance, the location of circuit breakers and relays is unimportant in making a load study. Breakers and relays are not shown if the primary function of the diagram is to provide information for such a study. On the other hand, determination of the stability of a system under transient conditions resulting from a fault depends on the speed with which relays and circuit breakers operate to isolate the faulted part of the system. Therefore, information about the circuit breakers may be of extreme importance. Sometimes one-line diagrams include information about the current and potential transformers which connect the relays to the system or which are installed for metering. The information found on a one-line diagram must be expected to vary according to the problem at hand and according to the practice of the particular company preparing the diagram.

The American National Standards Institute (ANSI) and the Institute of Electrical and Electronics Engineers have published a set of standard symbols for electrical diagrams.† Not all authors follow these symbols consistently, especially in indicating transformers. Figure 6.25 shows a few symbols which are commonly used. The basic symbol for a machine or rotating armature is a circle, but so many adaptations of the basic symbol are listed that every piece of rotating electric machinery in common use can be indicated. For anyone who is not working constantly with one-line diagrams, it is clearer to indicate a particular machine by the basic symbol followed by information on its type and rating.

† See Graphic Symbols for Electrical and Electronics Diagrams, IEEE Std 315-1975.

Figure 6.26 One-line diagram of an electric system.

It is important to know the location of points where a system is connected to ground in order to calculate the amount of current flowing when an unsymmetrical fault involving ground occurs. The standard symbol to designate a three-phase Y with the neutral solidly grounded is shown in Fig. 6-25. If a resistor or reactor is inserted between the neutral of the Y and ground to limit the flow of current to ground during a fault, the appropriate symbol for resistance or inductance may be added to the standard symbol for the grounded Y. Most transformer neutrals in transmission systems are solidly grounded. Generator neutrals are usually grounded through fairly high resistances and sometimes through inductance coils.

Figure 6.26 is the one-line diagram of a very simple power system. Two generators, one grounded through a reactor and one through a resistor, are connected to a bus and through a step-up transformer to a transmission line. Another generator, grounded through a reactor, is connected to a bus and through a transformer to the opposite end of the transmission line. A load is connected to each bus. On the diagram information about the loads, the ratings of the generators and transformers, and reactances of the different components of the circuit is often given.

6.12 IMPEDANCE AND REACTANCE DIAGRAMS

In order to calculate the performance of a system under load conditions or upon the occurrence of a fault, the one-line diagram is used to draw the single-phase equivalent circuit of the system. Figure 6.27 combines the equivalent circuits of

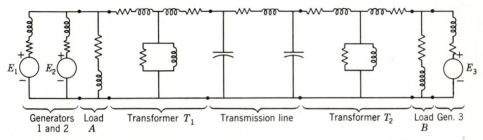

Figure 6.27 Impedance diagram corresponding to the one-line diagram of Fig. 6.26.

the various components shown in Fig. 6.26 to form the impedance diagram of the system. If a load study is to be made, the lagging loads A and B are represented by resistance and inductive reactance in series. The impedance diagram does not include the current-limiting impedances shown in the one-line diagram between the neutrals of the generators and ground because no current flows in the ground under balanced conditions and the neutrals of the generators are at the potential of the neutral of the system. Since the magnetizing current of a transformer is usually insignificant compared with the full-load current, the shunt admittance is usually omitted in the equivalent circuit of the transformer.

As previously mentioned, resistance is often omitted when making fault calculations even in digital-computer programs. Of course, omission of resistance introduces some error, but the results may be satisfactory since the inductive reactance of a system is much larger than its resistance. Resistance and inductive reactance do not add directly, and impedance is not far different from the inductive reactance if the resistance is small. Loads which do not involve rotating machinery have little effect on the total line current during a fault and are usually omitted. Synchronous motor loads, however, are always included in making fault calculations since their generated emfs contribute to the short-circuit current. The diagram should take induction motors into account by a generated emf in series with an inductive reactance if the diagram is to be used to determine the current immediately after the occurrence of a fault. Induction motors are ignored in computing the current a few cycles after the fault occurs because the current contributed by an induction motor dies out very quickly after the induction motor is short-circuited.

If we decide to simplify our calculation of fault current by omitting all static loads, all resistances, the magnetizing current of each transformer, and the capacitance of the transmission line, the impedance diagram reduces to the reactance diagram of Fig. 6.28. These simplifications apply to fault calculations only and not to load-flow studies, which are the subject of Chap. 8. If a computer is available, such simplification is not necessary.

The impedance and reactance diagrams discussed here are sometimes called the positive-sequence diagrams since they show impedances to balanced currents in a symmetrical three-phase system. The significance of this designation will become apparent when Chap. 11 is studied.

If data had been supplied with the one-line diagram we could mark values of reactance on Fig. 6.28. If ohmic values were to be shown, all would have to be

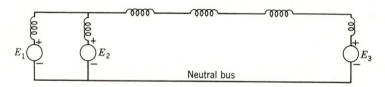

Figure 6.28 Reactance diagram adapted from Fig. 6.27 by omitting all loads, resistances, and shunt admittances.

referred to the same voltage level such as the transmission-line side of the transformer. As we have concluded, however, when bases are specified properly for the various parts of a circuit connected by a transformer the per-unit values of impedances determined in their own part of the system are the same when viewed from another part. Therefore it is necessary only to compute each impedance on the base of its own part of the circuit. The great advantage of using per-unit values is that no computations are necessary to refer an impedance from one side of a transformer to the other.

The following points should be kept in mind:

1. A base kilovolts and base kilovoltamperes is selected in one part of the system. The base values for a three-phase system are understood to be line-to-line kilovolts and three-phase kilovoltamperes or megavoltamperes.
2. For other parts of the system, that is, on other sides of transformers, the base kilovolts for each part is determined according to the line-to-line voltage ratios of the transformers. The base kilovoltamperes will be the same in all parts of the system. It will be helpful to mark the base kilovolts of each part of the system on the one-line diagram.
3. Impedance information available for three-phase transformers will usually be in per unit or percent on the base determined by the ratings.
4. For three single-phase transformers connected as a three-phase unit the three-phase ratings are determined from the single-phase rating of each individual transformer. Impedance in percent for the three-phase unit is the same as that for each individual transformer.
5. Per-unit impedance given on a base other than that determined for the part of the system in which the element is located must be changed to the proper base by Eq. (2.52).

Example 6.10 A 300 MVA, 20 kV three-phase generator has a subtransient reactance of 20%. The generator supplies a number of synchronous motors over a 64-km (40-mi) transmission line having transformers at both ends, as shown on the one-line diagram of Fig. 6.29. The motors, all rated 13.2 kV, are represented by just two equivalent motors. The neutral of one motor M_1 is grounded through reactance. The neutral of the second motor M_2 is not connected to ground (an unusual condition). Rated inputs to the motors are 200 MVA and 100 MVA for M_1 and M_2, respectively. For both motors $X'' = 20\%$. The three-phase transformer T_1 is rated 350 MVA, 230/20 kV

Figure 6.29 One-line diagram for Example 6.10.

with leakage reactance of 10%. Transformer T_2 is composed of three single-phase transformers each rated 127/13.2 kV, 100 MVA with leakage reactance of 10%. Series reactance of the transmission line is 0.5 Ω/km. Draw the reactance diagram with all reactances marked in per unit. Select the generator rating as base in the generator circuit.

SOLUTION The three-phase rating of transformer T_2 is

$$3 \times 100 = 300 \text{ kVA}$$

and its line-to-line voltage ratio is

$$\sqrt{3} \times 127/13.2 = 220/13.2 \text{ kV}$$

A base of 300 MVA, 20 kV in the generator circuit requires a 300 MVA base in all parts of the system and the following voltage bases:

In the transmission line: 230 kV (since T_1 is rated 230/20 kV)

In the motor circuit: $230\dfrac{13.2}{220} = 13.8 \text{ kV}$

These bases are shown in parentheses on the one-line diagram of Fig. 6.29. The reactances of the transformers converted to the proper base are

$$\text{Transformer } T_1: \quad X = 0.1 \times \frac{300}{350} = 0.0857 \text{ per unit}$$

$$\text{Transformer } T_2: \quad X = 0.1\left(\frac{13.2}{13.8}\right)^2 = 0.0915 \text{ per unit}$$

The base impedance of the transmission line is

$$\frac{(230)^2}{300} = 176.3 \ \Omega$$

and the reactance of the line is

$$\frac{0.5 \times 64}{176.3} = 0.1815 \text{ per unit}$$

$$\text{Reactance of motor } M_1 = 0.2\left(\frac{300}{200}\right)\left(\frac{13.2}{13.8}\right)^2 = 0.2745 \text{ per unit}$$

$$\text{Reactance of motor } M_2 = 0.2\left(\frac{300}{100}\right)\left(\frac{13.2}{13.8}\right)^2 = 0.5490 \text{ per unit}$$

Figure 6.30 is the required reactance diagram.

Example 6.11 If the motors M_1 and M_2 of Example 6.10 have inputs of 120 and 60 MW respectively at 13.2 kV, and both operate at unity power factor, find the voltage at the terminals of the generator.

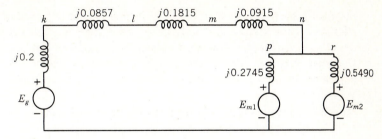

Figure 6.30 Reactance diagram for Example 6.10. Reactances are in per unit on the specified base.

SOLUTION Together the motors take 180 MW, or

$$\frac{180}{300} = 0.6 \text{ per unit}$$

Therefore with V and I at the motors in per unit

$$|V| \cdot |I| = 0.6 \text{ per unit}$$

and since

$$V = \frac{13.2}{13.8} = 0.9565\underline{/0°} \text{ per unit}$$

$$I = \frac{0.6}{0.9565} = 0.6273\underline{/0°} \text{ per unit}$$

At the generator

$$V = 0.9565 + 0.6273(j0.0915 + j0.1815 + j0.0857)$$

$$= 0.9565 + j0.2250 = 0.9826\underline{/13.2°} \text{ per unit}$$

The generator terminal voltage is

$$0.9826 \times 20 = 19.65 \text{ kV}$$

6.13 THE ADVANTAGES OF PER-UNIT COMPUTATIONS

Making computations for electric systems in terms of per-unit values simplifies the work greatly. A real appreciation of the value of the per-unit method comes through experience. Some of the advantages of the method are summarized briefly below.

1. Manufacturers usually specify the impedance of a piece of apparatus in percent or per unit on the base of the nameplate rating.
2. The per-unit impedances of machines of the same type and widely different rating usually lie within a narrow range, although the ohmic values differ

materially for machines of different ratings. For this reason, when the impedance is not known definitely, it is generally possible to select from tabulated average values a per-unit impedance which will be reasonably correct. Experience in working with per-unit values brings familiarity with the proper values of per-unit impedance for different types of apparatus.

3. When impedance in ohms is specified in an equivalent circuit, each impedance must be referred to the same circuit by multiplying it by the square of the ratio of the rated voltages of the two sides of the transformer connecting the reference circuit and the circuit containing the impedance. The per-unit impedance, once it is expressed on the proper base, is the same referred to either side of any transformer.

4. The way in which transformers are connected in three-phase circuits does not affect the per-unit impedances of the equivalent circuit, although the transformer connection does determine the relation between the voltage bases on the two sides of the transformer.

6.14 SUMMARY

The introduction in this chapter of the simplified equivalent circuits for the synchronous generator and transformer is of great importance for the remainder of our discussion throughout this book.

We have seen that the synchronous generator will deliver an increasing amount of reactive power to the system to which it is connected as its excitation is increased. Conversely, as its excitation is reduced it will furnish less reactive power and when underexcited will draw reactive power from the system. This analysis was made on the assumption of a generator supplying such a large system that the terminal voltage remained constant. In Chap. 8, we will extend this analysis to a generator supplying a system represented by its Thévenin equivalent.

Per-unit calculations will be used almost continuously throughout the chapters to follow. We have seen how the transformer is eliminated in the equivalent circuit by the use of per-unit calculations. It is important to remember that the square root of three does not enter the detailed per-unit computations because of the specification of a base line-to-line voltage and base line-to-neutral voltage related by the square root of three.

The concept of proper selection of base in the various parts of a circuit linked by transformers and the calculation of parameters in per unit on the base specified for the part of the circuit in which the parameters exist is fundamental in building an equivalent circuit from a one-line diagram.

PROBLEMS

6.1 Show the steps by which the sum of the three mmfs expressed in Eqs. (6.6) to (6.8) can be equated to the traveling wave of mmf given in Eq. (6.10).

6.2 Determine the highest speed at which two generators mounted on the same shaft can be driven so that the frequency of one generator is 60 Hz and the frequency of the other is 25 Hz. How many poles does each machine have?

6.3 The synchronous reactance of a generator is 1.0 per unit, and the leakage reactance of its armature is 0.1 per unit. The voltage to neutral of phase a at the bus of a large system to which the generator is connected is $1.0\underline{/0°}$ per unit, and the generator is delivering a current I_a equal to $1.0\underline{/-30°}$ per unit. Neglect resistance of the windings and find (a) the voltage drop in the machine due to armature reaction, (b) the no-load voltage to neutral E_g of phase a of the generator, and (c) the per-unit values of P and Q delivered to the bus.

6.4 Solve parts (b) and (c) of Prob. 6.3 for $I_a = 1.0\underline{/30}$ per unit and compare the results of these two problems.

6.5 For a certain armature current in a synchronous generator the mmf due to the field current is twice that due to armature reaction. Neglect saturation and find the ratio of the voltage E_r generated by the air-gap flux to the no-load voltage of the generator (a) when armature current I_a is in phase with E_r, (b) when I_a lags E_r by 90°, and (c) when I_a leads E_r by 90°.

6.6 A single-phase transformer is rated 440/220 V, 5.0 kVA. When the low-tension side is short-circuited and 35 V is applied to the high-tension side, rated current flows in the windings and the power input is 100 W. Find the resistance and reactance of the high- and low-tension windings if the power loss and ratio of reactance to resistance is the same in both windings.

6.7 A single-phase transformer rated 30 kVA, 1200/120 V is connected as an autotransformer to supply 1320 V from a 1200-V bus.

(a) Draw a diagram of the transformer connections showing the polarity marks on the windings and directions chosen as positive for current in each winding so that the currents will be in phase.

(b) Mark on the diagram the values of rated current in the windings and at the input and output.

(c) Determine the rated kilovoltamperes of the unit as an autotransformer.

(d) If the efficiency of the transformer connected for 1200/120-V operation at rated load unity power factor is 97%, determine its efficiency as an autotransformer with rated current in the windings and operating at rated voltage to supply a load at unity power factor.

6.8 Solve Prob. 6.7 if the transformer is to supply 1080 V from a 1200-V bus

6.9 A Δ-connected resistive load of 8000 kW is connected to the low-tension, Δ-connected side of a Y-Δ transformer rated 10,000 kVA, 138/13.8 kV. Find the load resistance in ohms in each phase as measured from line to neutral on the high-tension side of the transformer. Neglect transformer impedance and assume rated voltage is applied to the transformer primary.

6.10 Solve Prob. 6.7 if the same resistors are reconnected in Y.

6.11 Three transformers, each rated 5 kVA, 220 V on the secondary side, are connected Δ-Δ and have been supplying a 15 kW purely resistive load at 220 V. A change is made which reduces the load to 10 kW, still purely resistive. Someone suggests that, with two-thirds of the load, one transformer can be removed and the system can be operated open-Δ. Balanced three-phase voltages will still be supplied to the load since two of the line voltages (and thus also the third) will be unchanged.

To investigate further the suggestion

(a) Find each of the line currents (magnitude and angle) with the 10 kW load and the transformer between a and c removed. (Assume $V_{ab} = 220\underline{/0°}$ V, sequence $a\ b\ c$.)

(b) Find the kilovoltamperes supplied by each of the remaining transformers.

(c) What restriction must be placed on the load for open-Δ operation with these transformers?

(d) Think about why the individual transformer kilovoltampere values include a Q component when the load is purely resistive.

6.12 A transformer rated 200 MVA, 345Y/20.5Δ kV connects a load of rated 180 MVA, 22.5 kV, 0.8 power factor lag to a transmission line. Determine (*a*) the rating of each of three single-phase transformers which when properly connected will be equivalent to the three-phase transformer and (*b*) the complex impedance of the load in per unit in the impedance diagram if the base in the transmission line is 100 MVA, 345 kV.

6.13 A 120 MVA, 19.5 kV generator has $X_s = 1.5$ per unit and is connected to a transmission line by a transformer rated 150 MVA, 230Y/18Δ kV with $X = 0.1$ per unit. If the base to be used in the calculations is 100 MVA, 230 kV for the transmission line, find the per-unit values to be used for the transformer and generator reactances.

6.14 A transformer's three-phase rating is 5000 kVA, 115/13.2 kV, and its impedance is 0.007 + $j0.075$ per unit. The transformer is connected to a short transmission line whose impedance is $0.02 + j0.10$ per unit on a base of 10 MVA, 13.2 kV. The line supplies a three-phase load rated 3400 kW, 13.2 kV, with a lagging power factor of 0.85. If the high-tension voltage remains constant at 115 kV when the load at the end of the line is disconnected. find the voltage regulation at the load. Work in per unit and choose as base 10 MVA, 13.2 kV at the load.

6.15 The one-line diagram of an unloaded power system is shown in Fig. 6.31. Reactances of the two sections of transmission line are shown on the diagram. The generators and transformers are rated as follows:

Generator 1:   20 MVA, 13.8 kV, $X'' = 0.20$ per unit
Generator 2:   30 MVA, 18 kV, $X'' = 0.20$ per unit
Generator 3:   30 MVA, 20 kV, $X'' = 0.20$ per unit
Transformer T_1: 25 MVA, 220Y/13.8Δ kV, $X = 10\%$
Transformer T_2: Single-phase units each rated 10 MVA, 127/18 kV, $X = 10\%$
Transformer T_3: 35 MVA, 220Y/22Y kV, $X = 10\%$

Draw the impedance diagram with all reactances marked in per unit and with letters to indicate points corresponding to the one-line diagram. Choose a base of 50 MVA, 13.8 kV in the circuit of generator 1.

6.16 Draw the impedance diagram for the power system shown in Fig. 6.32. Mark impedances in per unit. Neglect resistance, and use a base of 50 kVA, 138 kV in the 40-Ω line. The ratings of the generators, motors, and transformers are:

Generator 1:   20 MVA, 18 kV, $X'' = 20\%$
Generator 2:   20 MVA, 18 kV, $X'' = 20\%$
Synchronous motor 3:   30 MVA, 13.8 kV, $X'' = 20\%$
Three-phase Y-Y transformers: 20 MVA, 138Y/20Y kV, $X = 10\%$
Three-phase Y-Δ transformers: 15 MVA, 138Y/13.8Δ kV, $X = 10\%$

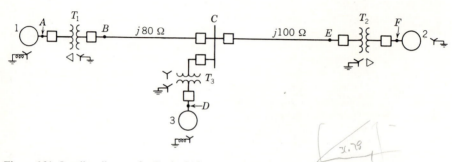

Figure 6.31 One-line diagram for Prob. 6.15.

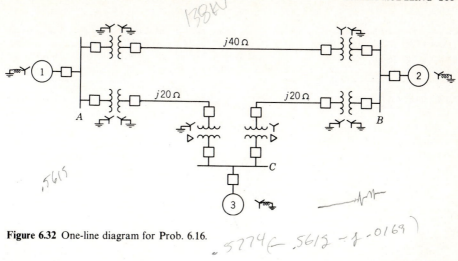

Figure 6.32 One-line diagram for Prob. 6.16.

6.17 If the voltage of bus *C* in Prob. 6.16 is 13.2 kV when the motor draws 24 MW at 0.8 power factor leading, calculate the voltages of buses *A* and *B*. Assume that the two generators divide the load equally. Give the answer in volts and in per unit on the base selected for Prob. 6.16. Find the voltages at *A* and *B* when the circuit breaker connecting generator 1 to bus *A* is open while the motor draws 12 MW at 13.2 kV with 0.8 power factor leading. All other circuit breakers remain closed.

SEVEN

NETWORK CALCULATIONS

The continued development of large, high-speed digital computers has brought about a change in the relative importance of various techniques in the solution of large networks. Digital-computer solutions depend upon network equations. So it is important that the power-system engineer understand the formulation of the equations from which, in obtaining a solution, the program that is followed by the computer is derived.

This chapter is not meant to be a comprehensive review of network equations but will serve to review and expand upon those methods of analysis upon which programs for computer solutions of power-system problems are very dependent.

Of particular importance in this chapter is the introduction of bus admittance and impedance matrices which will prove to be very useful in later work.

7.1 EQUIVALENCE OF SOURCES

A helpful procedure in some problems in network analysis is the substitution of a source of constant current in parallel with an impedance for a constant emf and series impedance. The two parts of Fig. 7.1 illustrate the circuits. Both sources with their associated impedances are connected to a two-terminal network having an input impedance Z_L. The load may be considered a passive network for the present; that is, any internal emfs in the load network are assumed to be short-circuited and any current sources opened.

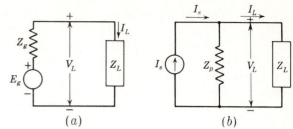

Figure 7.1 Circuits illustrating the equivalence of sources.

(a) (b)

For the circuit having the constant emf E_g and series impedance Z_g, the voltage across the load is

$$V_L = E_g - I_L Z_g \qquad (7.1)$$

where I_L is the load current. For the circuit having a source of constant current I_s with the shunt impedance Z_p, the voltage across the load is

$$V_L = (I_s - I_L)Z_p = I_s Z_p - I_L Z_p \qquad (7.2)$$

The two sources and their associated impedances will be equivalent if the voltage V_L is the same in both circuits. Of course, equal values of V_L will mean equal load currents I_L for identical loads.

Comparison of Eqs. (7.1) and (7.2) shows that V_L will be identical in both circuits and therefore that the emf and its series impedance can be interchanged with the current source and its shunt impedance provided

$$E_g = I_s Z_p \qquad (7.3)$$

and

$$Z_g = Z_p \qquad (7.4)$$

These relations show that a constant-current source and shunt impedance can be replaced by a constant emf and series impedance if the emf is equal to the product of the constant current and the shunt impedance and if the series impedance equals the shunt impedance. Conversely, a constant emf and series impedance can be replaced by a constant-current source and shunt impedance if the shunt impedance is identical to the series impedance and if the constant current is equal to the value of the emf divided by its series impedance.

We have shown the conditions for equivalence of sources connected to a passive network. By considering the principle of superposition we can show that the same provisions apply if the output is an active network, that is, if the output network includes voltage and current sources. To determine the contribution from the supply if the output network is active, the principle of superposition calls for shorting emfs in the output network and replacing current sources by open circuits while impedances remain intact. Thus the output is a passive network so far as the current component from the interchangeable sources is concerned. To determine the current components due to the sources in the load network, the emf of the supply source is shorted in one case and the current source is opened in the other case. Thus, only Z_g or its equivalent Z_p is connected across the input to the load to determine the effect of sources in the load network regardless of which type of source is the supply. So, in applying superposition, the components contributed by the sources in the load network are independent of the type of supply so long as the series impedance of the emf equals the shunt impedance of the constant-current source. Therefore, the same provisions for equivalence apply whether the load network is active or passive.

7.2 NODE EQUATIONS

The junctions formed when two or more pure elements (R, L, or C, or an ideal source of voltage or current) are connected to each other at their terminals are called *nodes*. Systematic formulation of equations determined at nodes of a circuit by applying Kirchhoff's current law is the basis of some excellent computer solutions of power-system problems. Usually it is convenient to consider only those nodes to which more than two elements are connected and to call these junction points *major nodes*.

In order to examine some features of node equations, we shall begin with the one-line diagram of a simple system shown in Fig. 7.2. Generators are connected through transformers to high-tension buses 1 and 3 and supply a synchronous motor load at bus 2. For purposes of analysis, all machines at any one bus are treated as a single machine and represented by a single emf and series reactance. The reactance diagram, with reactances specified in per unit, is shown in Fig. 7.3. Nodes are indicated by dots, but numbers are assigned only to major nodes. If the circuit is redrawn with the emfs and the impedances in series connecting them to the major nodes replaced by the equivalent current sources and shunt admittances, the result is the circuit of Fig. 7.4. Admittance values in per unit are shown instead of impedance values.

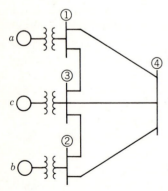

Figure 7.2 One-line diagram of a simple system.

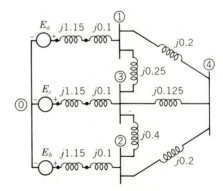

Figure 7.3 Reactance diagram for the system of Fig. 7.2. Reactance values are in per unit.

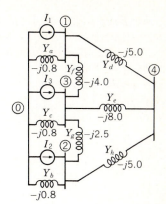

Figure 7.4 Circuit of Fig. 7.3 with current sources replacing the equivalent voltage sources. Values shown are admittances in per unit.

Single-subscript notation will be used to designate the voltage of each bus with respect to the neutral taken as the reference node 0. Applying Kirchhoff's current law at node 1 with current into the node from the source equated to current away from the node gives

$$I_1 = V_1 Y_a + (V_1 - V_3)Y_f + (V_1 - V_4)Y_d \tag{7.5}$$

and for node 4

$$0 = (V_4 - V_1)Y_d + (V_4 - V_2)Y_h + (V_4 - V_3)Y_e \tag{7.6}$$

Rearranging these equations yields

$$I_1 = V_1(Y_a + Y_f + Y_d) - V_3 Y_f - V_4 Y_d \tag{7.7}$$

$$0 = - V_1 Y_d - V_2 Y_h - V_3 Y_e + V_4(Y_d + Y_e + Y_h) \tag{7.8}$$

Similar equations can be formed for nodes 2 and 3, and the four equations can be solved simultaneously for the voltages V_1, V_2, V_3, and V_4. All branch currents can be found when these voltages are known, and so the required number of node equations is one less than the number of nodes in the network. A node equation formed for the reference node would yield no further information. In other words, the number of independent node equations is one less than the number of nodes.

We have not written the other two equations because we can already see how to formulate node equations in standard notation. In both Eqs. (7.7) and (7.8) it is apparent that the current flowing into the network from current sources connected to a node is equated to the sum of several products. At any node one product is the voltage of that node times the sum of all the admittances which terminate on the node. This product accounts for the current that flows away from the node if the voltage is zero at each other node. Each other product equals the *negative* of the voltage at another node times the admittance connected directly between the other node and the node at which the equation is formulated. For instance, at node 1 a product is $-V_3 Y_f$, which accounts for the current away from node 1 when all node voltages are zero except that at node 3.

The standard form for the four independent equations in matrix form is

$$
\begin{bmatrix} I_1 \\ I_2 \\ I_3 \\ I_4 \end{bmatrix} = \begin{bmatrix} Y_{11} & Y_{12} & Y_{13} & Y_{14} \\ Y_{21} & Y_{22} & Y_{23} & Y_{24} \\ Y_{31} & Y_{32} & Y_{33} & Y_{34} \\ Y_{41} & Y_{42} & Y_{43} & Y_{44} \end{bmatrix} \begin{bmatrix} V_1 \\ V_2 \\ V_3 \\ V_4 \end{bmatrix}
\tag{7.9}
$$

The symmetry of the equations in this form makes them easy to remember, and their extension to any number of nodes is apparent. The order of the Y subscripts is *effect-cause*; that is, the first subscript is that of the node at which the current is being expressed, and the second subscript is that of the voltage causing this component of current. The Y matrix is designated $\mathbf{Y}_{bus}$ and called the bus admittance matrix.† It is symmetrical around the principal diagonal. The admittances Y_{11}, Y_{22}, Y_{33}, and Y_{44} are called the *self-admittances* at the nodes, and each equals the sum of all the admittances terminating on the node identified by the repeated subscripts. The other admittances are the *mutual admittances* of the nodes, and each equals the negative of the sum of all admittances connected directly between the nodes identified by the double subscripts. For the network of Fig. 7.4 the mutual admittance Y_{13} equals $-Y_f$. Some authors call the self- and mutual admittances of the nodes the driving-point and transfer admittances of the nodes.

The general expression for the source current toward node k of a network having N independent nodes, that is, N buses other than the neutral, is

$$
I_k = \sum_{n=1}^{N} Y_{kn} V_n
\tag{7.10}
$$

One such equation must be written for each of the N buses at which the voltage of the network is unknown. If the voltage is fixed at any node, the equation is not written for that node. For instance, if both the magnitude and angle of the voltages at two of the high-tension buses of our example are fixed, only two equations are needed. Node equations would be written for the other two buses, the only ones at which the voltage would be unknown. A known emf and series impedance need not be replaced by the equivalent current source if one terminal of the emf element is connected to the reference node, for then the node which separates the emf and series impedance is one where voltage is known.

Example 7.1 Write in matrix form the node equations necessary to solve for the voltages of the numbered buses of Fig. 7.4. The network is equivalent to that of Fig. 7.3. The emfs shown in Fig. 7.3 are $E_a = 1.5\underline{/0°}$, $E_b = 1.5\underline{/-36.87°}$, and $E_c = 1.5\underline{/0°}$, all in per unit.

---

† Boldface type is used where one letter designates a matrix.

SOLUTION The current sources are

$$I_1 = I_3 = \frac{1.5\underline{/0^\circ}}{j1.25} = 1.2\ \underline{/-90^\circ} = 0 - j1.20 \text{ per unit}$$

$$I_2 = \frac{1.5\ \underline{/-36.87^\circ}}{j1.25} = 1.2\ \underline{/-126.87^\circ} = -0.72 - j0.96 \text{ per unit}$$

Self-admittances in per unit are

$$Y_{11} = -j5.0 - j4.0 - j0.8 = -j9.8$$

$$Y_{22} = -j5.0 - j2.5 - j0.8 = -j8.3$$

$$Y_{33} = -j4.0 - j2.5 - j8.0 - j0.8 = -j15.3$$

$$Y_{44} = -j5.0 - j5.0 - j8.0 = -j18.0$$

and the mutual admittances in per unit are

$$Y_{12} = Y_{21} = 0 \qquad Y_{23} = Y_{32} = +j2.5$$

$$Y_{13} = Y_{31} = +j4.0 \qquad Y_{24} = Y_{42} = +j5.0$$

$$Y_{14} = Y_{41} = +j5.0 \qquad Y_{34} = Y_{43} = +j8.0$$

The node equations in matrix form are

$$\begin{bmatrix} 0 & -j1.20 \\ -0.72 - j0.96 \\ 0 & -j1.20 \\ 0 \end{bmatrix} = \begin{bmatrix} -j9.8 & j0.0 & j4.0 & j5.0 \\ j0.0 & -j8.3 & j2.5 & j5.0 \\ j4.0 & j2.5 & -j15.3 & j8.0 \\ j5.0 & j5.0 & j8.0 & -j18.0 \end{bmatrix} \begin{bmatrix} V_1 \\ V_2 \\ V_3 \\ V_4 \end{bmatrix}$$

The square matrix above is recognized as the bus admittance matrix $\mathbf{Y}_{bus}$.

Example 7.2 Solve the node equations of the preceding example to find the bus voltages by inverting the bus admittance matrix.

SOLUTION Premultiplying both sides of the matrix equation of Example 7.1 by the inverse of the bus admittance matrix (determined by using a standard program on a digital computer) yields

$$\begin{bmatrix} j0.4774 & j0.3706 & j0.4020 & j0.4142 \\ j0.3706 & j0.4872 & j0.3922 & j0.4126 \\ j0.4020 & j0.3922 & j0.4558 & j0.4232 \\ j0.4142 & j0.4126 & j0.4232 & j0.4733 \end{bmatrix} \begin{bmatrix} 0 & -j1.20 \\ -0.72 - j0.96 \\ 0 & -j1.20 \\ 0 \end{bmatrix} = \begin{bmatrix} 1 & 0 & 0 & 0 \\ 0 & 1 & 0 & 0 \\ 0 & 0 & 1 & 0 \\ 0 & 0 & 0 & 1 \end{bmatrix} \begin{bmatrix} V_1 \\ V_2 \\ V_3 \\ V_4 \end{bmatrix}$$

The square matrix above obtained by inverting the bus admittance matrix is called the bus impedance matrix $\mathbf{Z}_{bus}$. Performing the indicated matrix multiplication yields

$$\begin{bmatrix} 1.4111 - j0.2668 \\ 1.3830 - j0.3508 \\ 1.4059 - j0.2824 \\ 1.4009 - j0.2971 \end{bmatrix} = \begin{bmatrix} V_1 \\ V_2 \\ V_3 \\ V_4 \end{bmatrix}$$

and so the node voltages are

$$V_1 = 1.4111 - j0.2668 = 1.436 \; \underline{/-10.71^\circ} \text{ per unit}$$

$$V_2 = 1.3830 - j0.3508 = 1.427 \; \underline{/-14.24^\circ} \text{ per unit}$$

$$V_3 = 1.4059 - j0.2824 = 1.434 \; \underline{/-11.36^\circ} \text{ per unit}$$

$$V_4 = 1.4009 - j0.2971 = 1.432 \; \underline{/-11.97^\circ} \text{ per unit}$$

7.3 MATRIX PARTITIONING

A useful method of matrix manipulation, called *partitioning*, consists in recognizing various parts of a matrix as submatrices which are treated as single elements in applying the usual rules of multiplication and addition. For instance, assume a 3×3 matrix $\mathbf{A}$, where

$$\mathbf{A} = \begin{bmatrix} a_{11} & a_{12} & a_{13} \\ a_{21} & a_{22} & a_{23} \\ \hline a_{31} & a_{32} & a_{33} \end{bmatrix} \tag{7.11}$$

The matrix is partitioned into four submatrices by the horizontal and vertical dashed lines. The matrix may be written

$$\mathbf{A} = \begin{bmatrix} \mathbf{D} & \mathbf{E} \\ \mathbf{F} & \mathbf{G} \end{bmatrix} \tag{7.12}$$

where the submatrices are

$$\mathbf{D} = \begin{bmatrix} a_{11} & a_{12} \\ a_{21} & a_{22} \end{bmatrix} \qquad \mathbf{E} = \begin{bmatrix} a_{13} \\ a_{23} \end{bmatrix}$$

$$\mathbf{F} = \begin{bmatrix} a_{31} & a_{32} \end{bmatrix} \qquad \mathbf{G} = a_{33}$$

To show the steps in matrix multiplication in terms of submatrices let us assume that $\mathbf{A}$ is to be postmultiplied by another matrix $\mathbf{B}$ to form the product

C, where

$$\mathbf{B} = \begin{bmatrix} b_{11} \\ b_{21} \\ \cdots \\ b_{31} \end{bmatrix} \tag{7.13}$$

With partitioning as indicated,

$$\mathbf{B} = \begin{bmatrix} \mathbf{H} \\ \mathbf{J} \end{bmatrix} \tag{7.14}$$

where the submatrices are

$$\mathbf{H} = \begin{bmatrix} b_{11} \\ b_{21} \end{bmatrix} \quad \text{and} \quad \mathbf{J} = b_{31}$$

Then the product is

$$\mathbf{C} = \mathbf{AB} = \begin{bmatrix} \mathbf{D} & \mathbf{E} \\ \mathbf{F} & \mathbf{G} \end{bmatrix} \begin{bmatrix} \mathbf{H} \\ \mathbf{J} \end{bmatrix} \tag{7.15}$$

The submatrices are treated as single elements to obtain

$$\mathbf{C} = \begin{bmatrix} \mathbf{DH} + \mathbf{EJ} \\ \mathbf{FH} + \mathbf{GJ} \end{bmatrix} \tag{7.16}$$

The product is finally determined by performing the indicated multiplication and addition of the submatrices.

If **C** is composed of the submatrices **M** and **N** so that

$$\mathbf{C} = \begin{bmatrix} \mathbf{M} \\ \mathbf{N} \end{bmatrix} \tag{7.17}$$

comparison with Eq. (7.16) shows

$$\mathbf{M} = \mathbf{DH} + \mathbf{EJ} \tag{7.18}$$

$$\mathbf{N} = \mathbf{FH} + \mathbf{GJ} \tag{7.19}$$

If we wish to find only the submatrix **N**, partitioning shows that

$$\mathbf{N} = \begin{bmatrix} a_{31} & a_{32} \end{bmatrix} \begin{bmatrix} b_{11} \\ b_{21} \end{bmatrix} + a_{33} b_{31}$$

$$= a_{31} b_{11} + a_{32} b_{21} + a_{33} b_{31} \tag{7.20}$$

The matrices to be multiplied must be compatible originally. Each vertical partitioning line between columns r and $r + 1$ of the first factor requires a horizontal partitioning line between rows r and $r + 1$ of the second factor in order for the submatrices to be multiplied. Horizontal partitioning lines may be drawn between any rows of the first factor, and vertical lines between any columns of the second, or omitted in either or both. An example that applies matrix partitioning appears at the end of the next section.

7.4 NODE ELIMINATION BY MATRIX ALGEBRA

Nodes may be eliminated by matrix manipulation of the standard node equations. However, only those nodes at which current does not enter or leave the network can be eliminated.

The standard node equations in matrix notation are expressed as

$$\mathbf{I} = \mathbf{Y}_{\text{bus}}\mathbf{V} \tag{7.21}$$

where $\mathbf{I}$ and $\mathbf{V}$ are column matrices and $\mathbf{Y}_{\text{bus}}$ is a symmetrical square matrix. The column matrices must be so arranged that elements associated with nodes to be eliminated are in the lower rows of the matrices. Elements of the square admittance matrix are located correspondingly. The column matrices are partitioned so that the elements associated with nodes to be eliminated are separated from the other elements. The admittance matrix is partitioned so that elements identified only with nodes to be eliminated are separated from the other elements by horizontal and vertical lines. When partitioned according to these rules, Eq. (7.21) becomes

$$\begin{bmatrix} \mathbf{I}_A \\ \mathbf{I}_X \end{bmatrix} = \begin{bmatrix} \mathbf{K} & \mathbf{L} \\ \mathbf{L}^T & \mathbf{M} \end{bmatrix} \begin{bmatrix} \mathbf{V}_A \\ \mathbf{V}_X \end{bmatrix} \tag{7.22}$$

where $\mathbf{I}_X$ is the submatrix composed of the currents entering the nodes to be eliminated and $\mathbf{V}_X$ is the submatrix composed of the voltages of these nodes. Of course, every element in $\mathbf{I}_X$ is zero, for the nodes could not be eliminated otherwise. The self- and mutual admittances composing $\mathbf{K}$ are those identified only with nodes to be retained. $\mathbf{M}$ is composed of the self- and mutual admittances identified only with nodes to be eliminated. It is a square matrix whose order is equal to the number of nodes to be eliminated. $\mathbf{L}$ and its transpose $\mathbf{L}^T$ are composed of only those mutual admittances common to a node to be retained and to one to be eliminated.

Performing the multiplication indicated in Eq. (7.22) gives

$$\mathbf{I}_A = \mathbf{K}\mathbf{V}_A + \mathbf{L}\mathbf{V}_X \tag{7.23}$$

and

$$\mathbf{I}_X = \mathbf{L}^T\mathbf{V}_A + \mathbf{M}\mathbf{V}_X \tag{7.24}$$

Since all elements of $\mathbf{I}_X$ are zero, subtracting $\mathbf{L}^T\mathbf{V}_A$ from both sides of Eq. (7.24) and multiplying both sides by the inverse of $\mathbf{M}$ (denoted by $\mathbf{M}^{-1}$) yields

$$-\mathbf{M}^{-1}\mathbf{L}^T\mathbf{V}_A = \mathbf{V}_X \tag{7.25}$$

This expression for $\mathbf{V}_X$ substituted in Eq. (7.23) gives

$$\mathbf{I}_A = \mathbf{K}\mathbf{V}_A - \mathbf{L}\mathbf{M}^{-1}\mathbf{L}^T\mathbf{V}_A \tag{7.26}$$

which is a node equation having the admittance matrix

$$\mathbf{Y}_{\text{bus}} = \mathbf{K} - \mathbf{L}\mathbf{M}^{-1}\mathbf{L}^T \tag{7.27}$$

This admittance matrix enables us to construct the circuit with the unwanted nodes eliminated, as we shall see in the following example.

Example 7.3 If the generator and transformer at bus 3 are removed from the circuit of Fig. 7.3, eliminate nodes 3 and 4 by the matrix-algebra procedure just described, find the equivalent circuit with these nodes eliminated, and find the complex power transferred into or out of the network at nodes 1 and 2. Also find the voltage at node 1.

SOLUTION The bus admittance matrix of the circuit partitioned for elimination of nodes 3 and 4 is

$$
\mathbf{Y}_{bus} = \begin{bmatrix} \mathbf{K} & \mathbf{L} \\ \mathbf{L}^T & \mathbf{M} \end{bmatrix} = \begin{bmatrix} -j9.8 & 0.0 & \vdots & j4.0 & j5.0 \\ 0.0 & -j8.3 & \vdots & j2.5 & j5.0 \\ \cdots & \cdots & & \cdots & \cdots \\ j4.0 & j2.5 & \vdots & -j14.5 & j8.0 \\ j5.0 & j5.0 & \vdots & j8.0 & -j18.0 \end{bmatrix}
$$

The inverse of the submatrix in the lower right position is

$$
\mathbf{M}^{-1} = \frac{1}{-197} \begin{bmatrix} -j18.0 & -j8.0 \\ -j8.0 & -j14.5 \end{bmatrix} = \begin{bmatrix} j0.0914 & j0.0406 \\ j0.0406 & j0.0736 \end{bmatrix}
$$

Then

$$
\mathbf{LM}^{-1}\mathbf{L}^T = \begin{bmatrix} j4.0 & j5.0 \\ j2.5 & j5.0 \end{bmatrix} \begin{bmatrix} j0.0914 & j0.0406 \\ j0.0406 & j0.0736 \end{bmatrix} \begin{bmatrix} j4.0 & j2.5 \\ j5.0 & j5.0 \end{bmatrix}
$$

$$
= - \begin{bmatrix} j4.9264 & j4.0736 \\ j4.0736 & j3.4264 \end{bmatrix}
$$

$$
\mathbf{Y}_{bus} = \mathbf{K} - \mathbf{LM}^{-1}\mathbf{L}^T = \begin{bmatrix} -j9.8 & 0.0 \\ 0.0 & -j8.3 \end{bmatrix} - \mathbf{LM}^{-1}\mathbf{L}^T
$$

$$
\mathbf{Y}_{bus} = \begin{bmatrix} -j4.8736 & j4.0736 \\ j4.0736 & -j4.8736 \end{bmatrix}
$$

Examination of the matrix shows us that the admittance between the two remaining buses 1 and 2 is $-j4.0736$, the reciprocal of which is the per-unit impedance between these buses. The admittance between each of these buses and the reference bus is

$$
-j4.8736 - (-j4.0736) = -j0.800 \text{ per unit}
$$

The resulting circuit is shown in Fig. 7.5a. When the current sources are converted to their equivalent emf sources the circuit, with impedances in per unit, is that of Fig. 7.5b. Then the current is

$$
I = \frac{1.5\underline{/0°} - 1.5\underline{/-36.87°}}{j(1.25 + 1.25 + 0.2455)} = \frac{1.5 - 1.2 + j0.9}{j(2.7455)}
$$

$$
= 0.3278 - j0.1093 = 0.3455 \underline{/-18.44} \text{ per unit}
$$

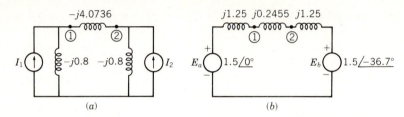

Figure 7.5 Circuit of Fig. 7.3 without the source at node 3 (a) with the equivalent current sources and (b) with the original voltage sources at nodes 1 and 2.

Power *out* of source a is

$$1.5\underline{/0°} \times 0.3455\underline{/18.44°} = 0.492 + j0.164 \text{ per unit}$$

And power *into* source b is

$$1.5\underline{/-36.87°} \times 0.3455\underline{/18.44°} = 0.492 - j0.164 \text{ per unit}$$

Note that the reactive voltamperes in the circuit equal

$$(0.3455)^2 \times 2.7455 = 0.328 = 0.164 + 0.164$$

The voltage at node 1 is

$$1.50 - j1.25(0.3278 - j0.1093) = 1.363 - j0.410 \text{ per unit}$$

In the simple circuit of this example node elimination could have been accomplished by Y-Δ transformations and by working with series and parallel combinations of impedances. The matrix partitioning method is a general method which is thereby more suitable for computer solutions. However, for the elimination of a large number of nodes, the matrix **M** whose inverse must be found will be large.

Inverting a matrix is avoided by eliminating one node at a time, and the process is very simple. The node to be eliminated must be the highest numbered node, and renumbering may be required. The matrix **M** becomes a single element and $\mathbf{M}^{-1}$ is the reciprocal of the element. The original admittance matrix partitioned into submatrices **K**, **L**, $\mathbf{L}^T$, and **M** is

$$\mathbf{Y}_{\text{bus}} = \begin{bmatrix} Y_{11} & \cdots & Y_{1j} & \cdots & Y_{1n} \\ \vdots & & \vdots & & \vdots \\ Y_{k1} & \cdots & Y_{kj} & \cdots & Y_{kn} \\ \vdots & & \vdots & & \vdots \\ Y_{n1} & \cdots & Y_{nj} & \cdots & Y_{nn} \end{bmatrix} \begin{matrix} \\ \\ \mathbf{L} \\ \\ \end{matrix} \qquad (7.28)$$

$$\underbrace{\phantom{Y_{11} \cdots Y_{1j}}}_{\mathbf{K}} \qquad \underbrace{}_{\mathbf{L}^T} \quad \underbrace{}_{\mathbf{M}}$$

the reduced $(n - 1) \times (n - 1)$ matrix will be, according to Eq. (7.27),

$$\mathbf{Y}_{\text{bus}} = \begin{bmatrix} Y_{11} & \cdots & Y_{1j} & \cdots \\ \vdots & & \vdots & \\ Y_{k1} & \cdots & Y_{kj} & \cdots \\ \vdots & & \vdots & \end{bmatrix} - \frac{1}{Y_{nn}} \begin{bmatrix} Y_{1n} \\ \vdots \\ Y_{kn} \\ \vdots \end{bmatrix} [Y_{n1} \cdots Y_{nj} \cdots] \qquad (7.29)$$

and when the indicated manipulation of the matrices is accomplished, the element in row k and column j of the resulting $(n - 1) \times (n - 1)$ matrix will be

$$Y_{kj \text{ (new)}} = Y_{kj \text{ (orig)}} - \frac{Y_{kn} Y_{nj}}{Y_{nn}} \qquad (7.30)$$

Each element in the original matrix **K** must be modified. When Eq. (7.28) is compared to Eq. (7.30) we can see how to proceed. We multiply the element in the last column and the same row as the element being modified by the element in the last row and the same column as the element being modified. We then divide this product by Y_{nn} and subtract the result from the element being modified. The following example illustrates the simple procedure.

Example 7.4 Perform the node elimination of Example 7.3 by first removing node 4 and then by removing node 3.

SOLUTION As in Example 7.3, the original matrix now partitioned for removal of one node is

$$\mathbf{Y}_{\text{bus}} = \begin{bmatrix} -j9.8 & 0.0 & j4.0 & \vdots & j5.0 \\ 0.0 & -j8.3 & j2.5 & \vdots & j5.0 \\ j4.0 & \boxed{j2.5} & -j14.5 & \vdots & \boxed{j8.0} \\ \hdashline j5.0 & \boxed{j5.0} & j8.0 & \vdots & -j18.0 \end{bmatrix}$$

To modify the element $j2.5$ in row 3, column 2 subtract from it the product of the elements enclosed by rectangles and divided by the element in the lower right corner. We find the modified element

$$Y_{32} = j2.5 - \frac{j8.0 \times j5.0}{-j18.0} = j4.7222$$

Similarly the new element in row 1, column 1 is

$$Y_{11} = -j9.8 - \frac{j5.0 \times j5.0}{-j18.0} = -j8.4111$$

Other elements are found in the same manner to yield

$$\mathbf{Y}_{\text{bus}} = \begin{bmatrix} -j8.4111 & j1.3889 & j6.2222 \\ j1.3889 & -j6.9111 & j4.7222 \\ j6.2222 & j4.7222 & -j10.9444 \end{bmatrix}$$

Reducing the above matrix to remove node 3 yields

$$\mathbf{Y}_{\text{bus}} = \begin{bmatrix} -j4.8736 & j4.0736 \\ j4.0736 & -j4.8736 \end{bmatrix}$$

which is identical to the matrix found by the matrix-partitioning method where two nodes were removed at the same time.

7.5 THE BUS ADMITTANCE AND IMPEDANCE MATRICES

In Example 7.2, we inverted the bus admittance matrix $\mathbf{Y}_{\text{bus}}$ and called the resultant matrix the bus impedance matrix $\mathbf{Z}_{\text{bus}}$. By definition

$$\mathbf{Z}_{\text{bus}} = \mathbf{Y}_{\text{bus}}^{-1} \tag{7.31}$$

and for a network of three independent nodes

$$\mathbf{Z}_{\text{bus}} = \begin{bmatrix} Z_{11} & Z_{12} & Z_{13} \\ Z_{21} & Z_{22} & Z_{23} \\ Z_{31} & Z_{32} & Z_{33} \end{bmatrix} \tag{7.32}$$

Since $\mathbf{Y}_{\text{bus}}$ is symmetrical around the principal diagonal, $\mathbf{Z}_{\text{bus}}$ must be symmetrical in the same manner.

The impedance elements of $\mathbf{Z}_{\text{bus}}$ on the principal diagonal are called *driving-point impedance of the nodes*, and the off-diagonal elements are called the *transfer impedances of the nodes*.

The bus admittance matrix need not be determined in order to obtain $\mathbf{Z}_{\text{bus}}$, and in another section of this chapter we shall see how $\mathbf{Z}_{\text{bus}}$ may be formulated directly.

The bus impedance matrix is important and very useful in making fault calculations as we shall see later. In order to understand the physical significance of the various impedances in the matrix we shall compare them with the node admittances. We can easily do so by looking at the equations at a particular node. For instance, starting with the node equations expressed as

$$\mathbf{I} = \mathbf{Y}_{\text{bus}} \mathbf{V} \tag{7.33}$$

we have at node 2 of the three independent nodes

$$I_2 = Y_{21} V_1 + Y_{22} V_2 + Y_{23} V_3 \tag{7.34}$$

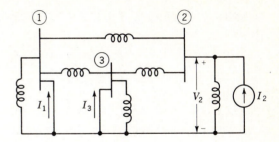

Figure 7.6 Circuit for measuring Y_{22}, Y_{12}, and Y_{32}.

If V_1 and V_3 are reduced to zero by shorting nodes 1 and 3 to the reference node and current I_2 is injected at node 2, the self-admittance at node 2 is

$$Y_{22} = \left.\frac{I_2}{V_2}\right|_{V_1 = V_3 = 0} \tag{7.35}$$

Thus, the self-admittance of a particular node could be measured by shorting all other nodes to the reference node and then finding the ratio of the current injected at the node to the voltage resulting at that node. Figure 7.6 illustrates the method for a three-node reactive network. The result is obviously equivalent to adding all the admittances directly connected to the node, as has been our procedure up to now.

Figure 7.6 also serves to illustrate mutual admittance. At node 1 the equation obtained by expanding Eq. (7.33) is

$$I_1 = Y_{11} V_1 + Y_{12} V_2 + Y_{13} V_3 \tag{7.36}$$

from which we see that

$$Y_{12} = \left.\frac{I_1}{V_2}\right|_{V_1 = V_3 = 0} \tag{7.37}$$

Thus the mutual admittance is measured by shorting all nodes except node 2 to the reference node and injecting a current I_2 at node 2, as shown in Fig. 7.6. Then Y_{12} is the ratio of the negative of the current leaving the network in the short circuit at node 1 to the voltage V_2. The negative of the current *leaving* node 1 is used since I_1 is defined as the current *entering* the network. The resultant admittance is the negative of the admittance directly connected beween nodes 1 and 2, as we would expect.

We have made this detailed examination of the node admittances in order to differentiate them clearly from the impedances of the bus impedance matrix.

We solve Eq. (7.33) by premultiplying both sides of the equation by $\mathbf{Y}_{bus}^{-1} = \mathbf{Z}_{bus}$ to yield

$$\mathbf{V} = \mathbf{Z}_{bus} \mathbf{I} \tag{7.38}$$

and we must remember when dealing with $\mathbf{Z}_{bus}$ that $\mathbf{V}$ and $\mathbf{I}$ are column matrices of the node voltages and the currents entering the nodes from current sources, respectively. Expanding Eq. (7.38) for a network of three independent nodes yields

$$V_1 = Z_{11}I_1 + Z_{12}I_2 + Z_{13}I_3 \tag{7.39}$$

$$V_2 = Z_{21}I_1 + Z_{22}I_2 + Z_{23}I_3 \tag{7.40}$$

$$V_3 = Z_{31}I_1 + Z_{32}I_2 + Z_{33}I_3 \tag{7.41}$$

From Eq. (7.40) we see that the driving-point impedance Z_{22} is determined by open-circuiting the current sources at nodes 1 and 3 and injecting the current I_2 at node 2. Then

$$Z_{22} = \frac{V_2}{I_2}\bigg|_{I_1=I_3=0} \tag{7.42}$$

Figure 7.7 shows the circuit described. Since Z_{22} is defined by opening the current sources connected to the other nodes whereas Y_{22} was found with the other nodes shorted, we must not expect any reciprocal relation between these two quantities.

The circuit of Fig. 7.7 also enables us to measure some transfer impedances, for we see from Eq. (7.39) that with current sources I_1 and I_3 open-circuited

$$Z_{12} = \frac{V_1}{I_2}\bigg|_{I_1=I_3=0}$$

and from Eq. (7.41)

$$Z_{32} = \frac{V_3}{I_2}\bigg|_{I_1=I_3=0}$$

Thus we can measure the transfer impedances Z_{12} and Z_{32} by injecting current at node 2 and finding the ratios of V_1 and V_3 to I_2 with the sources open at all nodes except node 2. We note that a mutual admittance is measured with all but one node short-circuited and that a transfer impedance is measured with all sources open-circuited except one.

Equation (7.39) tells us that if we inject current into node 1 with current sources 2 and 3 open, the only impedance through which I_1 flows is Z_{11}. Under

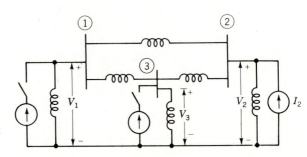

Figure 7.7 Circuit for measuring Z_{22}, Z_{12}, and Z_{32}.

the same conditions Eqs. (7.40) and (7.41) show that I_1 is causing voltages at buses 2 and 3 expressed by

$$V_2 = I_1 Z_{21} \quad \text{and} \quad V_3 = I_1 Z_{31} \tag{7.43}$$

We cannot set up a physically realizable passive circuit with these coupling impedances, but it is important to realize the implications of the preceding discussion, for $\mathbf{Z}_{\text{bus}}$ is sometimes used in load-flow studies and is extremely valuable in fault calculations, as we shall see later.

Example 7.5 A capacitor having a reactance of 5.0 per unit is connected to node 4 of the circuit of Examples 7.1 and 7.2. The emf's E_a, E_b, and E_c remain the same as in those examples. Find the current drawn by the capacitor.

SOLUTION The Thévenin equivalent of the circuit behind node 4 has an emf of

$$E_{th} = 1.432 \, \underline{/-11.97°}$$

which is the voltage at node 4 before the capacitor is connected and is the voltage V_4 found in Example 7.2.

To find the Thévenin impedance the emfs are short-circuited or the current sources are open-circuited, and the impedance between node 4 and the reference node must be determined. From $\mathbf{V} = \mathbf{Z}_{\text{bus}} \mathbf{I}$ we have at node 4

$$V_4 = Z_{41} I_1 + Z_{42} I_2 + Z_{43} I_3 + Z_{44} I_4$$

With emfs short-circuited (or with the emfs and their series impedances replaced by the equivalent current sources and shunt admittances with the current sources open) no current is entering the circuit from sources at nodes 1, 2, and 3. The ratio of a voltage applied at node 4 to the current this voltage will cause to flow in the network is Z_{44}, and this impedance is known since $\mathbf{Z}_{\text{bus}}$ was calculated in Example 7.2. By referring to that example we find

$$Z_{th} = Z_{44} = j0.4733$$

The current drawn by the capacitor is

$$I_C = \frac{1.432 \, \underline{/-11.97°}}{j0.4733 - j5.0} = 0.316 \underline{/78.03°} \text{ per unit}$$

Example 7.6 If a current of $-0.316 \underline{/78.03°}$ per unit is injected into the network at node 4 of Examples 7.1, 7.2, and 7.5, find the resulting voltages at nodes 1, 2, 3, and 4.

SOLUTION With original emfs short-circuited, the voltages at the nodes due only to the injected current will be calculated by making use of the bus impedance network which was found in Example 7.2. The required impedances are in column 4 of $\mathbf{Z}_{bus}$. From $\mathbf{V} = \mathbf{Z}_{bus}\mathbf{I}$, the voltages with all emfs shorted are

$$V_1 = I_4 Z_{14} = -0.316\underline{/78.03°} \times 0.4142\underline{/90°} = 0.1309\underline{/-11.97°}$$

$$V_2 = I_4 Z_{24} = -0.316\underline{/78.03°} \times 0.4126\underline{/90°} = 0.1304\underline{/-11.97°}$$

$$V_3 = I_4 Z_{34} = -0.316\underline{/78.03°} \times 0.4232\underline{/90°} = 0.1337\underline{/-11.97°}$$

$$V_4 = I_4 Z_{44} = -0.316\underline{/78.03°} \times 0.4733\underline{/90°} = 0.1496\underline{/-11.97°}$$

By superposition, the resulting voltages are determined by adding the voltages caused by the injected current with emfs shorted to the node voltages found in Example 7.2. The new node voltages are

$$V_1 = 1.436\underline{/-10.71°} + 0.1309\underline{/-11.97°} = 1.567\underline{/-10.81°} \text{ per unit}$$

$$V_2 = 1.427\underline{/-14.2°} + 0.1304\underline{/-11.97°} = 1.557\underline{/-14.04°} \text{ per unit}$$

$$V_3 = 1.434\underline{/-11.4°} + 0.1337\underline{/-11.97°} = 1.568\underline{/-11.41°} \text{ per unit}$$

$$V_4 = 1.432\underline{/-11.97°} + 0.1496\underline{/-11.97°} = 1.582\underline{/-11.97°} \text{ per unit}$$

Since the changes in voltages due to the injected current are all at the same angle and this angle differs little from the angles of the original voltages, an approximation will give satisfactory answers. The change in voltage magnitude at a bus is about equal to the product of the magnitude of the per-unit current and the magnitude of the appropriate impedance. These values added to the original voltage magnitudes give the magnitudes of the new voltages very closely. This approximation is valid because the network is purely reactive, but it provides a good estimate where the reactance is considerably larger than the resistance, as is usually the case.

The last two examples illustrate the importance of the bus impedance matrix and incidentally show how adding a capacitor at a bus will cause a rise in bus voltages. The assumption that the angles of voltage and current sources remain constant after connecting capacitors at a bus is not entirely valid if we are considering operation of a power system. We shall consider capacitors again in Chap. 8 and see an example using a computer load-flow program to calculate the effect of capacitors.

7.6 MODIFICATION OF AN EXISTING BUS IMPEDANCE MATRIX

Since $\mathbf{Z}_{bus}$ is such an important tool in power-system analysis we shall now examine how an existing $\mathbf{Z}_{bus}$ may be modified to add new buses or connect new lines to established buses. Of course we could create a new $\mathbf{Y}_{bus}$ and invert it, but direct methods of modifying $\mathbf{Z}_{bus}$ are available and are very much simpler than a matrix inversion even for a small number of buses. Also when we know how to modify $\mathbf{Z}_{\text{bus}}$ we can see how to build it directly.†

We recognize several types of modifications involving the addition of a branch having impedance Z_b to a network whose original $\mathbf{Z}_{\text{bus}}$ is known and is identified as $\mathbf{Z}_{\text{orig}}$, an $n \times n$ matrix.

In our analysis existing buses will be indentified by numbers or the letters h, i, j, and k. The letter p will designate a new bus to be added to the network to convert $\mathbf{Z}_{\text{orig}}$ to an $(n + 1) \times (n + 1)$ matrix. Four cases will be considered.

CASE 1: *Adding Z_b from a new bus p to the reference bus* The addition of the new bus p connected to the reference bus through Z_b without a connection to any of the buses of the original network cannot alter the original bus voltages when a current I_p is injected at the new bus. The voltage V_p at the new bus is equal to $I_p Z_b$. Then

$$
\begin{bmatrix} V_1 \\ V_2 \\ \vdots \\ V_n \\ \text{---} \\ V_p \end{bmatrix}
=
\left[
\begin{array}{ccccc:c}
 & & & & & 0 \\
 & & & & & 0 \\
 & & \mathbf{Z}_{\text{orig}} & & & \vdots \\
 & & & & & 0 \\
\hdashline
0 & 0 & \cdots & 0 & & Z_b
\end{array}
\right]
\underbrace{}_{\mathbf{Z}_{\text{bus(new)}}}
\begin{bmatrix} I_1 \\ I_2 \\ \vdots \\ I_n \\ \text{---} \\ I_p \end{bmatrix}
\tag{7.44}
$$

We note that the column matrix of currents multiplied by the new $\mathbf{Z}_{\text{bus}}$ will not alter the voltages of the original network and will result in the correct voltage at the new bus p.

CASE 2: *Adding Z_b from a new bus p to an existing bus k* The addition of a new bus p connected through Z_b to an existing bus k with I_p injected at bus p will cause the current entering the original network at bus k to become the sum of I_k which is injected at bus k plus the current I_p coming through Z_b as shown in Fig. 7.8.

The current I_p flowing into bus k will increase the original V_k by the voltage $I_p Z_{kk}$; that is

$$V_{k(\text{new})} = V_{k(\text{orig})} + I_p Z_{kk} \tag{7.45}$$

and V_p will be larger than the new V_k by the voltage $I_p Z_b$. So

$$V_p = V_{k(\text{orig})} + I_p Z_{kk} + I_p Z_b \tag{7.46}$$

and

$$V_p = \underbrace{I_1 Z_{k1} + I_2 Z_{k2} + \cdots + I_n Z_{kn}}_{V_{k(\text{orig})}} + I_p(Z_{kk} + Z_b) \tag{7.47}$$

† See H. E. Brown, *Solutions of Large Networks by Matrix Methods*, John Wiley & Sons, Inc., New York, 1975.

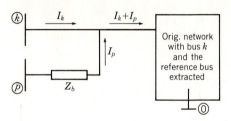

Figure 7.8 Addition of new bus p connected through impedance Z_b to existing bus k.

We now see that the new row which must be added to $\mathbf{Z}_{\text{orig}}$ in order to find V_p is

$$Z_{k1} \quad Z_{k2} \quad \cdots \quad Z_{kn} \quad (Z_{kk} + Z_b)$$

Since $\mathbf{Z}_{\text{bus}}$ must be a square matrix around the principal diagonal we must add a new column which is the transpose of the new row. The new column accounts for the increase of all bus voltages due to I_p. The matrix equation is

$$
\begin{bmatrix}
V_1 \\
V_2 \\
\vdots \\
V_n \\
\hdashline
V_p
\end{bmatrix}
=
\left[
\begin{array}{cccc:c}
 & & & & Z_{1k} \\
 & \mathbf{Z}_{\text{orig}} & & & Z_{2k} \\
 & & & & \vdots \\
 & & & & Z_{nk} \\
\hdashline
Z_{k1} & Z_{k2} & \cdots & Z_{kn} & Z_{kk} + Z_b
\end{array}
\right]
\underbrace{}_{\mathbf{Z}_{\text{bus(new)}}}
\begin{bmatrix}
I_1 \\
I_2 \\
\vdots \\
I_n \\
\hdashline
I_p
\end{bmatrix}
\tag{7.48}
$$

Note that the first n elements of the new row are the elements of row k of $\mathbf{Z}_{\text{orig}}$ and the first n elements of the new column are the elements of column k of $\mathbf{Z}_{\text{orig}}$.

CASE 3: *Adding Z_b from an existing bus k to the reference bus* To see how to alter $\mathbf{Z}_{(\text{orig})}$ by connecting an impedance Z_b from an existing bus k to the reference bus we shall add a new bus p connected through Z_b to bus k. Then we short-circuit bus p to the reference bus by letting V_p equal zero to yield the same matrix equation as Eq. (7.48) *except* that V_p is zero. So for the modification we proceed to create a new row and new column exactly the same as in case 2 but we then eliminate the $(n + 1)$ row and $(n + 1)$ column, which is possible because of the zero in the column matrix of voltages. We use the method developed in Eqs. (7.28) to (7.30) to find each element Z_{hi} in the new matrix where

$$Z_{hi(\text{new})} = Z_{hi(\text{orig})} - \frac{Z_{h(n+1)}Z_{(n+1)i}}{Z_{kk} + Z_b} \tag{7.49}$$

CASE 4: *Adding Z_b between two existing buses j and k* To add a branch impedance Z_b between already established buses j and k we examine Fig. 7.9 which shows these buses extracted from the original network. The current I_b is

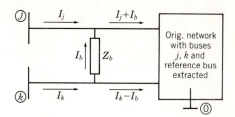

Figure 7.9 Addition of impedance Z_b between existing buses j and k.

shown flowing through Z_b from bus k to bus j. We now write some equations for node voltages.

$$V_1 = Z_{11}I_1 + \cdots + Z_{1j}(I_j + I_b) + Z_{1k}(I_k - I_b) + \cdots \qquad (7.50)$$

and upon rearranging

$$V_1 = Z_{11}I_1 + \cdots + Z_{1j}I_j + Z_{1k}I_k + \cdots + I_b(Z_{1j} - Z_{1k}) \qquad (7.51)$$

Similarly

$$V_j = Z_{j1}I_1 + \cdots + Z_{jj}I_j + Z_{jk}I_k + \cdots + I_b(Z_{jj} - Z_{jk}) \qquad (7.52)$$

$$V_k = Z_{k1}I_1 + \cdots + Z_{kj}I_j + Z_{kk}I_k + \cdots + I_b(Z_{kj} - Z_{kk}) \qquad (7.53)$$

We need one more equation since I_b is unknown. So we write

$$V_k - V_j = I_b Z_b \qquad (7.54)$$

or

$$0 = I_b Z_b + V_j - V_k \qquad (7.55)$$

and substituting the expressions for V_j and V_k given by Eqs. (7.52) and (7.53) in Eq. (7.55) we obtain

$$0 = I_b Z_b + (Z_{j1} - Z_{k1})I_1 + \cdots + (Z_{jj} - Z_{kj})I_j + \cdots$$
$$+ (Z_{jk} - Z_{kk})I_k + \cdots + (Z_{jj} + Z_{kk} - 2Z_{jk})I_b \qquad (7.56)$$

Collecting the coefficients of I_b and naming their sum Z_{bb} yields

$$Z_{bb} = Z_b + Z_{jj} + Z_{kk} - 2Z_{jk} \qquad (7.57)$$

By examining Eqs. (7.51) to (7.53) and (7.56) we can write the matrix equation

$$
\begin{bmatrix} V_1 \\ \vdots \\ V_j \\ V_k \\ \vdots \\ V_n \\ \hline 0 \end{bmatrix} =
\left[
\begin{array}{ccc|c}
 & & & Z_{1j} - Z_{1k} \\
 & & & \vdots \\
 & \mathbf{Z}_{\text{orig}} & & Z_{jj} - Z_{jk} \\
 & & & Z_{kj} - Z_{kk} \\
 & & & \vdots \\
 & & & Z_{nj} - Z_{nk} \\
 \hline
 (Z_{j1} - Z_{k1}) & \cdots & (Z_{kj} - Z_{kk}) & \cdots \quad Z_{bb}
\end{array}
\right]
\begin{bmatrix} I_1 \\ \vdots \\ I_j \\ I_k \\ \vdots \\ I_n \\ \hline I_b \end{bmatrix} \qquad (7.58)
$$

The new column is column j minus column k of $\mathbf{Z}_{\text{orig}}$ with Z_{bb} in the $(n+1)$ row. The new row is the transpose of the new column.

Eliminating the $(n+1)$ row and $(n+1)$ column of the square matrix of Eq. (7.58) in the same manner as previously we see that each element Z_{hi} in the new matrix is

$$Z_{hi(\text{new})} = Z_{hi(\text{orig})} - \frac{Z_{h(n+1)}Z_{(n+1)i}}{Z_b + Z_{jj} + Z_{kk} - 2Z_{jk}} \tag{7.59}$$

We need not consider the case of introducing two new buses connected by Z_b because we could always connect one of these new buses through an impedance to an existing bus or to the reference bus before adding the second new bus.

Example 7.7 Modify the bus impedance matrix of Example 7.2 to account for the connection of a capacitor having a reactance of 5.0 per unit between bus 4 and the reference bus of the circuit of Fig. 7.4. Then find V_4 using the impedances of the new matrix and the current sources of Example 7.2. Compare this value of V_4 with that found in Example 7.6.

SOLUTION We use Eq. (7.48) and recognize that $\mathbf{Z}_{\text{orig}}$ is the 4×4 matrix of Example 7.2, that subscript $k = 4$, and that $Z_b = -j5.0$ per unit to find

$$
\begin{bmatrix} V_1 \\ V_2 \\ V_3 \\ V_4 \\ \hdashline 0 \end{bmatrix} =
\left[\begin{array}{cccc:c}
 & & & & j0.4142 \\
 & & \mathbf{Z}_{\text{orig}} & & j0.4126 \\
 & & & & j0.4232 \\
 & & & & j0.4733 \\ \hdashline
j0.4142 & j0.4126 & j0.4232 & j0.4733 & -j4.5267
\end{array} \right]
\begin{bmatrix} I_1 \\ I_2 \\ I_3 \\ I_4 \\ \hdashline I_b \end{bmatrix}
$$

The terms in the fifth row and column were obtained by repeating the fourth row and column of $\mathbf{Z}_{\text{orig}}$ and noting that

$$Z_{55} = Z_{44} + Z_b = j0.4733 - j5.0 = -j4.5267$$

Then eliminating the fifth row and column we obtain for $\mathbf{Z}_{\text{bus(new)}}$ from Eq. (7.49)

$$Z_{11} = j0.4774 - \frac{j0.4142 \times j0.4142}{-j4.5267} = j0.5153$$

$$Z_{24} = j0.4126 - \frac{j0.4733 \times j0.4126}{-j4.5267} = j0.4557$$

and other elements in a similar manner to give

$$\mathbf{Z}_{\text{bus (new)}} =
\begin{bmatrix}
j0.5153 & j0.4084 & j0.4407 & j0.4575 \\
j0.4084 & j0.5248 & j0.4308 & j0.4557 \\
j0.4407 & j0.4308 & j0.4954 & j0.4674 \\
j0.4575 & j0.4557 & j0.4674 & j0.5228
\end{bmatrix}
$$

The column matrix of currents by which the new Z_{bus} is multiplied to obtain the new bus voltages is the same as in Example 7.2. So

$$V_4 = j0.4575(-j1.20) + j0.4557(-0.72 - j0.96) + j0.4674(-j1.20)$$

$$= 1.5474 - j0.3281 = 1.582\underline{/-11.97°} \text{ per unit}$$

as found in Example 7.6.

7.7 DIRECT DETERMINATION OF A BUS IMPEDANCE MATRIX

We have seen how to determine Z_{bus} by first finding Y_{bus} and inverting it. However, formulation of Z_{bus} directly is a straightforward process on the computer and simpler than inverting Y_{bus} for a large network.

To begin we have a list of the impedances showing the buses to which they are connected. We start by writing the equation for one bus connected through an impedance Z_a to the reference bus as

$$V_1 = I_1 Z_a$$

and this can be considered as a matrix equation where each of the three matrices has one row and one column. Now we might add a new bus connected to the first bus or to the reference bus. For instance, if the second bus is connected to the reference but through Z_b we have the matrix equation

$$\begin{bmatrix} V_1 \\ V_2 \end{bmatrix} = \begin{bmatrix} Z_a & 0 \\ 0 & Z_b \end{bmatrix} \begin{bmatrix} I_1 \\ I_2 \end{bmatrix} \tag{7.60}$$

and we proceed to modify our matrix by adding other buses following the procedures described in Sec. 7.6. Usually the buses of a network must be renumbered to agree with the order in which they are to be added to Z_{bus} as it is built up.

Example 7.8 Determine Z_{bus} for the network shown in Fig. 7.10 where impedances are shown in per unit. Preserve all three nodes.

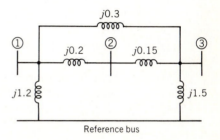

Figure 7.10 Network for Example 7.8.

SOLUTION We start by establishing bus 1 with its impedance to the reference bus and write

$$V_1 = j1.2I_1$$

We then have the 1×1 bus impedance matrix

$$\mathbf{Z}_{\text{bus}} = j1.2$$

To establish bus 2 with its impedance to bus 1 we follow Eq. (7.48) to write

$$\mathbf{Z}_{\text{bus (new)}} = \begin{bmatrix} j1.2 & j1.2 \\ j1.2 & j1.4 \end{bmatrix}$$

The term $j1.4$ above is the sum of $j1.2$ and $j0.2$. The elements $j1.2$ in the new row and column are the repetition of the elements of row 1 and column 1 of the matrix being modified.

Bus 3 with the impedance connecting it to bus 1 is established by writing

$$\begin{bmatrix} j1.2 & j1.2 & j1.2 \\ j1.2 & j1.4 & j1.2 \\ j1.2 & j1.2 & j1.5 \end{bmatrix}$$

Since node 1 is the node to which the new node 3 is being connected, the term $j1.5$ above is the sum of Z_{11} of the matrix being modified and the impedance Z_b of the branch being connected to bus 1 from bus 3. The other elements of the new row and column are the repetition of the row 1 and column 1 of the matrix being modified since the new node is being connected to bus 1.

If we now decide to add the impedance $Z_b = j1.5$ from node 3 to the reference bus, we follow Eq. (7.48) to connect a new bus 4 through Z_b and obtain the impedance matrix

$$\begin{bmatrix} j1.2 & j1.2 & j1.2 & \vdots & j1.2 \\ j1.2 & j1.4 & j1.2 & \vdots & j1.2 \\ j1.2 & j1.2 & j1.5 & \vdots & j1.5 \\ \hdashline j1.2 & j1.2 & j1.5 & \vdots & j3.0 \end{bmatrix}$$

where $j3.0$ above is the sum of $Z_{33} + Z_b$. The other elements in the new row and column are the repetition of row 3 and column 3 of the matrix being modified since bus 3 is the one which we are connecting to the reference bus through Z_b.

We now eliminate row 4 and column 4. Some of the elements of the new matrix from Eq. (7.49) are

$$Z_{11} = j1.2 - \frac{j1.2 \times j1.2}{j3.0} = j0.72$$

$$Z_{22} = j1.4 - \frac{j1.2 \times j1.2}{j3.0} = j0.92$$

$$Z_{23} = Z_{32} = j1.2 - \frac{j1.2 \times j1.5}{j3.0} = j0.60$$

When all the elements are determined we have

$$\mathbf{Z}_{\text{bus (new)}} = \begin{bmatrix} j0.72 & j0.72 & j0.60 \\ j0.72 & j0.92 & j0.60 \\ j0.60 & j0.60 & j0.75 \end{bmatrix}$$

Finally we add the impedance $Z_b = j0.15$ between buses 2 and 3. If we let j and k in Eq. (7.58) equal 2 and 3, respectively, we obtain the elements for row 4 and column 4.

$$Z_{14} = Z_{12} - Z_{13} = j0.72 - j0.60 = j0.12$$

$$Z_{24} = Z_{22} - Z_{23} = j0.92 - j0.60 = j0.32$$

$$Z_{34} = Z_{32} - Z_{33} = j0.60 - j0.75 = -j0.15$$

and from Eq. (7.57)

$$Z_{44} = Z_b + Z_{22} + Z_{33} - 2Z_{23}$$
$$= j0.15 + j0.92 + j0.75 - 2(j0.60) = j0.62$$

So we write

$$\begin{bmatrix} j0.72 & j0.72 & j0.60 & j0.12 \\ j0.72 & j0.92 & j0.60 & j0.32 \\ j0.60 & j0.60 & j0.75 & -j0.15 \\ j0.12 & j0.32 & -j0.15 & j0.62 \end{bmatrix}$$

and from Eq. (7.59) we find

$$\mathbf{Z}_{\text{bus (new)}} = \begin{bmatrix} j0.6968 & j0.6581 & j0.6290 \\ j0.6581 & j0.7548 & j0.6774 \\ j0.6290 & j0.6774 & j0.7137 \end{bmatrix}$$

which is the bus impedance matrix to be determined.

The procedure is simple for a computer which first must determine the types of modification involved as each impedance is added. However, the operations must follow a sequence such that we avoid connecting an impedance between two new buses.

As a matter of interest we can check the impedance values of $\mathbf{Z}_{\text{bus}}$ by the network calculations of Sec. 7.5.

Example 7.9 Find Z_{11} of the circuit of Example 7.8 by determining the impedance measured between node 1 and the reference bus when currents injected at nodes 2 and 3 are zero.

SOLUTION The equation corresponding to Eq. (7.42) is

$$Z_{11} = \frac{V_1}{I_1}\bigg|_{I_2 = I_3 = 0}$$

We recognize two parallel paths between nodes 1 and 3 of the circuit of Fig. 7.10 with the resulting impedance of

$$\frac{j0.3 \times j0.35}{j0.3 + j0.35} = j0.1615$$

This impedance in series with $j1.5$ is in parallel with $j1.2$ to yield

$$Z_{11} = \frac{j1.2(j1.5 + j0.1615)}{j(1.2 + 1.5 + 0.1615)} = j0.6968$$

which is identical with the value found in Example 7.8.

Although the network reduction method of Example 7.9 may appear to be simpler by comparison with other methods of forming $\mathbf{Z}_{\text{bus}}$ such is not the case because a different network reduction is required to evaluate each element of the matrix. In Example 7.9 the network reduction to find Z_{22}, for instance, is more difficult than that for finding Z_{11}. The digital computer could make a network reduction by node elimination but would have to repeat the process for each node.

7.8 SUMMARY

Equivalence of sources and node equations were reviewed briefly in this chapter to provide the essential background for understanding the bus admittance matrix which is the basis for most load-flow studies. Matrix partitioning was reviewed because of its usefulness in node-elimination methods.

The bus impedance matrix is preferred by some engineers for load-flow studies but finds its greatest value in fault calculations which we shall discuss later.

The modification of $\mathbf{Z}_{bus}$ was discussed to show the simplicity of accounting for the addition or removal of a transmission line without having to invert $\mathbf{Y}_{bus}$ each time a change is made. Direct formulation of $\mathbf{Z}_{bus}$ is a process which can be programmed in a straightforward manner.

PROBLEMS

7.1 Write the two node equations similar to Eqs. (7.5) and (7.6) required to find the voltages at nodes 1 and 2 of the circuit of Fig. 7.11 without changing the emf sources to current sources. Then write the equations in standard form after changing the emf sources to current sources.

7.2 Find the voltages at nodes 1 and 2 of the circuit of Fig. 7.11 by solving the equations determined in Prob. 7.1.

7.3 Eliminate nodes 3 and 4 of the network of Fig. 7.12 simultaneously by the method of partitioning employed in Example 7.3 to find the resulting 2×2 admittance matrix $\mathbf{Y}_{bus}$. Draw the circuit corresponding to the resulting matrix and show on the circuit the values of the parameters. Solve for V_1 and V_2 by matrix inversion.

7.4 Eliminate nodes 3 and 4 of the network of Fig. 7.12 to find the resulting 2×2 admittance matrix by eliminating node 4 first and then node 3 as in Example 7.4.

7.5 Modify $\mathbf{Z}_{bus}$ given in Example 7.2 for the circuit of Fig. 7.4 by adding a new node connected to bus 4 through an impedance of $j1.2$ per unit.

7.6 Modify $\mathbf{Z}_{bus}$ given in Example 7.2 by adding a branch having an impedance of $j1.2$ per unit between node 4 and the reference bus of the circuit of Fig. 7.4.

7.7 Determine the impedances in the first row of $\mathbf{Z}_{bus}$ for the circuit of Fig. 7.4 with the impedance connected between bus 3 and the reference bus removed by modifying the $\mathbf{Z}_{bus}$ found in Example 7.2. Then with the current sources connected only at buses 1 and 2 find the voltage at bus 1 and compare this value with that found in Example 7.3.

7.8 Modify $\mathbf{Z}_{bus}$ given in Example 7.2 by removing the impedance connected between nodes 2 and 3 of the network of Fig. 7.4.

Figure 7.11 Circuit for Probs. 7.1 and 7.2. Values shown are voltages and impedances in per unit.

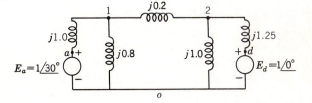

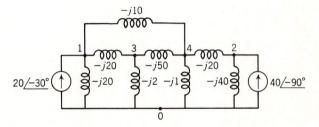

Figure 7.12 Circuit for Probs. 7.3 and 7.4. Values shown are currents and admittances in per unit.

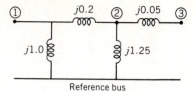

Figure 7.13 Circuit for Prob. 7.9. Values shown are reactances in per unit.

7.9 Find $\mathbf{Z}_{bus}$ for the network of Fig. 7.13 by the direct determination process discussed in Sec. 7.7.

7.10 For the reactance network of Fig. 7.14 find (*a*) $\mathbf{Z}_{bus}$ by direct formulation or by inversion of $\mathbf{Y}_{bus}$, (*b*) the voltage at each bus, (*c*) the current drawn by a capacitor having a reactance of 5.0 per unit connected from bus 3 to neutral, (*d*) the change in voltage at each bus when the capacitor is connected at bus 3, and (*e*) the voltage at each bus after connecting the capacitor. The magnitude and angle of each of the generated voltages may be assumed to remain constant.

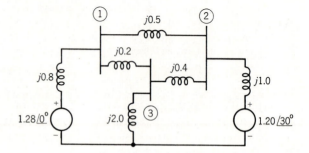

Figure 7.14 Circuit for Prob. 7.10. Voltages and impedances are in per unit.

EIGHT

LOAD-FLOW SOLUTIONS AND CONTROL

The great importance of load-flow studies in planning the future expansion of power systems as well as in determining the best operation of existing systems was discussed in Chap. 1. The principal information obtained from a load-flow study is the magnitude and phase angle of the voltage at each bus and the real and reactive power flowing in each line. However, much additional information of value is provided by the printout of the solution from computer programs used by the power companies. Most of these features will be brought out in our discussion of load-flow studies in this chapter which also treats the principles of the control of load flow.

We shall examine two of the methods upon which solutions to the load-flow problem are based. The great value of the digital computer in power-system design and operation will become apparent.

8.1 DATA FOR LOAD-FLOW STUDIES

Either the bus self- and mutual admittances which compose the bus admittance matrix Y_{bus} or the driving-point and transfer impedances which compose Z_{bus} may be used in solving the load-flow problem. We shall confine our study to methods using admittances. The starting point in obtaining the data which must be furnished to the computer is the one-line diagram of the system. Values of series impedances and shunt admittances of transmission lines are necessary so that the computer can determine all the Y_{bus} or Z_{bus} elements. Other essential information includes transformer ratings and impedances, shunt capacitor ratings, and transformer tap settings.

Operating conditions must always be selected for each study. At each bus except one the net real power into the network must be specified. The power drawn by a load is negative power input to the system. The other power inputs are from generators and positive or negative power entering over interconnections. In addition, at these buses either the net flow of reactive power into the network or the magnitude of the voltage must be specified; that is, at each bus a decision is required whether the voltage magnitude or the reactive-power flow is to be maintained constant. The usual case is to specify reactive power at load buses and voltage magnitude at generator buses, although sometimes reactive power is specified for the generators. In digital-computer programs provision is made for the calculation to consider voltage to be maintained constant at a bus only so long as the reactive-power generation remains within designated limits.

The one bus at which real-power flow is not specified, called the *swing bus*, is usually a bus to which a generator is connected. Obviously, the net power flow to the system cannot be fixed in advance at every bus because the loss in the system is not known until the study is complete. The generators at the swing bus supply the difference between the specified real power into the system at the other buses and the total system output plus losses. Both voltage magnitude and angle are specified at the swing bus. Real and reactive power at this bus are determined by the computer as part of the solution.

8.2 THE GAUSS-SEIDEL METHOD

The complexity of obtaining a formal solution for load flow in a power system arises because of the differences in the type of data specified for the different kinds of buses. Although the formulation of sufficient equations is not difficult, the closed form of solution is not practical. Digital solutions of the load-flow problems we shall consider at this time follow an iterative process by assigning estimated values to the unknown bus voltages and calculating a new value for each bus voltage from the estimated values at the other buses, the real power specified, and the specified reactive power or voltage magnitude. A new set of values for voltage is thus obtained for each bus and used to calculate still another set of bus voltages. Each calculation of a new set of voltages is called an *iteration*. The iterative process is repeated until the changes at each bus are less than a specified minimum value.

We shall examine first the solution based on expressing the voltage of a bus as a function of the real and reactive power delivered to a bus from generators or supplied to the load connected to the bus, the estimated or previously calculated voltages at the other buses, and the self- and mutual admittances of the nodes. The derivation of the fundamental equations starts with a node formulation of the network equations. We shall derive the equations for a four-bus system and write the general equations later. With the swing bus designated as number 1, computations start with bus 2. If P_2 and Q_2 are the scheduled real and reactive power entering the system at bus 2,

$$V_2 I_2^* = P_2 + jQ_2 \tag{8.1}$$

from which I_2 is expressed as

$$I_2 = \frac{P_2 - jQ_2}{V_2^*} \tag{8.2}$$

and in terms of self- and mutual admittances of the nodes, with generators and loads omitted since the current into each node is expressed as in Eq. (8.2),

$$\frac{P_2 - jQ_2}{V_2^*} = Y_{21} V_1 + Y_{22} V_2 + Y_{23} V_3 + Y_{24} V_4 \tag{8.3}$$

Solving for V_2 gives

$$V_2 = \frac{1}{Y_{22}} \left[\frac{P_2 - jQ_2}{V_2^*} - (Y_{21} V_1 + Y_{23} V_3 + Y_{24} V_4) \right] \tag{8.4}$$

Equation (8.4) gives a corrected value for V_2 based upon scheduled P_2 and Q_2 when the values estimated originally are substituted for the voltage expressions on the right side of the equation. The calculated value for V_2 and the estimated value for V_2^* will not agree. By substituting the conjugate of the calculated value of V_2 for V_2^* in Eq. (8.4) to calculate another value for V_2, agreement would be reached to a good degree of accuracy after several iterations and would be the correct value for V_2 with the estimated voltages and without regard to power at the other buses. This value would *not* be the solution for V_2 for the specific load-flow conditions, however, because the voltages upon which this calculation for V_2 depends are the estimated values of voltage at the other buses and the actual voltages are not yet known. Two successive calculations of V_2 (the second being like the first except for the correction of V_2^*) are recommended at each bus before proceeding to the next one.

As the corrected voltage is found at each bus, it is used in calculating the corrected voltage at the next. The process is repeated at each bus consecutively throughout the network (except at the swing bus) to complete the first iteration. Then the entire process is carried out again and again until the amount of correction in voltage at every bus is less than some predetermined precision index.

This process of solving linear algebraic equations is known as the *Gauss-Seidel iterative method*. If the same set of voltage values is used throughout a complete iteration (instead of immediately substituting each new value obtained to calculate the voltage at the next bus), the process is called the *Gauss iterative method*.

Convergence upon an erroneous solution may occur if the original voltages are widely different from the correct values. Erroneous convergence is usually avoided if the original values are of reasonable magnitude and do not differ too widely in phase. Any unwanted solution is usually detected easily by inspection of the results since the voltages of the system do not normally have a range in phase wider than 45° and the difference between nearby buses is less than about 10° and often very small.

For a total of N buses the calculated voltage at any bus k where P_k and Q_k are given is

$$V_k = \frac{1}{Y_{kk}} \left(\frac{P_k - jQ_k}{V_k^*} - \sum_{n=1}^{N} Y_{kn} V_n \right) \tag{8.5}$$

where $n \neq k$. The values for the voltages on the right side of the equation are the most recently calculated values for the corresponding buses (or the estimated voltage if no iteration has yet been made at that particular bus).

Experience with the Gauss-Seidel method of solution of power-flow problems has shown that an excessive number of iterations are required before the voltage corrections are within an acceptable precision index if the corrected voltage at a bus merely replaces the best previous value as the computations proceed from bus to bus. The number of iterations required is reduced considerably if the correction in voltage at each bus is multiplied by some constant that increases the amount of correction to bring the voltage closer to the value it is approaching. The multipliers that accomplish this improved convergence are called *acceleration factors*. The difference between the newly calculated voltage and the best previous voltage at the bus is multiplied by the appropriate acceleration factor to obtain a better correction to be added to the previous value. The acceleration factor for the real component of the correction may differ from that for the imaginary component. For any system, optimum values for acceleration factors exist, and poor choice of factors may result in less rapid convergence or make convergence impossible. An acceleration factor of 1.6 for both the real and imaginary components is usually a good choice. Studies may be made to determine the best choice for a particular system.

At a bus where voltage magnitude rather than reactive power is specified, the real and imaginary components of the voltage for each iteration are found by first computing a value for the reactive power. From Eq. (8.5)

$$P_k - jQ_k = \left(Y_{kk} V_k + \sum_{n=1}^{N} Y_{kn} V_n \right) V_k^* \tag{8.6}$$

where $n \neq k$. If we allow n to equal k

$$P_k - jQ_k = V_k^* \sum_{n=1}^{N} Y_{kn} V_n \tag{8.7}$$

$$Q_k = -\operatorname{Im} \left\{ V_k^* \sum_{n=1}^{N} Y_{kn} V_n \right\} \tag{8.8}$$

where Im means "imaginary part of."

Reactive power Q_k is evaluated by Eq. (8.8) for the best previous voltage values at the buses, and this value of Q_k is substituted in Eq. (8.5) to find a new V_k. The components of the new V_k are then multiplied by the ratio of the specified constant magnitude of V_k to the magnitude of the V_k found by Eq. (8.5). The result is the corrected complex voltage of the specified magnitude.

8.3 THE NEWTON-RAPHSON METHOD

Taylor's series expansion for a function of two or more variables is the basis for the Newton-Raphson method of solving the load-flow problem. Our study of the method will begin by a discussion of the solution of a problem involving only two equations and two variables. Then we shall see how to extend the analysis to the solution of load-flow equations.

Let us consider the equation of a function of two variables x_1 and x_2 equal to a constant K_1 expressed as

$$f_1(x_1, x_2) = K_1 \qquad (8.9)$$

and a second equation

$$f_2(x_1, x_2) = K_2 \qquad (8.10)$$

where K_1 and K_2 are constants.

We then estimate the solutions of these equations to be $x_1^{(0)}$ and $x_2^{(0)}$. The superscripts indicate that these values are initial estimates. We designate $\Delta x_1^{(0)}$ and $\Delta x_2^{(0)}$ as the values to be added to $x_1^{(0)}$ and $x_2^{(0)}$ to yield the correct solutions. So we can write

$$K_1 = f_1(x_1, x_2) = f_1(x_1^{(0)} + \Delta x_1^{(0)}, x_2^{(0)} + \Delta x_2^{(0)}) \qquad (8.11)$$

$$K_2 = f_2(x_1, x_2) = f_2(x_1^{(0)} + \Delta x_1^{(0)}, x_2^{(0)} + \Delta x_2^{(0)}) \qquad (8.12)$$

Our problem now is to solve for $\Delta x_1^{(0)}$ and $\Delta x_2^{(0)}$, which we shall do by expanding Eqs. (8.11) and (8.12) in Taylor's series to give

$$K_1 = f_1(x_1^{(0)}, x_2^{(0)}) + \Delta x_1^{(0)} \left. \frac{\partial f_1}{\partial x_1} \right|_{(0)} + \Delta x_2^{(0)} \left. \frac{\partial f_1}{\partial x_2} \right|_{(0)} + \cdots \qquad (8.13)$$

$$K_2 = f_2(x_1^{(0)}, x_2^{(0)}) + \Delta x_1^{(0)} \left. \frac{\partial f_2}{\partial x_1} \right|_{(0)} + \Delta x_2^{(0)} \left. \frac{\partial f_2}{\partial x_2} \right|_{(0)} + \cdots \qquad (8.14)$$

where the partial derivatives of order greater than 1 in the series of terms of the expansion have not been listed. The term $\partial f_1 / \partial x_1 |_{(0)}$ indicates that the partial derivative is evaluated for the values of $x_1^{(0)}$ and $x_2^{(0)}$. Other such terms are evaluated similarly.

If we neglect the partial derivatives of order greater than 1 we can rewrite Eqs. (8.13) and (8.14) in matrix form. We then have

$$\begin{bmatrix} K_1 - f_1(x_1^{(0)}, x_2^{(0)}) \\ K_2 - f_2(x_1^{(0)}, x_2^{(0)}) \end{bmatrix} = \begin{bmatrix} \dfrac{\partial f_1}{\partial x_1} & \dfrac{\partial f_1}{\partial x_2} \\ \dfrac{\partial f_2}{\partial x_1} & \dfrac{\partial f_2}{\partial x_2} \end{bmatrix} \begin{bmatrix} \Delta x_1^{(0)} \\ \Delta x_2^{(0)} \end{bmatrix} \qquad (8.15)$$

where the square matrix of partial derivatives is called the jacobian $\mathbf{J}$ or in this case $\mathbf{J}^{(0)}$ to indicate that the initial estimates $x_1^{(0)}$ and $x_2^{(0)}$ have been used to compute the numerical value of the partial derivatives. We note that $f_1(x_1^{(0)}, x_2^{(0)})$ is the calculated value of K_1 for the estimated values of $x_1^{(0)}$ and $x_2^{(0)}$ but this calculated value of K_1 is not the value specified by Eq. (8.9) unless our estimated values $x_1^{(0)}$ and $x_2^{(0)}$ are correct. If we designate as $\Delta K_1^{(0)}$ the specified value of K_1 minus the calculated value of K_1 and define $\Delta K_2^{(0)}$ similarly we have

$$\begin{bmatrix} \Delta K_1^{(0)} \\ \Delta K_2^{(0)} \end{bmatrix} = \mathbf{J}^{(0)} \begin{bmatrix} \Delta x_1^{(0)} \\ \Delta x_2^{(0)} \end{bmatrix} \qquad (8.16)$$

So by finding the inverse of the jacobian we can determine $\Delta x_1^{(0)}$ and $\Delta x_2^{(0)}$. However, since we truncated the series expansion, these values added to our initial guess do not determine for us the correct solution and we must try again by assuming new estimates $x_1^{(1)}$ and $x_2^{(1)}$ where

$$x_1^{(1)} = x_1^{(0)} + \Delta x_1^{(0)}$$

$$x_2^{(1)} = x_2^{(0)} + \Delta x_2^{(0)}$$

and repeat the process until the corrections become so small that they satisfy a chosen precision index.

To apply the Newton-Raphson method to the solution of load-flow equations we may choose to express bus voltages and line admittances in polar form or rectangular form. If we choose polar form and separate Eq. (8.7) into its real and imaginary components with

$$V_k = |V_k| \underline{/\delta_k} \qquad V_n = |V_n| \underline{/\delta_n} \qquad \text{and} \qquad Y_{kn} = |Y_{kn}| \underline{/\theta_{kn}}$$

we have

$$P_k - jQ_k = \sum_{n=1}^{N} |V_k V_n Y_{kn}| \underline{/\theta_{kn} + \delta_n - \delta_k} \tag{8.17}$$

So

$$P_k = \sum_{n=1}^{N} |V_k V_n Y_{kn}| \cos(\theta_{kn} + \delta_n - \delta_k) \tag{8.18}$$

and

$$Q_k = -\sum_{n=1}^{N} |V_k V_n Y_{kn}| \sin(\theta_{kn} + \delta_n - \delta_k) \tag{8.19}$$

As in the Gauss-Seidel method the swing bus is omitted from the iterative solution to determine voltages since both magnitude and angle of the voltage at the swing bus are specified. If we postpone consideration of voltage-controlled buses until later, we specify P and Q at all buses except the swing bus and estimate voltage magnitude and angle at all buses except the swing bus where voltage magnitude and angle are specified. The specified constant values of P and Q correspond to the K constants in Eq. (8.15). The estimated values of voltage magnitude and angle correspond to the estimated values for x_1 and x_2 in Eq. (8.15). We use these estimated values to calculate values of P_k and Q_k from Eqs. (8.18) and (8.19) and define

$$\Delta P_k = P_{k,\text{spec}} - P_{k,\text{calc}}$$

$$\Delta Q_k = Q_{k,\text{spec}} - Q_{k,\text{calc}}$$

which correspond to the ΔK values of Eq. (8.16).

The jacobian consists of the partial derivatives of P and Q with respect to each of the variables in Eqs. (8.18) and (8.19). The column matrix elements $\Delta \delta_k^{(0)}$

and $\Delta|V_k|^{(0)}$ correspond to $\Delta x_1^{(0)}$ and $\Delta x_2^{(0)}$ and are the corrections to be added to the original estimates $\delta_k^{(0)}$ and $|V_k|^{(0)}$ to obtain new values for computing $\Delta P_k^{(1)}$ and $\Delta Q_k^{(1)}$.

For the sake of simplicity we shall write the matrix equation for a system of only three buses. If the swing bus is number 1 we start our calculations at bus 2 since voltage magnitude and angle are specified at the swing bus. In matrix form

$$
\begin{bmatrix}
\Delta P_2 \\[4pt]
\Delta P_3 \\[4pt]
\text{----} \\[4pt]
\Delta Q_2 \\[4pt]
\Delta Q_3
\end{bmatrix}
=
\left[
\begin{array}{cc:cc}
\dfrac{\partial P_2}{\partial \delta_2} & \dfrac{\partial P_2}{\partial \delta_3} & \dfrac{\partial P_2}{\partial |V_2|} & \dfrac{\partial P_2}{\partial |V_3|} \\[10pt]
\dfrac{\partial P_3}{\partial \delta_2} & \dfrac{\partial P_3}{\partial \delta_3} & \dfrac{\partial P_3}{\partial |V_2|} & \dfrac{\partial P_3}{\partial |V_3|} \\[10pt]
\hdashline
\dfrac{\partial Q_2}{\partial \delta_2} & \dfrac{\partial Q_2}{\partial \delta_3} & \dfrac{\partial Q_2}{\partial |V_2|} & \dfrac{\partial Q_2}{\partial |V_3|} \\[10pt]
\dfrac{\partial Q_3}{\partial \delta_2} & \dfrac{\partial Q_3}{\partial \delta_3} & \dfrac{\partial Q_3}{\partial |V_2|} & \dfrac{\partial Q_3}{\partial |V_3|}
\end{array}
\right]
\begin{bmatrix}
\Delta \delta_2 \\[4pt]
\Delta \delta_3 \\[4pt]
\text{----} \\[4pt]
\Delta V_2 \\[4pt]
\Delta V_3
\end{bmatrix}
\tag{8.20}
$$

The superscripts which would indicate the number of the iteration are omitted in Eq. (8.20) because, of course, they change with each iteration. The elements of the jacobian are found by taking the partial derivatives of the expressions for P_k and Q_k and substituting therein the voltages assumed for the first iteration or calculated in the last previous iteration. The jacobian has been partitioned to emphasize the different general types of partial derivatives appearing in each submatrix. For instance, from Eq. (8.18) we find

$$
\frac{\partial P_k}{\partial \delta_n} = -\,|V_k V_n Y_{kn}|\sin\,(\theta_{kn} + \delta_n - \delta_k) \tag{8.21}
$$

where $n \neq k$ and

$$
\frac{\partial P_k}{\partial \delta_k} = \sum_{\substack{n=1 \\ n \neq k}}^{N} |V_k V_n Y_{kn}|\sin\,(\theta_{kn} + \delta_n - \delta_k) \tag{8.22}
$$

In the above summation $n \neq k$, which is apparent since δ_k drops out of Eq. (8.17) when $n = k$. Similar general forms of the partial derivatives may be found from Eqs. (8.18) and (8.19) for calculating the elements in the other submatrices.

Equation (8.20) and similar equations involving more buses are solved by inverting the jacobian. The values found for $\Delta \delta_k$ and $\Delta|V_k|$ are added to the previous values of voltage magnitude and angle to obtain new values for $P_{k,\,\text{calc}}^{(1)}$ and $Q_{k,\,\text{calc}}^{(1)}$ for starting the next iteration. The process is repeated until the precision index applied to the quantities in either column matrix is satisfied. To achieve convergence, however, the initial estimates of voltage must be reasonable, but this is seldom a problem in power-system work.

Voltage controlled buses are taken into account easily. Since the voltage magnitude is constant at such a bus we omit in the jacobian the column of partial differentials with respect to voltage magnitude of the bus. We are not interested at this point in the value of Q at the bus so we omit the row of partial differentials of Q for the voltage controlled bus. The value of Q at the bus can be determined after convergence by Eq. (8.19).†

The Newton-Raphson method, as noted previously, may also be used when equations are expressed in rectangular form. We chose to develop the equations in polar form because the jacobian provides interesting information which is lacking in the rectangular form. For instance, the dependence of P_k on δ_k and of Q_k on $|V_k|$ is seen immediately in the jacobian in polar form. Later in Sec. 8.10 we shall see how voltage regulating transformers in transmission lines principally affect the transfer of Q in a system while phase-shifting transformers principally affect the transfer of P.

Example 8.1 Figure 8.1 shows the one-line diagram of a very simple power system. Generators are connected at buses 1 and 3. Loads are indicated at buses 2, 4, and 5. Base values for the system are 100 MVA, 138 kV in the high-tension lines considered here. Table 8.1 gives impedances for the six lines which are identified by the buses on which they terminate. The charging megavars listed in the table account for the distributed capacitance of the lines and will be neglected in this example but discussed in Sec. 8.4 and

---

† For a more detailed explanation of the Newton-Raphson method and for excellent numerical examples carried through to a converging solution for both Gauss-Seidel and Newton-Raphson methods, see G. W. Stagg and A. H. El-Abiad, *Computer Methods in Power System Analysis*, chaps. 7 and 8, McGraw-Hill Book Company, New York, 1968.

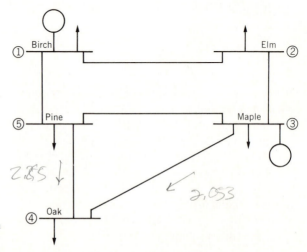

Figure 8.1 One-line diagram for Example 8.1.

Table 8.1

total resistance

| Line, bus to bus | Length | | R | X | R | X | Charging |
	km	mi	Ω	Ω	per unit	per unit	Mvar†
1–2	64.4	40	8	32	0.042	0.168	4.1
1–5	48.3	30	6	24	0.031	0.126	3.1
2–3	48.3	30	6	24	0.031	0.126	3.1
3–4	128.7	80	16	64	0.084	0.336	8.2
3–5	80.5	50	10	40	0.053	0.210	5.1
4–5	96.5	60	12	48	0.063	0.252	6.1

† At 138 kV.

100 MVA

135, MVA

included in the computer runs for the system. Table 8.2 lists values of P, Q, and V at each bus. Since values of P and Q in Eqs. (8.18) and (8.19) are positive for real power and inductive reactive voltamperes *input* to the network at each bus, net values of P and Q for these equations are negative at buses 2, 4, and 5. Generated Q is not specified where voltage magnitude is constant. In the voltage column the values for the load buses are the original estimates. Listed values of voltage magnitude and angle are to be held constant at the swing bus, and the listed voltage magnitude is to remain constant at bus 3. A load-flow study is to be made by the Newton-Raphson method using the polar form of the equations for P and Q. Determine the number of rows and columns in the jacobian. Calculate $\Delta P_2^{(0)}$ and the value of the second element in the first row of the jacobian using the specified values or initial estimates of the voltages.

SOLUTION Since the swing bus does not require a row and column of the jacobian an 8×8 matrix would be necessary if P and Q are specified for the remaining four buses. However, voltage magnitude is specified (held constant) at bus 3, and the jacobian will be a 7×7 matrix.

Table 8.2

| Bus | Generation | | Load | | V, per unit | Remarks |
	P, MW	Q, Mvar	P, MW	Q, Mvar		
1	....	....	65	30	1.04∠0°	Swing bus
2	0	0	115	60	1.00∠0°	Load bus (inductive)
3	180	....	70	40	1.02∠0°	Voltage magnitude constant
4	0	0	70	30	1.00∠0°	Load bus (inductive)
5	0	0	85	40	1.00∠0°	Load bus (inductive)

1.04 (handwritten note at bus 3 V)
.99 *98* *42* (handwritten notes at bus 4)
250MVA (handwritten note at left of bus 3)

6.588

In order to calculate $P_2^{(0)}$ for the estimated-voltage and fixed-voltage values of Table 8.2 we need only the admittances

$$Y_{21} = -\frac{1}{0.042 + j0.168} = 5.7747\underline{/104.04°}$$

$$Y_{23} = -\frac{1}{0.031 + j0.126} = 7.7067\underline{/103.82°}$$

and Y_{22}, which (since no other admittances terminate on bus 2) is expressed by

$$Y_{22} = -Y_{21} - Y_{23} = -|Y_{21}|\underline{/\theta_{21}} - |Y_{23}|\underline{/\theta_{23}}$$

From Eq. (8.18) since Y_{24} and Y_{25} are zero and since the initial values $\delta_1^{(0)} = \delta_2^{(0)} = \delta_3^{(0)} = 0$

$$\begin{aligned}
P_{2,\,calc}^{(0)} &= |V_2 V_1 Y_{21}| \cos \theta_{21} - |V_2 V_2 Y_{21}| \cos \theta_{21} \\
&\quad - |V_2 V_2 Y_{23}| \cos \theta_{23} + |V_2 V_3 Y_{23}| \cos \theta_{23} \\
&= (1.0 \times 1.04 - 1.0 \times 1.0)|Y_{21}| \cos \theta_{21} \\
&\quad - (1.0 \times 1.0 - 1.0 \times 1.02)|Y_{23}| \cos \theta_{23} \\
&= 0.04 \times 5.7747 \cos 104.04° + 0.02 \times 7.7067 \cos 103.82° \\
&= -0.0560 - 0.0368 = -0.0928 \text{ per unit}
\end{aligned}$$

Scheduled power into the network at bus 2 is

$$-\frac{115}{100} = -1.15 \text{ per unit}$$

So

$$\Delta P_2^{(0)} = -1.15 - (-0.0928) = -1.0572 \text{ per unit}$$

To find $\partial P_2/\partial\delta_3$ we use Eq. (8.21) and obtain

$$\frac{\partial P_2}{\partial\delta_3} = -|V_2 V_3 Y_{23}| \sin (\theta_{23} + \delta_3 - \delta_2)$$

$$= -1.0 \times 1.02 \times 7.067 \sin 103.82 = -7.6333 \text{ per unit}$$

The Newton-Raphson method is summarized in the following steps:

1. Determine values of $P_{k,\,calc}$ and $Q_{k,\,calc}$ flowing into the system at every bus for the specified or estimated values of voltage magnitudes and angles for the first iteration or the most recently determined voltages for subsequent iterations.
2. Calculate ΔP at every bus.
3. Calculate values for the jacobian using estimated or specified values of voltage magnitude and angle in the equations for partial derivatives determined by differentiation of Eqs. (8.18) and (8.19).

4. Invert the jacobian and calculate voltage corrections $\Delta\delta_k$ and $\Delta|V_k|$ at every bus.
5. Calculate new values of δ_k and $|V_k|$ by adding $\Delta\delta_k$ and $\Delta|V_k|$ to the previous values.
6. Return to step 1 and repeat the process using the most recently determined values of voltage magnitudes and angles until either all values of ΔP and ΔQ or all values of $\Delta\delta$ and $\Delta|V|$ are less than a chosen precision index.

P and Q at the swing bus and Q at voltage controlled buses can be determined from Eqs. (8.18) and (8.19). Line flow can be determined from the differences in bus voltages.

The number of iterations required by the Newton-Raphson method using bus admittances is practically independent of the number of buses. The time for the Gauss-Seidel method (bus admittances) increases almost directly with the number of buses. On the other hand, computing the elements of the jacobian is time-consuming, and the time per iteration is considerably longer for the Newton-Raphson method. The advantage of shorter computer time for a solution of the same accuracy is in favor of the Newton-Raphson method for all but very small systems.

8.4 DIGITAL-COMPUTER STUDIES OF LOAD FLOW

Power companies use very elaborate programs for making load-flow studies. A typical program is capable of handling systems of more than 2000 buses, 3000 lines, and 500 transformers. Of course programs can be expanded to even greater size provided the available computer facilities are sufficiently large.

Data supplied to the computer must include the numerical values given in Tables 8.1 and 8.2 and an indication of whether a bus is a swing bus, a regulated bus where voltage magnitude is held constant by generation of reactive power Q, or a bus with fixed P and Q. Where values are not to be held constant the quantities given in the tables are interpreted as initial estimates. Limits of P and Q generation usually must be specified as well as the limits of line kilovolt-amperes. Unless otherwise specified, programs usually assume a base of 100 MVA.

Total line charging in megavars specified for each line accounts for shunt capacitance and equals $\sqrt{3}$ times the *rated* line voltage in kilovolts times I_{chg}, as defined by Eqs. (4.24) and (4.25), divided by 10^3. This equals $\omega C_n|V|^2$ where $|V|$ is the rated line-to-line voltage in kilovolts, and C_n is line-to-neutral capacitance in farads for the entire length of the line. The program creates a nominal-π representation of the line by dividing equally between the two ends of the line the capacitance computed from the given value of charging megavars. For a long line, the computer could be programmed to compute the equivalent π for capacitance distributed evenly along the line.

8.5 INFORMATION OBTAINED IN A LOAD-FLOW STUDY

The information which is obtained from digital solutions of load flow is an indication of the great contribution digital computers have made to the power-system engineer's ability to obtain operating information about systems not yet built and to analyze the effects of changes on existing systems. The following discussion is not meant to list all the information obtainable but should provide some insight into the great importance of digital computers in power-system engineering.

The printout of results provided by the computer consists of a number of tabulations. Usually the most important information to be considered first is the table which lists each bus number and name, bus-voltage magnitude in per unit and phase angle, generation and load at each bus in megawatts and megavars, line charging, and megavars of static capacitors or reactors on the bus. Accompanying the bus information is the flow of megawatts and megavars from that bus over each transmission line connected to the bus. The totals of system generation and loads are listed in megawatts and megavars. The tabulation described is shown in Fig. 8.2 for the system of five buses of Example 8.1.

In the operation of power systems any appreciable drop in voltage on the primary of a transformer caused by a change of load may make it desirable to change the tap setting on transformers provided with adjustable taps in order to maintain proper voltage at the load. Where a tap-changing transformer has been specified to keep the voltage at a bus within designated tolerance limits, the voltage is examined before convergence is complete. If the voltage is not within the limits specified, the program causes the computer to perform a new set of iterations with a one-step change in the appropriate tap setting. The process is repeated as many times as necessary to cause the solution to conform to the desired conditions. The tap setting is listed in the tabulated results.

A system may be divided into areas, or one study may include the systems of several companies each designated as a different area. The computer program will examine the flow between areas, and deviations from the prescribed flow will be overcome by causing the appropriate change in generation of a selected generator in each area. In actual system operation interchange of power between areas is monitored to determine whether a given area is producing that amount of power which will result in the desired interchange.

Among other information that may be obtained is a listing of all buses where the voltage magnitude is above or below 1.05 or 0.95, respectively, or other limits that may be specified. A list of line loadings in megavoltamperes can be obtained. The printout will also list the total megawatt ($|I|^2 R$) and megavar ($|I|^2 X$) losses in the system and both P and Q mismatch at each bus. Mismatch is an indication of the preciseness of the solution and is the difference between P (and also usually Q) entering and leaving each bus.

SAVE THE TREES POWER COMPANY - LOAD FLOW STUDY — DATE 04/26/79 TIME 00,00.12

REPORT OF POWER FLOW CALCULATIONS FOR AREA 1, — 3 ITERATIONS, SWING BUS IS 1,

BUS	NAME	VOLTS	ANGLE	GENERATION MW	GENERATION MVAR	LOAD MW	LOAD MVAR	CAP/REAC MVAR	TO BUS	NAME	LINE FLOW MW	LINE FLOW MVAR	TAP
1	BIRCH	1.040	0.0	234.7	100.1	65.0	30.0		2	ELM	73.98	31.55	
									5	PINE	95.68	38.56	
2	ELM	0.961	-6.3	0.0	0.0	115.0	60.0		1	BIRCH	-71.41	-25.39	
									3	MAPLE	-43.59	-34.60	
3	MAPLE	1.020	-3.7	180.0	110.3R	70.0	40.0		2	ELM	44.59	35.65	
									4	OAK	40.46	18.06	
									5	PINE	24.95	16.58	
4	OAK	0.920	-10.9	0.0	0.0	70.0	30.0		3	MAPLE	-38.74	-18.90	
									5	PINE	-31.25	-11.09	
5	PINE	0.968	-6.2	0.0	0.0	85.0	40.0		1	BIRCH	-92.59	-29.14	
									3	MAPLE	-24.44	-19.62	
									4	OAK	32.03	8.77	
	AREA TOTALS			414.7	210.4	405.0	200.0	0.0					

SOLUTION TIME 0.10 SECONDS.
TOTAL TIME 0.54 SECONDS.

Figure 8.2 Digital-computer solution of load-flow for the system of Example 8.1. Base is 100 MVA.

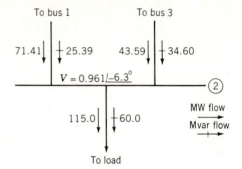

Figure 8.3 Flow of P and Q at bus 2 for the system of Example 8.1. Numbers beside the arrows show the flow of P and Q in megawatts and megavars. The bus voltage is shown in per unit.

8.6 NUMERICAL RESULTS

The load-flow study for the system described in Example 8.1 for which Fig. 8.2 is the printout was run on a program attributed to the Philadelphia Electric Company and subsequently modified. Three Newton-Raphson iterations were required. Similar size studies run using other programs also required three Newton-Raphson iterations but required 22 Gauss-Seidel iterations with the same precision index. Figure 8.2 may be examined for more information than just the tabulated results. For instance, the megawatt loss in any of the lines can be found by comparing the values of P and Q at the two ends of the line. As an example we see that 95.68 MW flow from bus 1 into line 1–5 and 92.59 MW flow into bus 5 from the line. Evidently the $|I|^2 R$ loss in the line is 3.09 MW. On another page of the printout, not reproduced here, $|I|^2 R$ losses of the system are listed as 9.67 MW.

Information provided by Fig. 8.2 can be displayed on a diagram showing the entire system. Figure 8.3 shows a portion of such a diagram at bus 2.

8.7 CONTROL OF POWER INTO A NETWORK

We studied some characteristics of synchronous machines in Chap. 6. We developed the principle that an overexcited generator supplies reactive power Q to a system and an underexcited generator absorbs reactive power Q from a system. Thus, we understand how the exciter controls the flow of reactive power between the generator and the system.

Now we turn our attention to real power P. Assume that a generator supplying a large system is delivering power under stable conditions so that a certain angle δ exists between V_t, the voltage at the system bus, and E_g, the generated voltage of the machine. If E_g leads V_t we have the phasor diagram of Fig. 8.4a which is identical to Fig. 6.9a. If the power input to the generator is increased by a larger opening of the valves through which steam (or water) enters a turbine while $|E_g|$ is constant, the rotor speed will start to increase and the angle between E_g and V_t will increase. Increasing δ results in a larger I_a and lower θ, as

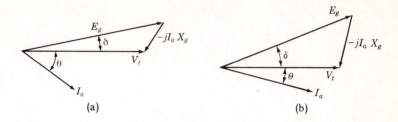

Figure 8.4 Phasor diagrams of a generator having constant values of $|E_g|$ and $|V_t|$ for (a) angle δ small and (b) angle δ larger to show the increase in power delivered as δ increases.

may be seen by comparing Figs. 8.4a and 8.4b. The generator will therefore deliver more power to the network, and the input from the prime mover will again equal the output to the network if losses are disregarded. Equilibrium will be reestablished at the speed corresponding to the frequency of the infinite bus with a larger δ. Figure 8.4b is drawn for the same dc field excitation and therefore the same $|E_g|$ as Fig. 8.4a, but the power output equal to $|V_t| \cdot |I_a| \cos \theta$ is greater for the condition of Fig. 8.4b, and the increase in δ has caused the generator to deliver the additional power to the network.

The dependence of power on the power angle is also shown by an equation giving $P + jQ$ supplied by a generator in terms of δ. If

$$V_t = |V_t| \underline{/0^\circ} \qquad \text{and} \qquad E_g = |E_g| \underline{/\delta}$$

where V_t and E_g are expressed in volts to neutral or in per unit, then

$$I_a = \frac{|E_g| \underline{/\delta} - |V_t|}{jX_g} \tag{8.23}$$

and

$$I_a^* = \frac{|E_g| \underline{/-\delta} - |V_t|}{-jX_g} \tag{8.24}$$

Therefore

$$P + jQ = V_t I_a^*$$

$$= \frac{|V_t| \cdot |E_g| \underline{/-\delta} - |V_t|^2}{-jX_g}$$

$$= \frac{|V_t| \cdot |E_g| \underline{/90 - \delta} - |V_t|^2 \underline{/90^\circ}}{X_g} \tag{8.25}$$

The real part of Eq. (8.25) is

$$P = \frac{|V_t| \cdot |E_g|}{X_g} \cos (90 - \delta) = \frac{|V_t| \cdot |E_g|}{X_g} \sin \delta \tag{8.26}$$

and the imaginary part of Eq. (9.13) is

$$Q = \frac{|V_t| \cdot |E_g|}{X_g} \sin(90 - \delta) - \frac{|V_t|^2}{X_g}$$

$$= \frac{|V_t|}{X_g} (|E_g| \cos \delta - |V_t|) \tag{8.27}$$

When volts rather than per-unit values are substituted for V_t and E_g in Eqs. (8.26) and (8.27), we must be careful to note that V_t and E_g are line-to-neutral voltages and P and Q will be per-phase quantities. However, line-to-line voltage values substituted for V_t and E_g will yield total three-phase values for P and Q. The per-unit P and Q of Eqs. (8.26) and (8.27) are multiplied by base three-phase megavoltamperes or base megavoltamperes per phase depending on whether total three-phase power or power per phase is wanted.

Equation (8.26) shows very clearly the dependence of power transferred to the network on the power angle δ if $|E_g|$ and $|V_t|$ are constant. However, if P and V_t are constant, Eq. (8.26) shows that δ must decrease if $|E_g|$ is increased by increasing the dc field excitation. In Eq. (8.27) with P constant both an increase in $|E_g|$ and a decrease in δ mean that Q will increase if already positive or decrease in magnitude and perhaps become positive if Q is negative before increasing the field excitation. This agrees with the conclusions reached in Sec. 6.4.

Equation (8.26) can be interpreted as the power transferred from one bus in a network to another bus through a reactance X connecting the two buses. If the bus voltages are V_1 and V_2 and δ is the angle by which V_1 leads V_2,

$$P = \frac{|V_1| \cdot |V_2|}{X} \sin \delta \tag{8.28}$$

Similarly from Eq. (8.27), Q received at bus 2 is

$$Q = \frac{|V_2|}{X} (|V_1| \cos \delta - |V_2|) \tag{8.29}$$

The equations derived in Sec. 5.8 to develop circle diagrams are more general than Eqs. (8.28) and (8.29) because they account for resistance and capacitance. Equations (5.59) and (5.60), however, are identical with Eqs. (8.28) and (8.29) if the only line parameter considered is the inductance.

From Eqs. (8.28) and (8.29) we see that an increase in δ causes a larger change in P than in Q when δ is small. This difference is explained when we recognize that $\sin \delta$ changes widely but $\cos \delta$ changes by only a small amount with a change in δ when δ is less than 10 or 15°.

8.8 THE SPECIFICATION OF BUS VOLTAGES

In Chap. 6 and Sec. 8.7 we have been considering the synchronous generator from the viewpoint of supplying power to an infinite bus. We have examined the effect of generator excitation and power angle when the terminal voltage of the generator remains constant. In load-flow studies on a digital computer, however, we found that it was necessary to specify voltage magnitude or reactive power at every bus except the swing bus, where voltage was specified by both magnitude and angle. Although the computer can easily tell us the results over the entire system of specifying various voltage magnitudes at particular buses, it may be helpful to look at what happens in a very simple case.

Usually it is at the buses on which there is generation that the voltage magnitude is specified when a load-flow study is made on a computer. At such buses real power P supplied by the generator is also specified. The reactive power Q is then determined by the computer in solving the problem. Therefore, our purpose at this point is to examine the effect of the magnitude of the specified bus voltage on the value of Q supplied by the generator to the power network.

Figure 8.5 shows a generator represented by its equivalent circuit with the relatively small resistance neglected in order to simplify our analysis. The power system is represented by its Thévenin equivalent voltage E_{th} in series with the Thévenin impedance X_{th}, where again resistance has been neglected. Any local load on the bus is included in the Thévenin equivalent. For constant power delivered by the generator the component of I in phase with E_{th} must remain constant. The voltage specified at the bus is $|V_t|$, and

$$V_t = E_{th} + jIX_{th} \qquad (8.30)$$

The phasor diagram for the circuit of Fig. 8.5 is shown in Fig. 8.6 for three different phase angles between E_{th} and I. In all three cases, however, the component of I in phase with E_{th} is constant.

Figure 8.6 shows that larger magnitudes of bus voltage V_t with constant power input to the bus require a larger $|E_g|$, and, of course, the larger $|E_g|$ is obtained by increasing the excitation of the dc field winding of the generator. Increasing the bus voltage by increasing $|E_g|$ causes the current to become more lagging, as we see from Fig. 8.6 and as we expect from our discussion of the synchronous generator. When we are making a load-flow study, increasing the voltage specified at a generator bus means that the generator feeding the bus will

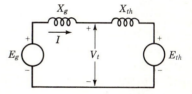

Figure 8.5 Generator with internal voltage E_g connected to a power system represented by its Thévenin equivalent.

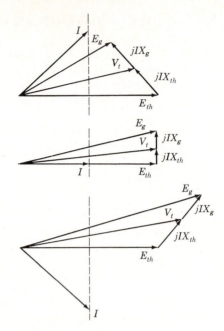

Figure 8.6 Phasor diagrams of a generator supplying the same power to a system at three different values of bus voltage V_t.

increase its output of reactive power to the bus. From the standpoint of operation of the system we are controlling bus voltage and Q generation by adjusting the generator excitation.

Since we have represented the system by its Thévenin equivalent, we are assuming that all E_g and E_m values in the system remain constant in magnitude and angle. This assumption is not strictly true under actual operating conditions. When a change is made in the excitation of one generator, other changes may be made elsewhere in the system. An example which would require changing E_g of generators or motors at other buses is the specification to hold voltage constant at these buses. The computer program takes care of all such conditions that are imposed. However, our assumption of constant E_g and E_m values in the system except where we are making a change is very useful to illustrate the effect of a change of voltage magnitude at a particular bus.

Example 8.2 A generator is supplying a large system which can be represented by its Thévenin equivalent circuit, consisting of a generator with a voltage E_{th} in series with $Z_{th} = j0.2$ per unit. The voltage at the generator terminals is $V_t = 0.97\underline{/0°}$ per unit when delivering a current of $0.8 - j0.2$ per unit. Synchronous reactance of the generator is 1.0 per unit. Find P and Q into the system at the generator terminals and compute E_g (a) for the conditions described above and (b) if $|V_t| = 1.0$ per unit when the generator is delivering the same power P to the system. Assume that the system is so large that E_{th} is not affected by the change in $|V_t|$. The bus at the generator terminals is not an infinite bus, however, because Z_{th} is not zero.

SOLUTION (a) From the generator into the system

$$P + jQ = 0.97(0.8 + j0.2) = 0.776 + j0.194 \text{ per unit}$$

and

$$E_g = 0.97 + j1(0.8 - j0.2)$$
$$= 1.17 + j0.8 = 1.42\underline{/34.4°} \text{ per unit}$$

(b) To find P and Q when $|V_t| = 1.0$ we must find the phase angle of V_t, as follows:

$$E_{th} = 0.97 - j0.2(0.8 - j0.2)$$
$$= 0.93 - j0.16 = 0.944 \underline{/-9.76°} \text{ per unit}$$

The phase angle of V_t is determined by finding the angle δ between V_t and E_{th} for $|V_t| = 1.0$ and $P = 0.776$. By Eq. (8.28)

$$\frac{1.0 \times 0.944}{0.2} \sin \delta = 0.776$$

$$\delta = 9.46°$$

(by which V_t leads E_{th}). Therefore,

$$V_t = 1.0 \underline{/-9.76°} + 9.46° = 1.0 \underline{/-0.3°} = 1.0 - j0.005$$

$$I = \frac{1.0 - j0.005 - (0.93 - j0.16)}{j0.2}$$

$$= \frac{0.07 + j0.155}{j0.2} = 0.775 - j0.350$$

$$= 0.850 \underline{/-24.3°}$$

and

$$E_g = 1.0 - j0.005 + j1(0.775 - j0.350)$$
$$= 1.350 + j0.770 = 1.55\underline{/29.7°} \text{ per unit}$$

At the generator terminals into the system

$$P + jQ = 1.0\underline{/-0.3°} \times 0.850\underline{/24.3°}$$
$$= 0.850\underline{/24.0°} = 0.776 + j0.346 \text{ per unit}$$

This example verifies our reasoning that specifying a higher terminal voltage at a system bus to which a generator is connected results in a larger reactive power supplied to the system by the generator and requires larger generated voltage obtained by increasing the dc field excitation of the generator. In this example Q increased from 0.194 to 0.346 per unit and $|E_g|$ had to be increased from 1.42 to 1.55 per unit.

8.9 CAPACITOR BANKS

Another very important method of controlling bus voltage is by shunt capacitor banks at the buses at both transmission and distribution levels along lines or at substations and loads. Essentially capacitors are a means of supplying vars at the point of installation. Capacitor banks may be permanently connected, but as regulators of voltage they may be switched on and off the system as changes in load demand. Switching may be manually or automatically controlled either by time clocks or in response to voltage or reactive-power requirements. When they are in parallel with a load having a lagging power factor, the capacitors are the source of some or perhaps all of the reactive power of the load. Thus, capacitors reduce the line current necessary to supply the load and reduce the voltage drop in the line as the power factor is improved. Since capacitors lower the reactive requirement from generators, more real-power output is available. The reader may wish to review the effect of power factor on voltage regulation by referring to Fig. 5.5.

In the load-flow computer program, voltage magnitude can be specified only if there is a source of reactive-power generation. Therefore at load buses at which there are no generators, capacitor banks must be assumed, and the computer will specify the value of Q required.

If capacitors are applied at a particular node, the increase in voltage at the node can be determined by Thévenin's theorem. Figure 8.7 shows the system represented by its Thévenin equivalent at the node where capacitors will be applied by closing the switch. The resistance in the equivalent circuit is indicated but is always much smaller than the inductive reactance. With the switch open the node voltage V_t is equal to the Thévenin voltage E_{th}. When the switch is closed, the current drawn by the capacitor is

$$I_C = \frac{E_{th}}{Z_{th} - jX_C} \qquad (8.31)$$

The phasor diagram is shown in Fig. 8.8. The increase in V_t caused by adding the capacitor is very nearly equal to $|I_C| X_{th}$ if we assume that E_{th} remains un-

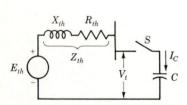

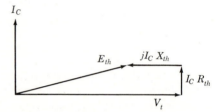

Figure 8.7 Circuit showing a capacitor to be connected through switch S to a system represented by its Thévenin equivalent.

Figure 8.8 Phasor diagram of the circuit of Fig. 8.7 with the capacitor connected. Before connection of the capacitor $V_t = E_{th}$.

changed since E_{th} and V_t are identical before adding the capacitor. This phasor diagram serves to explain the increase in voltage at the bus where the capacitor is installed. Example 7.6 was introduced as a part of our study of Z_{bus}; and this example should be reviewed because it shows how the change in voltage magnitude due to the added capacitor can be calculated at all the buses of a system where there are no regulated buses and loads are represented by impedances.

Here again we have made the assumption that E_g and E_m values in the system remain constant. As described in Sec. 8.8, the assumption is not strictly true but provides a good estimate of the increase of bus voltages due to adding capacitors except at buses where voltage is held constant. If the capacitors are added to a load bus which is remote from any generation, the estimate is quite good for nearby buses.

The digital load-flow printout of Fig. 8.2 shows a voltage at bus 4 of 0.920 per unit. The same load-flow program can be used to determine the amount of reactive power which must be supplied by capacitors at this bus to raise the voltage to any specified value. The procedure is to designate bus 4 as a regulated bus to be held at the specified voltage and with a generator at the bus to supply only reactive power. If generator losses are neglected, such a generator is equivalent to a lossless, unloaded, and overexcited synchronous motor and is known as a *synchronous condenser*. The computer determines the necessary amount of reactive power, which may be supplied to the system by either static capacitors or a synchronous condenser.

When the voltage of bus 4 is specified to be 0.950 per unit, the required reactive-power generation is found to be 15.3 kvar. This reactive-power input at bus 4 also raises the voltage of bus 5 from 0.968 to 0.976 per unit. At bus 2, the only other unregulated bus, the voltage is unchanged because of its separation from bus 4 by regulated buses 1 and 3.

The flow of real and reactive power determined by the computer in the lines connected to bus 4 with and without the added capacitors is shown in Fig. 8.9.

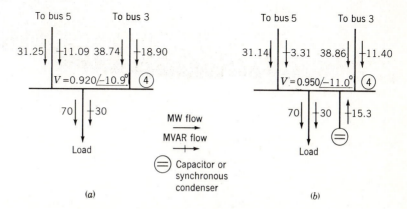

(a)

(b)

Figure 8.9 Flow of P and Q at bus 4 of the system of Example 8.1 (a) as found in the original load-flow study and (b) with capacitors added at the bus to raise the voltage to 0.950 per unit.

The voltage drops on the lines from buses 3 and 5 to bus 4 are reduced by supplying reactive power at bus 4 because the reactive power flowing in these lines is reduced. The increase of voltage at bus 4 obtained by applying capacitors at the bus results in causing the reactive power reaching bus 4 over the two transmission lines to be apportioned between the lines to accomplish the required voltage drop in each.

8.10 CONTROL BY TRANSFORMERS

Transformers provide an additional means of control of the flow of both real and reactive power. Our usual concept of the function of transformers in a power system is that of changing from one voltage level to another, as when a transformer converts the voltage of a generator to the transmission-line voltage. However, transformers which provide a small adjustment of voltage magnitude, usually in the range of $\pm 10\%$, and others which shift the phase angle of the line voltages are important components of a power system. Some transformers regulate both the magnitude and phase angle.

Almost all transformers provide taps on windings to adjust the ratio of transformation by changing taps when the transformer is deenergized. A change in tap can be made while the transformer is energized, and such transformers are called *load-tap-changing* (LTC) *transformers* or *tap-changing-under-load* (TCUL) *transformers*. The tap changing is automatic and operated by motors which respond to relays set to hold the voltage at the prescribed level. Special circuits allow the change to be made without interrupting the current.

A type of transformer designed for small adjustments of voltage rather than for changing voltage levels is called a *regulating transformer*. Figure 8.10 shows a

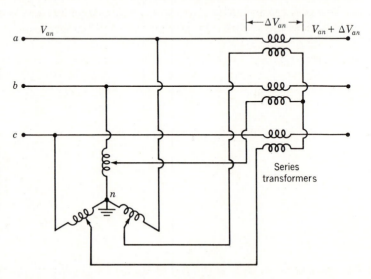

Figure 8.10 Regulating transformer for control of voltage magnitude.

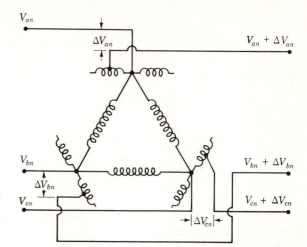

Figure 8.11 Regulating transformer for control of phase angle. Windings drawn parallel to each other are on the same iron core.

regulating transformer for control of voltage magnitude, and Fig. 8.11 shows a regulating transformer for phase-angle control. The phasor diagram of Fig. 8.12 helps to explain the shift in phase angle. Each of the three windings to which taps are made is on the same magnetic core as the phase winding whose voltage is $90°$ out of phase with the voltage from neutral to the point connected to the center of the tapped winding. For instance, the voltage to neutral V_{an} is increased by a component ΔV_{an} which is in phase or $180°$ out of phase with V_{bc}. Figure 8.12 shows how the three line voltages are shifted in phase angle with very little change in magnitude.

The procedure to determine $\mathbf{Y}_{bus}$ and $\mathbf{Z}_{bus}$ in per unit for a network containing a regulating transformer is the same as the procedure to account for any transformer whose turns ratio is other than the ratio used to select the ratio of base voltages on the two sides of the transformer. Such a transformer, which we shall now investigate, is said to have an off-nominal turns ratio.

If we have two buses connected by a transformer, and if the ratio of the line-to-line voltages of the transformer is the same as the ratio of the base

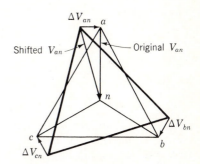

Figure 8.12 Phasor diagram for the regulating transformer shown in Fig. 8.11.

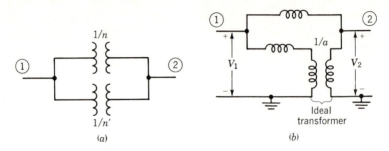

Figure 8.13 Transformers with differing turns ratios connected in parallel: (*a*) the one-line diagram; (*b*) the reactance diagram in per unit. The turns ratio $1/a$ is equal to n/n'.

voltages of the two buses, the equivalent circuit (with the magnetizing current neglected) is simply the transformer impedance in per unit on the chosen base connected between the buses. Figure 8.13*a* is a one-line diagram of two transformers in parallel. Let us assume that one of them has the voltage ratio $1/n$, which is also the ratio of base voltages on the two sides of the transformer, and that the voltage ratio of the other is $1/n'$. The equivalent circuit is then that of Fig. 8.13*b*. We need the ideal (no impedance) transformer with the ratio $1/a$ in the per-unit reactance diagram to take care of the off-nominal turns ratio of the second transformer because base voltages were determined by the turns ratio of the first transformer.

If we have a regulating transformer (rather than the LTC, which changes voltage level as well as providing the tap-changing feature), Fig. 8.13*b* may be interpreted as two transmission lines in parallel with a regulating transformer in one line.

Evidently our problem is to find the node admittances of Fig. 8.14, which is a more detailed representation of the LTC with a turns ratio of $1/n'$ or of the regulating transformer with the transformation ratio $1/a$. The admittance Y in the figure is the reciprocal of the per-unit impedance of the transformer. Since the admittance Y is shown on the side of the ideal transformer nearest node 1, the tap-changing side (or the side corresponding to n') is nearest node 2. This designation is important in using the equations which are to be derived. If we are considering a transformer with off-nominal turns ratio, a is the ratio n'/n. If we have a regulating transformer, a may be real or imaginary, such as 1.02 for a 2% boost in magnitude or $\varepsilon^{j\pi/60}$ for a 3° shift per phase.

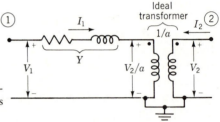

Figure 8.14 More detailed per-unit reactance diagram of the transformer of Fig. 8.13*b*, whose turns ratio is $1/a$.

Figure 8.14 has been labeled to show currents I_1 and I_2 entering the two nodes, and the voltages are V_1 and V_2 referred to the reference node. The complex expression for power into the ideal transformer in the direction from node 1 is

$$S_1 = \frac{V_2}{a} I_1^* \tag{8.32}$$

and into the transformer from node 2

$$S_2 = V_2 I_2^* \tag{8.33}$$

Since we are assuming that we have an ideal transformer with no losses, the power into the ideal transformer from node 1 must equal the power out of the transformer from node 2, and so

$$\frac{V_2}{a} I_1^* = -V_2 I_2^* \tag{8.34}$$

and

$$I_1 = -a^* I_2 \tag{8.35}$$

The current I_1 can be expressed by

$$I_1 = \left(V_1 - \frac{V_2}{a} \right) Y \tag{8.36}$$

or

$$I_1 = V_1 Y - V_2 \frac{Y}{a} \tag{8.37}$$

Substituting $-a^* I_2$ for I_1 and solving for I_2 yields

$$I_2 = -V_1 \frac{Y}{a^*} + V_2 \frac{Y}{aa^*} \tag{8.38}$$

Comparing Eqs. (8.37) and (8.38), we have, since $aa^* = |a|^2$, the node admittances

$$Y_{11} = Y \qquad Y_{22} = \frac{Y}{|a|^2}$$

$$Y_{12} = -\frac{Y}{a} \qquad Y_{21} = -\frac{Y}{a^*} \tag{8.39}$$

The equivalent π corresponding to these values of node admittances can be found only if a is real, so that $Y_{21} = Y_{12}$. If the transformer is changing magnitude, not phase shifting, the circuit is that of Fig. 8.15. This circuit cannot be realized if Y has a real component, which would require a negative resistance in the circuit. The important factor, however, is that we can now account for magnitude, phase shifting, and off-nominal-turns-ratio transformers in calculations to obtain $\mathbf{Y}_{\text{bus}}$ and $\mathbf{Z}_{\text{bus}}$.

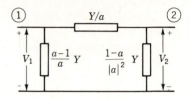

Figure 8.15 Circuit having the node admittances of Eqs. (8.39) providing a is real.

Example 8.3 Two transformers are connected in parallel to supply an impedance to neutral per phase of $0.8 + j0.6$ per unit at a voltage of $V_2 = 1.0 \underline{/0°}$ per unit. Transformer T_a has a voltage ratio equal to the ratio of the base voltages on the two sides of the transformer. This transformer has an impedance of $j0.1$ per unit on the appropriate base. The second transformer T_b has a step-up toward the load of 1.05 times that of T_a (secondary windings on 1.05 tap), and its impedance is $j0.1$ per unit on the base of the circuit on its low-tension side. Figure 8.16 shows the equivalent circuit with transformer T_b represented by its impedance and an ideal transformer. Find the complex power transmitted to the load through each transformer.

SOLUTION

$$I_2 = -\frac{1.0}{0.8 + j0.6} = -0.8 + j0.6$$

$$a = 1.05$$

To determine the current in each transformer we need to find V_1 from the equation

$$I_2 = V_1 Y_{21} + V_2 Y_{22}$$

where the node admittances are those of the parallel combination of the two transformers. For transformer T_a alone

$$Y_{21} = -\frac{1}{j0.1} = j10$$

$$Y_{22} = \frac{1}{j0.1} = -j10$$

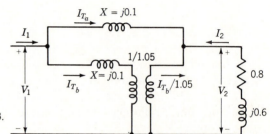

Figure 8.16 Circuit for Example 8.3. Values in per unit.

For transformer T_b alone

$$Y_{21} = -\frac{1/j0.1}{1.05} = j9.52$$

$$Y_{22} = \frac{1/j0.1}{|1.05|^2} = -j9.07$$

For the two transformers in parallel

$$Y_{21} = j10 + j9.52 = j19.52$$

$$Y_{22} = -j10 - j9.07 = -j19.07$$

Then, from the node equation for I_2,

$$-0.8 + j0.6 = V_1(j19.52) - j19.07 \times 1.0$$

$$V_1 = 1.008 + j0.041$$

$$V_1 - V_2 = 0.008 + j0.041$$

Therefore

$$I_{T_a} = (V_1 - V_2)(-j10) = 0.41 - j0.08$$

From Eq. (8.35) the current into bus 2 from the transformer T_b is I_{T_b}/a^*, and from Fig. 8.16 this current is $-(I_{T_a} + I_2)$, giving

$$\frac{I_{T_b}}{a^*} = -I_2 - I_{T_a} = 0.8 - j0.6 - (0.41 - j0.08) = 0.39 - j0.52$$

The complex powers are

$$S_{T_a} = V_2 I_{T_a}^* = 0.41 + j0.08 \text{ per unit}$$

$$S_{T_b} = V_2\left(\frac{I_{T_b}}{a^*}\right)^* = 0.39 + j0.52 \text{ per unit}$$

An approximate solution to this problem is found by recognizing that Fig. 8.17 with switch S closed is also an equivalent circuit for the problem if the voltage ΔV, which is in the branch of the circuit equivalent to transformer T_b, is

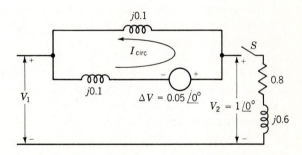

Figure 8.17 A circuit equivalent when switch S is closed to that of Fig. 8.16.

equal to $a - 1$ in per unit. In other words, if T_a is providing a voltage ratio 5% higher than T_b, a equals 1.05 and ΔV equals 0.05 per unit. To the extent that we can say that the current set up by ΔV circulates around the loop indicated by I_{circ} with switch S open and that with S closed none of that current goes through the load impedance because it is much larger than the transformer impedance, we can use the superposition principle. Then

$$I_{circ} = \frac{0.05}{j0.2} = -j0.25 \text{ per unit}$$

With ΔV short-circuited the current in each path is half of the load current, or $0.4 - j0.3$. Then superimposing the circulating current gives

$$I_{T_a} = 0.4 - j0.3 - (-j0.25) = 0.4 - j0.05$$

$$I_{T_b} = 0.4 - j0.3 + (-j0.25) = 0.4 - j0.55$$

so that

$$S_{T_a} = 0.40 + j0.05 \text{ per unit}$$

and

$$S_{T_b} = 0.40 + j0.55 \text{ per unit}$$

These values, although approximate, are so nearly equal to values originally found that the method is often used because of its simplicity.

This example shows that the transformer with the higher tap setting is supplying most of the reactive power to the load. The real power is dividing equally between the transformers. Since both transformers have the same impedance, they would share both the real and reactive power equally if they had the same turns ratio. In that case each would be represented by the same per-unit reactance of $j0.1$ between the two buses and would carry equal current. When two transformers are in parallel, we can vary the distribution of reactive power between the transformers by adjusting the voltage-magnitude ratios. When two paralleled transformers of equal kilovoltamperes do not share the kilovoltamperes equally because their impedances differ, the kilovoltamperes may be more nearly equalized by adjustment of the voltage-magnitude ratios through tap changing.

If a particular transmission line in a system is carrying too small or too large a reactive power, a regulating transformer to adjust voltage magnitude can be provided at one end of the line to make the line transmit a larger or smaller reactive power. We can investigate this by means of the automatic tap-changing feature in the load-flow program on a digital computer. For instance, we can raise the voltage at bus 4 of Example 8.1 by inserting a magnitude-regulating transformer in the line from bus 5 to bus 4 at bus 4, and we tell the computer to consider this as an LTC with a tap setting to hold the bus voltage at about 0.950 per unit. There is a definite step between tap settings, and the voltage will not

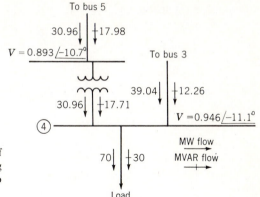

Figure 8.18 Flow of P and Q at bus 4 of the system of Fig. 8.1 when a regulating transformer in line 5-4 at bus 4 raises V_4 to 0.946 per unit.

necessarily be exactly 0.950 per unit. The results achieved at bus 4 are shown in the one-line diagram of Fig. 8.18. A per-unit reactance of 0.08 was assumed for the LTC.

When the voltage of bus 4 is raised by the LTC in line 5-4, the voltage drop over line 3-4 must be less and we expect this to be accomplished by a reduced flow of reactive power through the line with little change in real power. By comparing Fig. 8.18 with Fig. 8.9a we see that the Q flowing into bus 4 through line 3-4 is reduced from 18.90 to 12.26 Mvar without much change in P. To supply the 30 Mvar required by the load 17.71 Mvar now flows into bus 4 through the line from bus 5 and the LTC. The increased megavars on the line cause the voltage on the low-tension side of the LTC to be quite low, but the transformer steps up the voltage to 0.946 per unit at bus 4 by selecting the proper tap setting.

The voltage at bus 5 dropped from the original value of 0.968 per unit to 0.962 per unit. As a comparison, we noted in Sec. 8.9 that the voltage at bus 5 increased to 0.976 per unit when capacitors were added to bus 4. The reason for the decrease in the voltage at bus 5 in the present case is that the increased reactive power supplied to bus 4 from bus 5 caused an increase in the reactive power which had to be fed to bus 5 from the voltage-regulated buses 1 and 3.

To determine the effect of phase-shifting transformers we need only let a be complex with a magnitude of unity in Eqs. (8.39).

Example 8.4 Repeat Example 8.3 except that T_b includes both a transformer having the same turns ratio as T_a and a regulating transformer with a phase shift of 3° ($a = \varepsilon^{j\pi/60} = 1.0\underline{/3°}$). The impedance of the two components of T_b is $j0.1$ per unit on the base of T_a.

SOLUTION For transformer T_a alone, as in Example 8.3,

$$Y_{21} = j10 \qquad Y_{22} = -j10$$

and for transformer T_b

$$Y_{21} = \frac{j10}{a^*} = 10\underline{/93°}$$

$$Y_{22} = \frac{-j10}{|1.0\underline{/3°}|^2} = -j10$$

Combining the transformers in parallel gives

$$Y_{21} = 10\underline{/90°} + 10\underline{/93°} = -0.523 + j20.0$$

$$Y_{22} = -j10 - j10 = -j20$$

Following the procedure of Example 8.3, we have

$$-0.8 + j0.6 = V_1(-0.523 + j20) + (-j20)(1.0)$$

$$V_1 = \frac{-0.8 + j20.6}{-0.523 + j20} = \frac{0.418 - j10.77 + j16.0 + 412.0}{400}$$

$$= 1.03 + j0.013$$

$$V_1 - V_2 = 0.03 + j0.013$$

$$I_{T_a} = (0.03 + j0.013)(-j10) = 0.13 - j0.30$$

$$\frac{I_{T_b}}{a^*} = 0.8 - j0.6 - (0.13 - j0.30) = 0.67 - j0.30$$

$$S_{T_a} = 0.13 + j0.30 \text{ per unit}$$

$$S_{T_b} = 0.67 + j0.30 \text{ per unit}$$

As in Example 8.3, we can obtain an approximate solution of the problem by inserting a voltage source ΔV in series with the impedance of transformer T_b. The proper per-unit voltage is

$$a - 1 = 1.0\underline{/3°} - 1.0\underline{/0°} = (2 \sin 1.5°)\underline{/91.5°} = 0.0524\underline{/91.5°}$$

$$I_{\text{circ}} = \frac{0.0524\underline{/91.5°}}{0.2\underline{/90°}} = 0.262 + j0.0069$$

$$I_{T_a} = 0.4 - j0.3 - (0.262 + j0.007) = 0.138 - j0.307$$

$$I_{T_b} = 0.4 - j0.3 + (0.262 + j0.007) = 0.662 - j0.293$$

So

$$S_{T_a} = 0.138 + j0.307 \text{ per unit}$$

$$S_{T_b} = 0.662 + j0.293 \text{ per unit}$$

Again the approximate values are close to the values found previously.

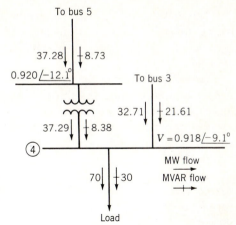

Figure 8.19 Flow of P and Q at bus 4 of the system of Fig. 8.1 when a regulating transformer in line 5-4 at bus 4 causes a phase shift of 3° across its terminals.

The example shows that the phase-shifting transformer is useful to control the amount of real power flow but has less effect on the flow of reactive power. Both Examples 8.3 and 8.4 are illustrative of two transmission lines in parallel with a regulating transformer in one of the lines. For instance, Eqs. (8.39) would apply to a transmission line having a regulating or off-nominal-turns-ratio transformer at one end and with shunt admittance and the impedance of the transformer neglected or included in the series impedance of the line. In that case, Y of Eqs. (8.39) would be the reciprocal of the series impedance of the line in per unit. In a load-flow study on the digital computer a transformer at the end of a line may be taken into account by adding a bus so that the transformer is directly connected to buses on both sides.

Figure 8.19 shows the flow of real and reactive power and the voltage at bus 4 of the system of Example 8.1 when a phase-shifting transformer is placed in line 5-4 at bus 4. Input data to the computer specified a shift of 3° across the transformer. The result was a shift in real power from line 3-4 to line 5-4, which would be expected from our discussion of transformers or transmission lines in parallel. In this case the two lines being compared are not in parallel and there is a significant change (about half as great as the change in P) in reactive power in the lines into bus 4. This change in Q is consistent with Eq. (8.29) even though we are not neglecting resistance and is explained by the reduction in δ between buses 3 and 4 which increases Q over that line.

8.11 SUMMARY

In addition to discussing how load-flow studies are made on a computer this chapter has presented some methods of controlling voltage and the flow of power from the standpoint of understanding how this control is accomplished. The load-flow study on a computer is the best way to obtain quantitative answers for the effect of specific control operations.

The analysis of the effect of excitation of the synchronous generator connected to a bus having a constant voltage as discussed in Chap. 6 has been extended to a generator supplying a system represented by its Thévenin equivalent.

When we looked at the application of capacitors at a load, we saw that the reactive power furnished by the capacitors caused the voltage at the load to rise. Since increasing the excitation of the synchronous generator causes input of reactive power to the system, the effect is the same as the addition of capacitors and will cause the voltage at the generator bus to rise unless the system is very large.

Since voltage magnitude and generator real power delivered are usually specified for a load-flow study, we examined how generator excitation must be varied to meet the specified bus voltage for constant P from the generator. Finally we derived expressions for P and Q from the generator in terms of $|V_t|$, $|E_g|$, and power angle δ to show the dependence of real power on δ.

The results of paralleling two transformers when the voltage-magnitude ratios were different or when one provided a phase shift were examined. Equations (8.39) provide us with the equations for the node admittances of the equivalent circuits of such transformers. Examples were provided to show that LTC transformers controlling voltage magnitude and regulating transformers of the magnitude- and phase-shifting types could control the flow of real and reactive power on transmission lines.

PROBLEMS

8.1 Evaluate $\Delta P_4^{(0)}$ for continuing Example 8.1.

8.2 Determine the value of the element $(\partial P_3/\partial \delta_4)$ in the third column and second row of the jacobian for the first iteration in continuing Example 8.1.

8.3 Evaluate for the first iteration the element in the third column and third row of the jacobian of Example 8.1.

8.4 Evaluate for the first iteration the element in the sixth column and third row of the jacobian of Example 8.1.

8.5 Draw a diagram similar to Fig. 8.3 for bus 3 of the system of Example 8.1 from the information provided by the printout of load flow in Fig. 8.2. What is the apparent megawatt and megavar mismatch at this bus?

8.6 Reproduce Fig. 8.20 and on it indicate for Example 8.1 the values of

(a) P and Q leaving bus 5 on line 5-4.

(b) Q supplied by the fixed capacitance of the nominal π of line 5-4 at bus 5. (Note that this value of Q varies as $|V_5|^2$.)

(c) P and Q at both ends of the series part of the nominal π of the line.

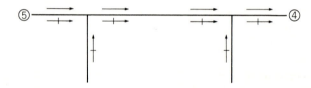

Figure 8.20 Diagram for Prob. 8.6.

(*d*) Q supplied by the fixed capacitance of the nominal π of line 5-4 at bus 4.

(*e*) P and Q into bus 4 on line 5-4.

8.7 As part of the load-flow solution of Example 8.1 the computer listed a total line loss of 9.67 MW. How does this compare with the sum of the losses which can be found from the load-flow listings of each individual line?

8.8 The effect of field excitation discussed in Sec. 6.4 can now be calculated. Consider a generator having a synchronous reactance of 1.0 per unit and connected to a large system. Resistance may be neglected. If the bus voltage is $1.0\underline{/0°}$ per unit and the generator is supplying to the bus a current of 0.8 per unit at 0.8 power-factor lagging, find the magnitude and angle of the no-load voltage E_g of the generator and P and Q delivered to the bus. Then find the angle δ between E_g and the bus voltage, the current I_a, and Q delivered to the bus by the generator if the power output of the generator remains constant but the excitation of the generator is (*a*) decreased so that $|E_g|$ is 15% smaller and (*b*) increased so that $|E_g|$ is 15% larger. What is the percent change in Q upon reducing and increasing $|E_g|$? Do the results of this problem agree with the conclusions reached in Sec. 6.4?

8.9 A power system to which a generator is to be connected at a certain bus may be represented by the Thévenin voltage $E_{th} = 0.9\underline{/0°}$ per unit in series with $Z_{th} = 0.25\underline{/90°}$ per unit. When connected to the system, E_g of the generator is $1.4\underline{/30°}$ per unit. Synchronous reactance of the generator on the system base is 1.0 per unit. (*a*) Find the bus voltage V_t and P and Q transferred to the system at the bus; (*b*) if the bus voltage is to be raised to $|V_t| = 1.0$ per unit for the same P transferred to the system, find the value of E_g required and the value of Q transferred to the system at the bus. Assume all other system emfs are unchanged in magnitude and angle; that is, E_{th} and Z_{th} are constant.

8.10 In Prob. 7.10 voltages at the three buses were calculated before and after connecting a capacitor from neutral to bus 3. Determine P and Q entering or leaving bus 3 over transmission lines, through the reactance connected between the bus and neutral, and from the capacitor before and after the capacitor is connected. Assume generated voltages remain constant in magnitude and angle. Draw diagrams similar to those of Fig. 8.9 to show the values calculated.

8.11 Figure 8.9 shows that 15.3 Mvar must be supplied by a capacitor bank at bus 4 of the 60-Hz system of Example 8.1 to raise the bus voltage to 0.950 per unit. If the base voltage is 138 kV, find the capacitance in each phase if the capacitors are (*a*) connected in Y and (*b*) connected in Δ.

8.12 Two buses *a* and *b* are connected to each other through impedances $X_1 = 0.1$ and $X_2 = 0.2$ per unit in parallel. Bus *b* is a load bus supplying a current $I = 1.0\underline{/-30°}$ per unit. The per-unit bus voltage V_b is $1.0\underline{/0°}$. Find P and Q into bus *b* through each of the parallel branches (*a*) in the circuit described, (*b*) if a regulating transformer is connected at bus *b* in the line of higher reactance to give a boost of 3% in voltage magnitude toward the load ($a = 1.03$), and (*c*) if the regulating transformer advances the phase 2° ($a = \varepsilon^{j\pi/90}$). Use the circulating-current method for parts (*b*) and (*c*), and assume that V_a is adjusted for each part of the problem so that V_b remains constant. Figure 8.21 is the one-line diagram showing buses *a* and *b* of the system with the regulating transformer in place. Neglect the impedance of the transformer.

8.13 Two reactances $X_1 = 0.08$ and $X_2 = 0.12$ per unit are in parallel between two buses *a* and *b* in a power system. If $V_a = 1.05\underline{/10°}$ and $V_b = 1.0\underline{/0°}$ per unit, what should be the turns ratio of the regulating transformer to be inserted in series with X_2 at bus *b* so that no vars flow into bus *b* from the branch whose reactance is X_1? Use the circulating-current method, and neglect the reactance of the regulating transformer. P and Q of the load and V_b remain constant.

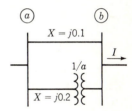

Figure 8.21 Circuit for Prob. 8.12.

8.14 Two transformers each rated 115Y/13.2Δ kV operate in parallel to supply a load of 35 MVA, 13.2 kV at 0.8 power-factor lag. Transformer 1 is rated 20 MVA with $X = 0.09$ per unit, and transformer 2 is rated 15 MVA with $X = 0.07$ per unit. Find the magnitude of the current in per unit through each transformer, the megavoltampere output of each transformer, and the megavoltamperes to which the total load must be limited so that neither transformer is overloaded. If the taps on transformer 1 are set at 111 kV to give a 3.6% boost in voltage toward the low-tension side of that transformer compared to transformer 2, which remains on the 115 kV tap, find the megavoltampere output of each transformer for the original 35 MVA total load and the maximum megavoltamperes of the total load which will not overload the transformers. Use a base of 35 MVA, 13.2 kV on the low-tension side. The circulating-current method is satisfactory for this problem.

8.15 If the impedance of the load on bus b of the circuit described in Prob. 8.12 is $0.866 + j0.5$ per unit, and if V_a is $1.04\underline{/0°}$ per unit (the voltage V_b and the load current no longer specified), find V_b for the conditions described in parts (a), (b), and (c) of Prob. 8.12. Also find P and Q into bus b through each of the parallel branches in all three cases. Equations (8.39) should be used in this problem, and the load impedance can be included in Y_{22} of the node admittance equations of the complete circuit.

NINE

ECONOMIC OPERATION OF POWER SYSTEMS

An engineer is always concerned with the cost of products and services. For a power system to return a profit on the capital invested, proper operation is very important. Rates fixed by regulatory bodies and the importance of conservation of fuel place extreme pressure on power companies to achieve maximum efficiency of operation and to improve efficiency continually in order to maintain a reasonable relation between cost of a kilowatthour to a consumer and the cost to the company of delivering a kilowatthour in the face of constantly rising prices for fuel, labor, supplies, and maintenance.

Engineers have been very successful in increasing the efficiency of boilers, turbines, and generators so continuously that each new unit added to the generating plants of a system operates more efficiently than any older unit on the system. In operating the system for any load condition the contribution from each plant and from each unit within a plant must be determined so that the cost of delivered power is a minimum. How the engineer has met and solved this challenging problem is the subject of this chapter.

An early method of attempting to minimize the cost of delivered power called for supplying power from only the most efficient plant at light loads. As load increased, power would be supplied by the most efficient plant until the point of maximum efficiency of that plant was reached. Then for further increase in load the next most efficient plant would start to feed power to the system, and a third plant would not be called upon until the point of maximum efficiency of the second plant was reached. Even with transmission losses neglected this method fails to minimize cost.

We shall study first the most economic distribution of the output of a plant between the generators, or units, within the plant. Since system generation is often expanded by adding units to existing plants, the various units within a plant usually have different characteristics. The method that will be developed is also applicable to economic scheduling of plant outputs for a given loading of the system without consideration of transmission losses. We shall proceed to develop a method of expressing transmission loss as a function of the outputs of the various plants. Then we shall determine how the output of each of the plants of a system is scheduled to achieve minimum cost of power delivered to the load.

9.1 DISTRIBUTION OF LOAD BETWEEN UNITS WITHIN A PLANT

To determine the economic distribution of load between the various units consisting of a turbine, generator, and steam supply the variable operating costs of the unit must be expressed in terms of the power output. Fuel cost is the principal factor in fossil-fuel plants, and cost of nuclear fuel can also be expressed as a function of output. Most of our electric energy will continue to come from fossil and nuclear fuels for many years until other energy sources are able to assume some of the task. We shall base our discussion on the economics of fuel cost with the realization that other costs which are a function of power output can be included in the expression for fuel cost. A typical input-output curve which is a plot of fuel input for a fossil-fuel plant in Btu per hour versus power output of the unit in megawatts is shown in Fig. 9.1. The ordinates of the curve are converted to dollars per hour by multiplying the fuel input by the cost of fuel in dollars per million Btu.

If a line is drawn through the origin to any point on the input-output curve, the reciprocal of the slope can be expressed in megawatts divided by input in millions of Btu per hour, or the ratio of energy output in megawatthours to fuel input measured in millions of Btu. This ratio is the fuel efficiency. Maximum efficiency occurs at that point where the slope of the line from the origin to a point on the curve is a minimum, that is, at the point where the line is tangent to

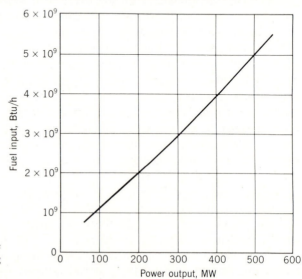

Figure 9.1 Input-output curve for a generating unit showing fuel input versus power output.

the curve. For the unit whose input-output curve is shown in Fig. 9.1, the maximum efficiency is at an output of approximately 280 MW, which requires an input of 2.8×10^9 Btu/h. The fuel requirement is 10.0×10^6 Btu/MWh. By comparison, when the output of the unit is 100 MW, the fuel requirement is 11.0×10^6 Btu/MWh.

Of course the fuel requirement for a given output is easily converted into dollars per megawatthour. As we shall see, the criterion for distribution of the load between any two units is based upon whether increasing the load on one unit as the load is decreased on the other unit by the same amount results in an increase or decrease in total cost. Thus, we are concerned with *incremental* cost, which is determined by the slopes of the input-output curves of the two units. If we express the ordinates of the input-output curve in dollars per hour and let

$$F_n = \text{input to unit } n, \text{ dollars per hour}$$

$$P_n = \text{output of unit } n, \text{ MW}$$

the incremental fuel cost of the unit in dollars per megawatthour is dF_n/dP_n.

Incremental fuel cost for a unit for any given power output is the limit of the ratio of the increase in cost of fuel input in dollars per hour to the corresponding increase in power output in megawatts as the increase in power output approaches zero. Approximately, the incremental fuel cost could be obtained by determining the increased cost of fuel for a definite time interval during which the power output is increased by a small amount. For instance, the approximate incremental cost at any particular output is the additional cost in dollars per hour to increase the output by 1 MW. Actually incremental cost is determined by measuring the slope of the input-output curve and multiplying by cost per Btu in the proper units. Since mills (tenths of a cent) per kilowatthour are equal to dollars per megawatthour, and since a kilowatt is a very small amount of power in comparison with the usual output of a unit of a steam plant, incremental fuel cost may be considered to be the cost of fuel in mills per hour to supply an additional kilowatt output.

A typical plot of incremental fuel cost versus power output is shown in Fig. 9.2. This figure is obtained by measuring the slope of the input-output curve of Fig. 9.1 for various outputs and applying a fuel cost of $1.30 per million Btu. However, the cost of fuel in terms of Btu is not very predictable, and the reader should not assume that cost figures throughout this chapter are applicable at any particular time. Figure 9.2 shows that incremental fuel cost is quite linear with respect to power output over an appreciable range. In analytical work the curve is usually approximated by one or two straight lines. The dashed line in the figure is a good representation of the curve. The equation of the line is

$$\frac{dF_n}{dP_n} = 0.0126P + 8.9$$

so that when the power output is 300 MW, the incremental cost determined by the linear approximation is $12.68 per megawatthour. This value is the approximate additional cost per hour of increasing the output by 1 MW and the saving

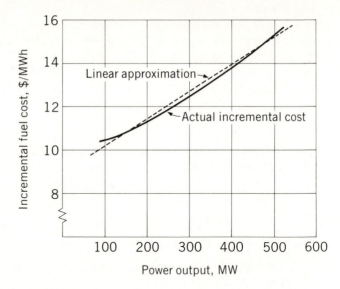

Figure 9.2 Incremental fuel cost versus power output for the unit whose input-output curve is shown in Fig. 9.1.

in cost per hour of reducing the output by 1 MW. The *actual* incremental cost at 300 MW is $12.50 per megawatthour, but this power output is near the point of maximum deviation between the actual value and the linear approximation of incremental cost. For greater accuracy two straight lines may be drawn to represent this curve in its upper and lower range.

We now have the background to understand the guiding principle for distributing the load among the units within a plant. For instance, suppose that the total output of a plant is supplied by two units and that the division of load between these units is such that the incremental fuel cost of one is higher than that of the other. Now suppose that some of the load is transferred from the unit with the higher incremental cost to the unit with the lower incremental cost. Reducing the load on the unit with the higher incremental cost will result in a greater reduction of cost than the increase in cost for adding that same amount of load to the unit with the lower incremental cost. The transfer of load from one to the other can be continued with a reduction in total fuel cost until the incremental fuel costs of the two units are equal. The same reasoning can be extended to a plant with more than two units. Thus the criterion for economical division of load between units within a plant is that all units must operate at the same incremental fuel cost. If the plant output is to be increased, the incremental cost at which each unit operates will rise but must remain the same for all.

The criterion which we have developed intuitively can be found mathematically. For a plant with K units, let

$$F_T = F_1 + F_2 + \cdots + F_K = \sum_{n=1}^{K} F_n \qquad (9.1)$$

$$P_R = P_1 + P_2 + \cdots + P_K = \sum_{n=1}^{K} P_n \qquad (9.2)$$

where F_T is the total fuel cost and P_R is the total power received by the plant bus and transferred to the power system. The fuel costs of the individual units are F_1, F_2, ..., F_K with corresponding outputs P_1, P_2, ..., P_K. Our objective is to obtain a minimum F_T for a given P_R, which requires that the total differential $dF_T = 0$. Since total fuel cost is dependent on the power output of each unit

$$dF_T = \frac{\partial F_T}{\partial P_1} dP_1 + \frac{\partial F_T}{\partial P_2} dP_2 + \cdots + \frac{\partial F_T}{\partial P_K} dP_K = 0 \tag{9.3}$$

With total fuel cost F_T dependent upon the various unit outputs, the requirement of constant P_R means that Eq. (9.2) is a constraint on the minimum value of F_T. The restriction that P_R remain constant requires that $dP_R = 0$, and so

$$dP_1 + dP_2 + \cdots + dP_K = 0 \tag{9.4}$$

Multiplying Eq. (9.4) by λ and subtracting the resulting equation from Eq. (9.3) yields, when terms are collected,

$$\left(\frac{\partial F_T}{\partial P_1} - \lambda\right) dP_1 + \left(\frac{\partial F_T}{\partial P_2} - \lambda\right) dP_2 + \cdots + \left(\frac{\partial F_T}{\partial P_K} - \lambda\right) dP_K = 0 \tag{9.5}$$

This equation is satisfied if each term is equal to zero. Each partial derivative becomes a full derivative since only the fuel cost of any one unit will vary if only the power output of that unit is varied. For example $\partial F_T / \partial P_K$ becomes dF_K / dP_K. Equation (9.5) is satisfied if

$$\frac{dF_1}{dP_1} = \lambda, \frac{dF_2}{dP_2} = \lambda, \ldots, \frac{dF_K}{dP_K} = \lambda \tag{9.6}$$

and so all units must operate at the same incremental fuel cost λ for minimum cost in dollars per hour. Thus we have proved mathematically the same criterion which we reached intuitively. The procedure is known as the method of lagrangian multipliers. We shall need this mathematical method when we consider the effect of transmission losses on the distribution of loads between several plants to achieve minimum fuel cost for a specified loading of a power system.

When the incremental fuel cost of each of the units in a plant is nearly linear with respect to power output over a range of operation under consideration, equations that represent incremental fuel costs as linear functions of power output will simplify the computations. A schedule for assigning loads to each unit in a plant can be prepared by assuming various values of λ, obtaining the corresponding outputs of each unit, and adding outputs to find plant load for each assumed λ. A curve of λ versus plant load establishes the value of λ at which each unit should operate for a given total plant load. If maximum and minimum loads are specified for each unit, some units will be unable to operate at the same incremental fuel cost as the other units and still remain within the limits specified for light and heavy loads.

Example 9.1 Incremental fuel costs in dollars per megawatthour for a plant consisting of two units are given by

$$\frac{dF_1}{dP_1} = 0.0080P_1 + 8.0 \qquad \frac{dF_2}{dF_2} = 0.0096P_2 + 6.4$$

Assume that both units are operating at all times, that total load varies from 250 to 1250 MW, and that maximum and minimum loads on each unit are to be 625 and 100 MW, respectively. Find the incremental fuel cost and the allocation of load between units for the minimum cost of various total loads.

SOLUTION At light loads unit 1 will have the higher incremental fuel cost and will operate at its lower limit of 100 MW for which dF_1/dP_1 is $8.8 per megawatthour. When the output of unit 2 is also 100 MW, dF_2/dP_2 is $7.36 per megawatthour. Therefore, as plant output increases, the additional load should come from unit 2 until dF_2/dP_2 equals $8.8 per megawatthour. Until that point is reached the incremental fuel cost λ of the plant is determined by unit 2 alone. When the plant load is 250 MW, unit 2 will supply 150 MW with dF_2/dP_2 equal to $7.84 per megawatthour. When dF_2/dP_2 equals $8.8 per megawatthour

$$0.0096P_2 + 6.4 = 8.8$$

$$P_2 = \frac{2.4}{0.0096} = 250 \text{ MW}$$

and the total plant output is 350 MW. From this point on the required output of each unit for economic load distribution is found by assuming various values of λ and calculating each unit's output and the total plant output. Results are shown in Table 9.1.

Figure 9.3 shows plant λ plotted versus plant output. We note in Table 9.1 that at $\lambda = 12.4$, unit 2 is operating at its upper limit, and that additional load must come from unit 1, which then determines the plant λ.

If we wish to know the distribution of load between the units for a plant output of 1000 MW, we could plot the output of each individual unit versus plant output as shown in Fig. 9.4 from which each unit's output can be read for any plant output. A digital computer could easily determine the correct

Table 9.1 Outputs of each unit and total output for various values of λ for Example 9.1

Plant λ, $/MWh	Unit 1 P_1, MW	Unit 2 P_2, MW	Plant $P_1 + P_2$ MW
7.84	100	150	250
8.8	100	250	350
9.6	200	333	533
10.4	300	417	717
11.2	400	500	900
12.0	500	583	1083
12.4	550	625	1175
13.0	625	625	1250

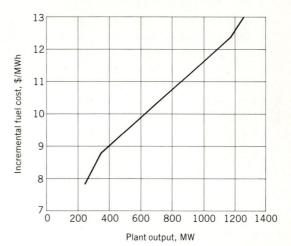

Figure 9.3 Incremental fuel cost versus plant output with total plant load economically distributed between units, as found in Example 9.1.

output of each of many units by requiring all unit incremental costs to be equal for any total plant output. For the two units of the example for a total output of 1000 MW

$$P_1 + P_2 = 1000$$

and

$$0.008P_1 + 8.0 = 0.0096(1000 - P_1) + 6.4$$

$$P_1 = 454.55 \text{ MW}$$

$$P_2 = 545.45 \text{ MW}$$

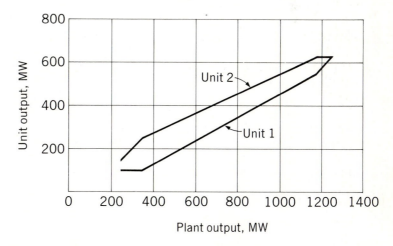

Figure 9.4 Output of each unit versus plant output for economical operation of the plant of Example 9.1.

and for each unit $\lambda = 11.636$. Such accuracy, however, is not necessary because of the uncertainty in determining costs and the use of an approximate equation in this example to express the incremental costs.

The savings effected by economic distribution of load rather than some arbitrary distribution can be found by integrating the expression for incremental fuel cost and comparing increases and decreases of cost for the units as load is shifted from the most economical allocation.

Example 9.2 Determine the saving in fuel cost in dollars per hour for the economic distribution of a total load of 900 MW between the two units of the plant described in Example 9.1 compared with equal distribution of the same total load.

SOLUTION Example 9.1 shows that unit 1 should supply 400 MW and unit 2 should supply 500 MW. If each unit supplies 450 MW, the increase in cost for unit 1 is

$$\int_{400}^{450} (0.008P_1 + 8) \, dP_1 = \left| 0.004P_1^2 + 8P_1 \right|_{400}^{450} = \$570 \text{ per hour}$$

Similarly, for unit 2

$$\int_{500}^{450} (0.0096P_2 + 6.4) \, dP_2 = \left| 0.0048P_2^2 + 6.4P \right|_{500}^{450} = -\$548 \text{ per hour}$$

The negative sign indicates a decrease in cost, as we expect for a decrease in output. The net increase in cost is $\$570 - \$548 = \$22$ per hour. The saving seems small, but this amount saved every hour for a year of continuous operation would reduce fuel cost by \$192,720 for the year.

The saving effected by economic distribution of load justifies devices for controlling the loading of each unit automatically. We shall consider automatic control of generation after investigating the coordination of transmission losses in the economic distribution of load between plants.

9.2 TRANSMISSION LOSS AS A FUNCTION OF PLANT GENERATION

In determining the economic distribution of load between plants we encounter the need to consider losses in the transmission lines. Although the incremental fuel cost at one plant bus may be lower than that of another plant for a given distribution of load between the plants, the plant with the lower incremental cost at its bus may be much farther from the load center. The losses in transmission from the plant having the lower incremental cost may be so great that economy may dictate lowering the load at the plant with the lower incremental cost and

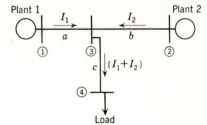

Figure 9.5 A simple system connecting two generating plants to one load.

increasing it at the plant with the higher incremental cost. To coordinate transmission loss in the problem of determining economic loading of plants, we need to express the total transmission loss of a system as a function of plant loadings.

Determining the transmission loss in a simple system consisting of two generating plants and one load will help us see the principles involved in expressing loss in terms of power output of the plants.† Figure 9.5 shows such a system. If R_a, R_b, and R_c are the resistances of lines a, b, and c, respectively, the total loss for the three-phase transmission system is

$$P_L = 3|I_1|^2 R_a + 3|I_2|^2 R_b + 3|I_1 + I_2|^2 R_c \tag{9.7}$$

If we assume that I_1 and I_2 are in phase,

$$|I_1 + I_2| = |I_1| + |I_2| \tag{9.8}$$

and upon simplification

$$P_L = 3|I_1|^2(R_a + R_c) + 3 \times 2|I_1||I_2|R_c + 3|I_2|^2(R_b + R_c) \tag{9.9}$$

If P_1 and P_2 are the three-phase power outputs of plants 1 and 2 at power factors of pf_1 and pf_2 and if V_1 and V_2 are the bus voltages at the plants,

$$|I_1| = \frac{P_1}{\sqrt{3}|V_1|pf_1} \quad \text{and} \quad |I_2| = \frac{P_2}{\sqrt{3}|V_2|pf_2} \tag{9.10}$$

Upon substitution in Eq. (9.9), we obtain

$$P_L = P_1^2 \frac{R_a + R_c}{|V_1|^2(pf_1)^2} + 2P_1 P_2 \frac{R_c}{|V_1||V_2|(pf_1)(pf_2)} + P_2^2 \frac{R_b + R_c}{|V_2|^2(pf_2)^2}$$

$$= P_1^2 B_{11} + 2P_1 P_2 B_{12} + P_2^2 B_{22} \tag{9.11}$$

---

† For a simple discussion of loss formulas in a manner similar to our approach, see D. C. Harker, A Primer on Loss Formulas, *Trans. AIEE*, vol. 77, pt. III, 1958, pp. 1434–1436.

where

$$B_{11} = \frac{R_a + R_c}{|V_1|^2 (pf_1)^2}$$

$$B_{12} = \frac{R_c}{|V_1||V_2|(pf_1)(pf_2)}$$

(9.12)

$$B_{22} = \frac{R_b + R_c}{|V_2|^2 (pf_2)^2}$$

The terms B_{11}, B_{12}, and B_{22} are called *loss coefficients* or *B coefficients*. If the voltages in Eq. (9.12) are line-to-line kilovolts with resistances in ohms, the units for the loss coefficients are reciprocal megawatts. Then, in Eq. (9.11), with three-phase powers P_1 and P_2 in megawatts, P_L will be in megawatts also. Of course, the computations may be made in per unit.

For the system for which they are derived and the assumption of I_1 and I_2 in phase, these coefficients yield the exact loss by Eq. (9.11) only for the particular values of P_1 and P_2 which result in the voltages and power factors used in Eqs. (9.12). The B coefficients are constant, as P_1 and P_2 vary, only insofar as bus voltages at the plants maintain constant magnitude and plant power factors remain constant. Fortunately, the use of constant values for the loss coefficients in Eq. (9.11) yields reasonably accurate results when the coefficients are calculated for some average operating condition and if extremely wide shifts of load between plants or in total load do not occur. In practice large systems are economically loaded by calculations based on several sets of loss coefficients depending on load conditions.†

Example 9.3 For the system whose one-line diagram is shown in Fig. 9.5, assume $I_1 = 1.0\underline{/0°}$ per unit and $I_2 = 0.8\underline{/0°}$ per unit. If the voltage at bus 3 is $V_3 = 1.0\underline{/0°}$ per unit, find the loss coefficients. Line impedances are 0.04 + j0.16 per unit, 0.03 + j0.12 per unit, and 0.02 + j0.08 per unit for sections *a*, *b*, and *c*, respectively.

SOLUTION Ordinarily load currents and bus voltages are available from load studies. For this problem bus voltages can be calculated from the data given:

$$V_1 = 1.0 + (1.0 + j0)(0.04 + j0.16) = 1.04 + j0.16 \text{ per unit}$$

$$V_2 = 1.0 + (0.8 + j0)(0.03 + j0.12) = 1.024 + j0.096 \text{ per unit}$$

Since all currents have phase angles of zero, the power factor at each source node is the cosine of the angle of the voltage at the node, and voltage

---

† For two of many sources of further information see L. K. Kirchmayer, *Economic Operation of Power Systems*, John Wiley and Sons, Inc., New York, 1958; W. S. Meyer and V. D. Albertson, Improved Loss Formula Computation by Optimally Ordered Elimination Techniques, *IEEE Trans. Power Appar. Syst.*, vol. 90, 1971, pp. 716–731. Also see the footnotes in Sec. 9.4 for a method used by a number of companies.

magnitude times power factor equals the real part of the complex expression for voltage. Therefore

$$B_{11} = \frac{0.04 + 0.02}{1.04^2} = 0.0554 \text{ per unit}$$

$$B_{12} = \frac{0.02}{1.024 \times 1.04} = 0.0188 \text{ per unit}$$

$$B_{22} = \frac{0.03 + 0.02}{1.024^2} = 0.0477 \text{ per unit}$$

Example 9.4 Calculate the transmission loss for Example 9.3 by the loss formula of Eq. (9.11), and check the result.

SOLUTION Source powers are ordinarily available from the load study for the operating condition used to calculate loss coefficients. For this problem P_1 and P_2 must be calculated.

$$P_1 = \text{Re } \{(1.0 + j0)(1.04 + j0.16)\} = 1.04 \text{ per unit}$$

$$P_2 = \text{Re } \{(0.8 + j0)(1.024 + j0.096)\} = 0.8192 \text{ per unit}$$

$$P_L = 1.04^2 \times 0.0554 + 2 \times 1.04 \times 0.8192 \times 0.0188 + 0.8192^2 \times 0.0477$$

$$= 0.06 + 0.032 + 0.032 = 0.124 \text{ per unit}$$

Adding the loss in each section computed by I^2R yields

$$P_L = 1.0^2 \times 0.04 + 1.8^2 \times 0.02 + 0.8^2 \times 0.03$$

$$= 0.04 + 0.0648 + 0.0192 = 0.124 \text{ per unit}$$

Exact agreement between methods is expected since the loss coefficients were determined for the condition for which loss was calculated. The amount of error introduced by using the same loss coefficients for two other operating conditions may be seen by examining the results shown in Table 9.2.

Table 9.2 Comparison of transmission loss calculated by loss coefficients and I^2R for data of Example 9.3 with several operating conditions
All quantities are in per unit

I_1	I_2	P_1	P_2	P_L by loss coefficients	P_L by I^2R	Conditions
1.0	0.8	1.040	0.819	0.124	0.124	Original case
0.5	0.4	0.510	0.405	0.030	0.031	P_1, P_2 reduced 50%
0.5	1.3	0.510	1.351	0.128	0.126	0.53 shift P_1 to P_2

The general form of the loss equation for any number of sources is

$$P_L = \sum_m \sum_n P_m B_{mn} P_n \tag{9.13}$$

where $\sum_m$ and $\sum_n$ indicate independent summations to include all sources. For instance, for three sources,

$$P_L = P_1^2 B_{11} + P_2^2 B_{22} + P_3^2 B_{33} + 2P_1 P_2 B_{12} + 2P_2 P_3 B_{23} + 2P_1 P_3 B_{13} \tag{9.14}$$

The matrix form of the transmission-loss equation is

$$\mathbf{P}_L = \mathbf{P}^T \mathbf{B} \mathbf{P} \tag{9.15}$$

where for a total of s sources

$$\mathbf{P} = \begin{bmatrix} P_1 \\ P_2 \\ \vdots \\ P_s \end{bmatrix} \quad \text{and} \quad \mathbf{B} = \begin{bmatrix} B_{11} & B_{12} & B_{13} & \cdots & B_{1s} \\ B_{21} & B_{22} & B_{23} & \cdots & B_{2s} \\ \vdots & \vdots & \vdots & & \vdots \\ B_{s1} & B_{s2} & B_{s3} & \cdots & B_{ss} \end{bmatrix}$$

9.3 DISTRIBUTION OF LOAD BETWEEN PLANTS

The method developed to express transmission loss in terms of plant outputs enables us to coordinate transmission loss in scheduling the output of each plant for maximum economy for a given system load. The mathematical treatment is similar to that of scheduling units within a plant except that we shall now include transmission loss as an additional constraint.

In the equation

$$F_T = F_1 + F_2 + \cdots + F_K = \sum_{n=1}^{K} F_n \tag{9.16}$$

F_T is now the total cost of all the fuel for the entire system and is the sum of the fuel costs of the individual plants $F_1, F_2, \ldots, F_K$. The total input to the network from all the plants is

$$P_T = P_1 + P_2 + \cdots + P_K = \sum_{n=1}^{K} P_n \tag{9.17}$$

where $P_1, P_2, \ldots, P_K$ are the individual plant inputs to the network. The total fuel cost of the system is a function of the power inputs. The constraining relation on the minimum value of F_T is

$$\sum_{n=1}^{K} P_n - P_L - P_R = 0 \tag{9.18}$$

where P_R is the total power received by the loads on the system and P_L is the transmission loss expressed as a function of the loss coefficients and the power input to the network from each plant. Since P_R is constant, $dP_R = 0$; therefore

$$\sum_{n=1}^{K} dP_n - dP_L = 0 \tag{9.19}$$

and since minimum cost means $dF_T = 0$,

$$dF_T = \sum_{n=1}^{K} \frac{\partial F_T}{\partial P_n} dP_n = 0 \tag{9.20}$$

Transmission loss P_L is dependent upon plant outputs, and dP_L is expressed by

$$dP_L = \sum_{n=1}^{K} \frac{\partial P_L}{\partial P_n} dP_n \tag{9.21}$$

Substituting dP_L from Eq. (9.21) in Eq. (9.19), multiplying by λ, and subtracting the result from Eq. (9.20) yield

$$\sum_{n=1}^{K} \left(\frac{\partial F_T}{\partial P_n} + \lambda \frac{\partial P_L}{\partial P_n} - \lambda \right) dP_n = 0 \tag{9.22}$$

This equation is satisfied provided that

$$\frac{\partial F_T}{\partial P_n} + \lambda \frac{\partial P_L}{\partial P_n} - \lambda = 0 \tag{9.23}$$

for every value of n. Rearranging Eq. (9.23) and recognizing that changing the output of only one plant can affect the cost at only that plant, we have

$$\frac{dF_n}{dP_n} \frac{1}{1 - \partial P_L/\partial P_n} = \lambda \tag{9.24}$$

or

$$\frac{dF_n}{dP_n} L_n = \lambda \tag{9.25}$$

where L_n is called the penalty factor of plant n and

$$L_n = \frac{1}{1 - \partial P_L/\partial P_n} \tag{9.26}$$

The multiplier λ is in dollars per megawatthour when fuel cost is in dollars per hour and power is in megawatts. The result is analogous to that for scheduling units within a plant. Minimum fuel cost is obtained when the incremental fuel cost of each plant multiplied by its penalty factor is the same for all plants in the system. The products are equal to λ, which is called the system λ and is approximately the cost in dollars per hour to increase the total delivered load by 1 MW. For a system of three plants, for instance,

$$\frac{dF_1}{dP_1} L_1 = \frac{dF_2}{dP_2} L_2 = \frac{dF_3}{dP_3} L_3 = \lambda \tag{9.27}$$

The transmission loss P_L is expressed by Eq. (9.13). For K plants partial differentiation with respect to P_n yields

$$\frac{\partial P_L}{\partial P_n} = \frac{\partial}{\partial P_n} \sum_{m=1}^{K} \sum_{n=1}^{K} P_m B_{mn} P_n = 2 \sum_{m=1}^{K} P_m B_{mn} \tag{9.28}$$

The simultaneous equations obtained by writing Eq. (9.24) for each plant of the system can be solved by assuming a value for λ. Economical loading for each plant is found for the assumed λ. By solving the equations for several values of λ, data are found for plotting generation at each plant against total generation. If transmission loss is calculated for each λ, plant outputs can be plotted against total received load. If fixed amounts of power are transferred over tie lines with other systems or received from hydro stations, the distribution of the remaining load among the other plants is affected by the changes in transmission loss brought about by the flow through these additional points of entry to the system. No new variables are introduced, but additional loss coefficients are required. For instance, a system having five steam plants, three hydro plants, and seven interconnections would require a 15×15 matrix of loss coefficients, but the only unknowns to be found for any given λ would be the five inputs to the system from the steam plants.

Example 9.5 A system consists of two plants connected by a transmission line. The only load is located at plant 2. When 200 MW is transmitted from plant 1 to plant 2 power loss in the line is 16 MW. Find the required generation for each plant and the power received by the load when λ for the system is \$12.50 per megawatthour. Assume that the incremental fuel costs can be approximated by the following equations:

$$\frac{dF_1}{dP_1} = 0.010P_1 + 8.5 \qquad \text{\$/MWh}$$

$$\frac{dF_2}{dP_2} = 0.015P_2 + 9.5 \qquad \text{\$/MWh}$$

SOLUTION For a two-plant system

$$P_L = P_1^2 B_{11} + 2P_1 P_2 B_{12} + P_2^2 B_{22}$$

Since all the load is at plant 2, varying P_2 cannot affect P_L. Therefore

$$B_{22} = 0 \qquad \text{and} \qquad B_{12} = 0$$

When $P_1 = 200$ MW, $P_L = 16$ MW; so

$$16 = 200^2 B_{11}$$

$$B_{11} = 0.0004 \text{ MW}^{-1}$$

and

$$\frac{\partial P_L}{\partial P_1} = 2P_1 B_{11} + 2P_2 B_{12} = 0.0008P_1$$

$$\frac{\partial P_2}{\partial P_1} = 2P_2 B_{22} + 2P_1 B_{12} = 0$$

Penalty factors are

$$L_1 = \frac{1}{1 - 0.0008P_1} \qquad \text{and} \qquad L_2 = 1.0$$

For $\lambda = 12.5$

$$\frac{0.010P_1 + 8.5}{1 - 0.0008P_1} = 12.5$$

$$P_1 = 200 \text{ MW}$$

$$0.015P_2 + 9.5 = 12.5$$

$$P_2 = 200 \text{ MW}$$

Economic load dispatching therefore requires equal division of load between the two plants for $\lambda = 12.5$. The power loss in transmission is

$$P_L = 0.0004 \times 200^2 = 16 \text{ MW}$$

and the delivered load is

$$P_R = P_1 + P_2 - P_L = 384 \text{ MW}$$

Example 9.6 For the system of Example 9.5 with 384 MW received by the load, find the savings in dollars per hour obtained by coordinating rather than neglecting the transmission loss in determining the load of the plants.

SOLUTION If transmission loss is neglected, the incremental fuel costs at the two plants are equated to give

$$0.010P_1 + 8.5 = 0.015P_2 + 9.5$$

The power delivered to the load is

$$P_1 + P_2 - 0.0004P_1^2 = 384$$

Solving these two equations for P_1 and P_2 gives the following values for plant generation with losses not coordinated:

$$P_1 = 290.7 \text{ MW} \qquad \text{and} \qquad P_2 = 127.1 \text{ MW}$$

The load on plant 1 is increased from 200 to 290.7 MW. The increase in fuel cost is

$$\int_{200}^{290.7} (0.010P_1 + 8.5)\, dP_1 = \left| \frac{0.010}{2} P_1^2 + 8.5P_1 \right|_{200}^{290.7}$$

$$= 222.53 + 770.95 = 993.48$$

The load on plant 2 is decreased from 200 to 127.1 MW. The decrease (negative increase) in cost for plant 2 is

$$-\int_{200}^{127.1} (0.015P_2 + 9.5)\, dP_2 = \left| \frac{0.015}{2} P_2^2 + 9.5P_2 \right|_{200}^{127.1}$$

$$= 178.84 + 692.55 = 871.39$$

The net savings by accounting for transmission loss in scheduling the received load of 384 MW is

$$993.48 - 871.39 = \$122.09 \text{ per hour}$$

9.4 A METHOD OF COMPUTING PENALTY FACTORS AND LOSS COEFFICIENTS

Expressing loss in a transmission system as a function of plant outputs in terms of B coefficients is the most widely used method of calculating $\partial P_L / \partial P_n$ for cost minimization. The simplicity of the loss equation in terms of B coefficients is the principal advantage of this method which has resulted in great savings in system operating costs. The rapid development of digital computers has made other methods attractive.

One method of evaluating $\partial P_L / \partial P_n$ from load-flow studies and used by a number of power companies as part of a program to determine B coefficients will be discussed briefly.† The method depends on the fact that

$$\frac{\partial P_L}{\partial P_n} = \sum_{j=1}^{N} \frac{\partial P_L}{\partial \theta_j} \frac{\partial \theta_j}{\partial P_n} \tag{9.29}$$

where θ_j is the phase angle of the voltage at node j in a system of N buses. If bus voltages are assumed constant, it can be shown that in terms of voltage phase angles,

$$\frac{\partial P_L}{\partial \theta_j} = 2 \sum_{k=1}^{N} |V_j| \cdot |V_k| G_{jk} \sin (\theta_k - \theta_j) \tag{9.30}$$

where G_{jk} is the real part of Y_{jk} of the bus admittance matrix. The difficulty with Eq. (9.29) is in expressing $\partial \theta_j / \partial P_n$; direct differentiation is impossible because voltage phase angles cannot be expressed in terms of plant generated powers.

The terms $\partial \theta_j / \partial P_n$ remain quite constant for changes in load levels and system generation schedules. Since the terms express a change in the voltage phase angles θ_j for a change in plant generation P_n with all other plant generation remaining constant, these terms can be approximated very closely through load-flow studies. For some typical load pattern the total received load is increased by increasing each individual load by the same small amount, say 5%. The change in total received power plus losses is provided by one plant n while other plant outputs are held constant. The changes in each voltage phase angle θ_j are determined, and the ratios of change in phase angle to change in plant input to the system $\Delta\theta_j / \Delta P_n$ are found for all values of j for plant n. The load-flow program of the digital computer is used and the process is repeated for

---

† E. F. Hill and W. D. Stevenson, Jr., An Improved Method of Determining Incremental Loss Factors from Power System Admittances and Voltages, *IEEE Trans. Power Appar. Syst.,* vol. PAS-87, June 1968, pp. 1419–1425.

every plant supplying the load change in turn. A set of A_{jn} coefficients is found, where

$$A_{jn} = \frac{\Delta \theta_j}{\Delta P_n} \tag{9.31}$$

Then incremental loss for plant n is given by

$$\frac{\partial P_L}{\partial P_n} = \sum_{j=1}^{N} \frac{\partial P_L}{\partial \theta_j} A_{jn} \tag{9.32}$$

The A_{jn} values are essentially constant regardless of the various combinations of generation scheduling and load levels. Thus once a matrix of A_{jn} coefficients has been determined, an on-line computer monitoring load flow can calculate plant penalty factors continually by solving Eqs. (9.30) and (9.32).

The usual practice, however, is to use the A_{jn} coefficients to calculate B coefficients by the equation.†

$$B_{mn} = \frac{1}{2} \sum_{i=1}^{N} \sum_{j=1}^{N} \frac{\partial^2 P_L}{\partial \theta_i \, \partial \theta_j} A_{im} A_{jn} \tag{9.33}$$

In determining the second partial derivative of P_L with respect to θ_i and θ_j Eq. (9.30) is used with the values of θ found in the load-flow solution of the typical load pattern used in determining the A_{jn} coefficients. The on-line computer will then calculate incremental loss $\partial P_L / \partial P_n$ from the B coefficients and control the system for economic load distribution, as described in the following section.

9.5 AUTOMATIC GENERATION CONTROL

Computer control of the output of each plant and of each unit within a plant is common practice in power-system operation. By continually monitoring all plant outputs and the power flowing in interconnections interchange of power with other systems is controlled. Most control systems are digital or a combination of digital and analog. In this section we shall consider one of a variety of ways in which computer control is accomplished.

In discussing control the term *area* means that part of an interconnected system in which one or more companies control their generation to absorb all their own load changes and maintain a prearranged net interchange of power with other areas for specified periods. Monitoring the flow of power on the tie lines between areas determines whether a particular area is absorbing satisfactorily all the load changes within its own boundaries. The function of the computer is to require the area to absorb its own load changes, to provide the agreed

---

† See E. F. Hill and W. D. Stevenson, Jr., A New Method of Determining Loss Coefficients, *IEEE Trans. Power Appar. Syst.*, vol. PAS-87, July 1968, pp. 1548–1552.

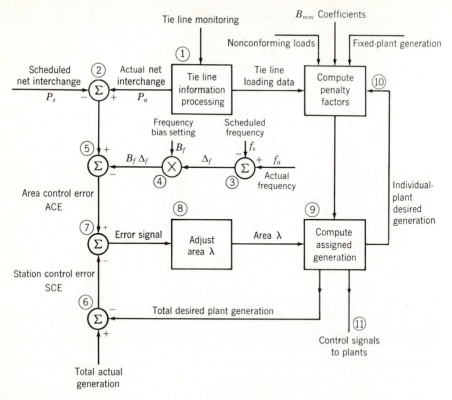

Figure 9.6 Block diagram to illustrate the operation of a computer controlling a particular area.

net interchange with neighboring areas, to determine the desired generation of each plant in the area for economic dispatch, and to cause the area to do its share to maintain the desired frequency of the interconnected system.

The block diagram of Fig. 9.6 indicates the flow of information in a computer controlling a particular area. The numbers enclosed by circles adjacent to the diagram identify positions on the diagram to simplify our discussion of the control operation. The larger circles on the diagram enclosing the symbols $\times$ or Σ indicate points of multiplication or algebraic summation of incoming signals.

At position 1 processing of information about power flow on tie lines to other control areas is indicated. The *actual net interchange* P_a is the algebraic sum of the power on the tie lines and is positive when net power is out of the area. The prearranged net interchange is called the *scheduled net interchange P_s*. At position 2 the scheduled net interchange is subtracted from the actual net interchange.† We shall discuss the condition where both actual and scheduled net interchange are out of the system and therefore positive.

---

† Subtraction of standard or reference value from actual value to obtain the error is the accepted convention of power-system engineers and is the negative of the definition of control error found in the literature of control theory.

Position 3 on the diagram indicates the subtraction of the scheduled frequency f_s (for instance 60 Hz) from the actual frequency f_a to obtain Δf, the frequency deviation. Position 4 on the diagram indicates that the *frequency bias setting* B_f, a factor with *negative sign*, is multiplied by Δf to obtain a value of megawatts called the *frequency bias* $B_f \Delta f$.

The frequency bias which is positive when the actual frequency is less than the scheduled frequency is subtracted from $P_a - P_s$ at position 5 to obtain the area control error (ACE), which may be positive or negative. As an equation

$$ACE = P_a - P_s - B_f(f_a - f_s) \tag{9.34}$$

A negative ACE means that the area is not generating enough power to send the desired amount out of the area. There is a deficiency in net power output. Without frequency bias the indicated deficiency would be less because there would be no positive offset $B\Delta f$ added to P_s (subtracted from P_a) when actual frequency is less than scheduled frequency and the ACE would be less. The area would produce sufficient generation to supply its own load and the prearranged interchange but would not provide the additional output to assist neighboring interconnected areas to raise the frequency.

Station control error (SCE) is the amount of actual generation of all the area plants minus the desired generation as indicated at position 6 of the diagram. This SCE is negative when desired generation is greater than existing generation.

The key to the whole control operation is the comparison of ACE and SCE. Their difference is an error signal, as indicated at position 7 of the diagram. If ACE and SCE are negative and equal, the deficiency in the output from the area equals the excess of the desired generation over the actual generation and no error signal is produced. However this excess of desired generation will cause a signal indicated at position 11 to the plants to increase their generation to reduce the magnitude of the SCE, and the resulting increase in output from the area will reduce the magnitude of the ACE at the same time.

If ACE is more negative than SCE, there will be an error signal to increase the λ of the area and this increase will in turn cause the desired plant generation to increase (position 9). Each plant will receive a signal to increase its output as determined by the principles of economic dispatch.

This discussion has considered specifically only the case of scheduled net interchange out of the area (positive scheduled net interchange) greater than actual net interchange with ACE equal to or more negative than SCE. The reader should be able to extend the discussion to the other possibilities by referring to Fig. 9.6.

Position 10 on the diagram indicates the computation of penalty factors for each plant. Here the B coefficients are stored to calculate $\partial P_1 / \partial P_n$. Non-conforming loads, plants where generation is not to be allowed to vary, and tie-line loading enter the calculation of penalty factors. The penalty factors are transmitted to the section (position 9) which calculates the individual plant generation to provide with economic dispatch the total desired plant generation.

One other point of importance (not indicated on Fig. 9.6) is the offset in scheduled net interchange of power that varies in proportion to the integral in cycles between actual and rated (60 Hz) frequency. The offset is in the direction to help in reducing the integrated difference to zero and thereby keep electric clocks accurate.

PROBLEMS

9.1 For a certain generating unit in a plant the fuel input in millions of Btu per hour expressed as a function of power output P in megawatts is

$$0.0001P^3 + 0.015P^2 + 3.0P + 90$$

(a) Determine the equation for incremental fuel cost in dollars per megawatthour as a function of power output in megawatts based on a fuel cost of $1.40 per million Btu.

(b) Find the equation for a good linear approximation of incremental fuel cost as a function of power output over a range of 20 to 120 MW.

(c) What is the average cost of fuel per megawatthour when the plant is delivering 100 MW?

(d) What is the approximate additional fuel cost per hour to raise the output of the plant from 100 to 101 MW?

9.2 The incremental fuel costs for two units of a plant are

$$\frac{dF_1}{dP_1} = 0.010P_1 + 11.0 \quad \text{and} \quad \frac{dF_2}{dP_2} = 0.012P_2 + 8.0$$

where F is in dollars per hour and P is in megawatts. If both units operate at all times and maximum and minimum loads on each unit are 625 and 100 MW, plot λ in dollars per megawatthour versus plant output in megawatts for lowest fuel cost as total load varies from 200 to 1250 MW.

9.3 Find the savings in dollars per hour for economical allocation of load between the units of Prob. 9.2 compared with their sharing the output equally when the total output is 750 MW.

9.4 A plant has two generators supplying the plant bus, and neither is to operate below 100 MW or above 625 MW. Incremental costs with P_1 and P_2 in megawatts are

$$\frac{dF_1}{dP_1} = 0.012P_1 + 8.0 \qquad \text{\$/MWh}$$

$$\frac{dF_2}{dP_2} = 0.018P_2 + 7.0 \qquad \text{\$/MWh}$$

For economic dispatch find the plant λ when $P_1 + P_2$ equals (a) 200 MW, (b) 500 MW, and (c) 1150 MW.

9.5 Calculate the power loss in the system of Example 9.3 by the loss coefficients of the example and by $|I|^2|R|$ for $I_1 = 1.5\underline{/0°}$ per unit and $I_2 = 1.2\underline{/0°}$ per unit. Assume $V_3 = 1.0\underline{/0°}$ per unit.

9.6 Find the loss coefficients that will give the true power loss for the system of Example 9.3 for $I_1 = 0.8\underline{/0°}$ per unit, $I_2 = 0.8\underline{/0°}$ per unit, and $V_3 = 1.0\underline{/0°}$ per unit.

9.7 A power system has only two generating plants, and power is being dispatched economically with $P_1 = 140$ MW and $P_2 = 250$ MW. The loss coefficients are:

$$B_{11} = 0.10 \times 10^{-2} \text{ MW}^{-1}$$

$$B_{12} = -0.01 \times 10^{-2} \text{ MW}^{-1}$$

$$B_{22} = 0.13 \times 10^{-2} \text{ MW}^{-1}$$

To raise the *total load* on the system by 1 MW will cost an additional $12 per hour. Find (*a*) the penalty factor for plant 1, and (*b*) the additional cost per hour to increase the output of *this plant* by 1 MW.

9.8 On a system consisting of two generating plants the incremental costs in dollars per megawatt-hour with P_1 and P_2 in megawatts are

$$\frac{dF_1}{dP_1} = 0.008P_1 + 8.0 \qquad \frac{dF_2}{dP_2} = 0.012P_2 + 9.0$$

The system is operating on economic dispatch with $P_1 = P_2 = 500$ MW and $\partial P_L/\partial P_2 = 0.2$. Find the penalty factor of plant 1.

9.9 A power system is operating on economic load dispatch with a system λ of $12.5 per megawatt-hour. If raising the output of plant 2 by 100 kW (while other outputs are kept constant) results in increased $|I|^2|R|$ losses of 12 kW for the system, what is the approximate additional cost per hour if the output of this plant is increased by 1 MW?

9.10 A power system is supplied by only two plants, both of which are operating on economic dispatch. At the bus of plant 1 the incremental cost is $11.0 per megawatthour, and at plant 2 is $10.0 per megawatthour. Which plant has the higher penalty factor? What is the penalty factor of plant 1 if the cost per hour of increasing the load on the system by 1 MW is $12.5?

9.11 Calculate the values listed below for the system of Example 9.5 with system $\lambda = \$13.5$ per megawatthour. Assume fuel costs at no load of $200 and $400 per hour for plants 1 and 2, respectively.

(*a*) P_1, P_2, and power delivered to the load for economic dispatch with transmission loss coordinated.

(*b*) P_1 and P_2 for the value of power delivered to the load found in part (*a*) but with transmission loss not coordinated. Transmission loss must be included, however, in determining the total power input to the system.

(*c*) Total fuel cost in dollars per hour for parts (*a*) and (*b*).

TEN

SYMMETRICAL THREE-PHASE FAULTS

When a fault occurs in a power network, the current flowing is determined by the internal emfs of the machines in the network, by their impedances, and by the impedances in the network between the machines and the fault. The current flowing in a synchronous machine immediately after the occurrence of a fault, that flowing a few cycles later, and the sustained, or steady-state, value of the fault current differ considerably because of the effect of the armature current on the flux that generates the voltage in the machine. The current changes relatively slowly from its initial value to its steady-state value. This chapter discusses the calculation of fault current at different periods and explains the changes in reactance and internal voltage of a synchronous machine as the current changes from its initial value upon the occurrence of a fault to its steady-state value. Description of a computer program for calculating fault currents will be postponed until we have discussed unsymmetrical faults since programs are not confined to three-phase faults.†

10.1 TRANSIENTS IN *RL* SERIES CIRCUITS

The selection of a circuit breaker for a power system depends not only upon the current the breaker is to carry under normal operating conditions but also upon the maximum current it may have to carry momentarily and the current it may have to interrupt at the voltage of the line in which it is placed.

---

† For a book devoted entirely to fault studies see P. M. Anderson, *Analysis of Faulted Power Systems*, Iowa State University Press, Ames, Iowa, 1973.

In order to approach the problem of calculating the initial current when a synchronous generator is short-circuited, consider what happens when an ac voltage is applied to a circuit containing constant values of resistance and inductance. Let the applied voltage be $V_{max} \sin (\omega t + \alpha)$, where t is zero at the time of applying the voltage. Then α determines the magnitude of the voltage when the circuit is closed. If the instantaneous voltage is zero and increasing in a positive direction when it is applied by closing a switch, α is zero. If the voltage is at its positive maximum instantaneous value, α is $\pi/2$. The differential equation is

$$V_{max} \sin (\omega t + \alpha) = Ri + L \frac{di}{dt} \tag{10.1}$$

The solution of this equation is

$$i = \frac{V_{max}}{|Z|} [\sin (\omega t + \alpha - \theta) - \varepsilon^{-Rt/L} \sin (\alpha - \theta)] \tag{10.2}$$

where $|Z|$ is $\sqrt{R^2 + (\omega L)^2}$ and θ is $\tan^{-1} (\omega L/R)$.

The first term of Eq. (10.2) varies sinusoidally with time. The second term is nonperiodic and decays exponentially with a time constant of L/R. This nonperiodic term is called the dc component of the current. We recognize the sinusoidal term as the steady-state value of the current in an RL circuit for the given applied voltage. If the value of the steady-state term is not zero when $t = 0$, the dc component appears in the solution in order to satisfy the physical condition of zero current at the instant of closing the switch. Note that the dc term does not exist if the circuit is closed at a point on the voltage wave such that $\alpha - \theta = 0$ or $\alpha - \theta = \pi$. Figure 10.1 shows the variation of current with time according to Eq. (10.2) when $\alpha - \theta = 0$. If the switch is closed at a point on the voltage wave such that $\alpha - \theta = \pm \pi/2$, the dc component has its maximum initial value, which is equal to the maximum value of the sinusoidal component. Figure 10.2 shows current versus time when $\alpha - \theta = -\pi/2$. The dc component may have any value from 0 to $V_{max}/|Z|$, depending on the instantaneous value of the voltage when the circuit is closed and on the power factor of the circuit. At the instant of applying the voltage, the dc and steady-state components always have the same magnitude but are opposite in sign in order to express the zero value of current then existing.

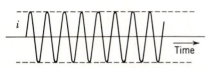

Figure 10.1 Current as a function of time in an RL circuit for $\alpha - \theta = 0$, where $\theta = \tan^{-1} (\omega L/R)$. The voltage is $V_{max} \sin (\omega t + \alpha)$ applied at $t = 0$.

Figure 10.2 Current as a function of time in an RL circuit for $\alpha - \theta = -\pi/2$, where $\theta = \tan^{-1} (\omega L/R)$. The voltage is $V_{max} \sin (\omega t + \alpha)$ applied at $t = 0$.

In Secs. 6.2 and 6.3 we discussed the principles of operation of a synchronous generator consisting of a rotating magnetic field which generates a voltage in an armature winding having resistance and reactance. The current flowing when a generator is short-circuited is similar to that flowing when an alternating voltage is suddenly applied to a resistance and an inductance in series. There are important differences, however, because the current in the armature affects the rotating field.

A good way to analyze the effect of a three-phase short circuit at the terminals of a previously unloaded alternator is to take an oscillogram of the current in one of the phases upon the occurrence of such a fault. Since the voltages generated in the phases of a three-phase machine are displaced 120 electrical degrees from each other, the short circuit occurs at different points on the voltage wave of each phase. For this reason the unidirectional or dc transient component of current is different in each phase. If the dc component of current is eliminated from the current of each phase, the resulting plot of each phase current versus time is that shown in Fig. 10.3. Comparison of Figs. 10.1 and 10.3 shows the difference between applying a voltage to the ordinary RL circuit and applying a short circuit to a synchronous machine. There is no dc component in either of these figures. In a synchronous machine the flux across the air gap of the machine is much larger at the instant the short circuit occurs than it is a few cycles later. The reduction of flux is caused by the mmf of the current in the armature. Our discussion in Sec. 6.2 concerned the effect of armature current, which is called *armature reaction*. The equivalent circuit developed in Sec. 6.3

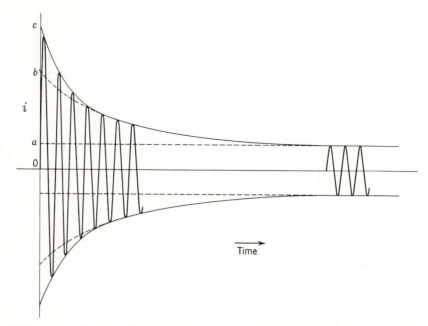

Figure 10.3 Current as a function of time for a synchronous generator short-circuited while running at no load. The unidirectional transient component of current has been eliminated in redrawing the oscillogram.

accounts for the reduction of flux due to armature reaction and applies to the steady-state condition after the dc transient has disappeared and after the amplitude of the wave shown in Fig. 10.3 has become constant. When a short circuit occurs at the terminals of a synchronous machine, time is required for the reduction in flux across the air gap. As the flux diminishes, the armature current decreases because the voltage generated by the air-gap flux determines the current which will flow through the resistance and leakage reactance of the armature winding.

10.2 SHORT-CIRCUIT CURRENTS AND THE REACTANCES OF SYNCHRONOUS MACHINES†

Certain terms that are valuable in the calculation of short-circuit current in a power system can be defined from Fig. 10.3. The reactances that we shall discuss are *direct-axis* reactances which we mentioned in Sec. 6.3. We recall that direct-axis reactance is used for computing voltage drops caused by that component of the armature current which is in quadrature (90° out of phase) with the voltage generated at no load. Since the resistance in a faulted circuit is small compared with the inductive reactance, current during a fault is always lagging by a large angle, and the so-called direct-axis reactance is required. In the discussion to follow, it should be remembered that the current shown in the oscillogram of Fig. 10.3 is that which flows in an alternator operating at no load before the fault occurs.

In Fig. 10.3 the distance *oa* is the maximum value of the sustained short-circuit current. This value of current times 0.707 is the rms value $|I|$ of the sustained, or steady-state, short-circuit current. The no-load voltage of the alternator $|E_g|$ divided by the steady-state current $|I|$ is called the *synchronous reactance* of the generator or the *direct-axis synchronous reactance* X_d since the power factor is low during the short circuit. The comparatively small resistance of the armature is neglected.

If the envelope of the current wave is extended back to zero time and the first few cycles where the decrement appears to be very rapid are neglected, the intercept is the distance *ob*. The rms value of the current represented by this intercept, or 0.707 times *ob* in amperes, is known as the *transient current* $|I'|$. A new machine reactance may now be defined. It is called the *transient reactance*, or in this particular case the *direct-axis transient reactance* X'_d and is equal to $|E_g|/|I'|$ for an alternator operating at no load before the fault. If the rapid decrement of the first few cycles is neglected, the point of intersection that the current envelope makes with the zero axis can be determined more accurately by plotting on semilogarithmic paper the *excess* of the current envelope over the

† For a more complete discussion see C. F. Wagner, "Machine Characteristics," in Central Station Engineers of the Westinghouse Electric Corporation, *Electrical Transmission and Distribution Reference Book*, 4th ed., Chap. 6, pp. 145–194, East Pittsburgh, Pa., 1964; and A. E. Fitzgerald, C. Kingsley, and A. Kusko, *Electric Machinery*, 3d ed., pp. 312–319, 479–492, McGraw-Hill Book Company, New York, 1971.

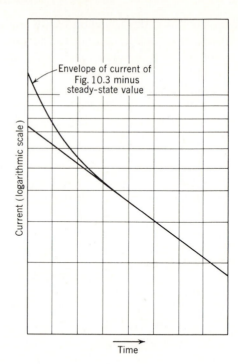

Figure 10.4 Excess of the current envelope of Fig. 10.3 over the sustained maximum current, plotted on semilogarithmic scales.

sustained value represented by *oa*, as shown in Fig. 10.4. The straight-line portion of this curve is extended to the zero-time axis, and the intercept is added to the maximum instantaneous value of the sustained current to obtain the maximum instantaneous value of transient current corresponding to *ob* in Fig. 10.3.

The rms value of the current determined by the intercept of the current envelope with zero time is called the *subtransient current* $|I''|$. In Fig. 10.3 the subtransient current is 0.707 times the ordinate *oc*. Subtransient current is often called the *initial symmetrical rms current*, which is more descriptive because it conveys the idea of neglecting the dc component and taking the rms value of the ac component of current immediately after the occurrence of the fault. *Direct-axis subtransient reactance* X_d'' for an alternator operating at no load before the occurrence of a three-phase fault at its terminals is $|E_g|/|I''|$.

The currents and reactances discussed above are defined by the following equations, which apply to an alternator operating at no load before the occurrence of a three-phase fault at its terminals:

$$|I| = \frac{oa}{\sqrt{2}} = \frac{|E_g|}{X_d} \tag{10.3}$$

$$|I'| = \frac{ob}{\sqrt{2}} = \frac{|E_g|}{X_d'} \tag{10.4}$$

$$|I''| = \frac{oc}{\sqrt{2}} = \frac{|E_g|}{X_d''} \tag{10.5}$$

where $|I|$ = steady-state current, rms value
$\quad\quad |I'|$ = transient current, rms value excluding dc component
$\quad\quad |I''|$ = subtransient current, rms value excluding dc component
$\quad\quad X_d$ = direct-axis synchronous reactance
$\quad\quad X'_d$ = direct-axis transient reactance
$\quad\quad X''_d$ = direct-axis subtransient reactance
$\quad\quad |E_g|$ = rms voltage from one terminal to neutral at no load
$\quad oa, ob, oc$ = intercepts shown in Fig. 10.3.

In analytical work the steady-state, transient, and subtransient currents may be expressed as phasors, usually with E_g as the reference.

The subtransient current $|I''|$ is much larger than the steady-state current $|I|$ because the decrease in flux across the air gap of the machine caused by the armature current, as described in Sec. 6.3, cannot take place immediately. So a larger voltage is induced in the armature windings just after the fault occurs than exists after steady state is reached. However, we account for the difference in induced voltage by using different reactances in series with the no-load voltage E_g to calculate currents for subtransient, transient, and steady-state conditions. We shall examine the transients for machines carrying a load in the next section.

Equations (10.3) to (10.5) indicate the method of determining fault current in a generator when its reactances are known. If the generator is unloaded when the fault occurs, the machine is represented by the no-load voltage to neutral in series with the proper reactance. The resistance is taken into account if greater accuracy is desired. If there is impedance external to the generator between its terminals and the short circuit, the external impedance must be included in the circuit.

Example 10.1 Two generators are connected in parallel to the low-voltage side of a three-phase Δ-Y transformer as shown in Fig. 10.5. Generator 1 is rated 50,000 kVA, 13.8 kV. Generator 2 is rated 25,000 kVA, 13.8 kV. Each generator has a subtransient reactance of 25%. The transformer is rated 75,000 kVA, 13.8Δ/69Y kV, with a reactance of 10%. Before the fault occurs, the voltage on the high-tension side of the transformer is 66 kV. The transformer is unloaded, and there is no circulating current between the generators. Find the subtransient current in each generator when a three-phase short circuit occurs on the high-tension side of the transformer.

SOLUTION Select as base in the high-tension circuit 69 kV, 75,000 kVA. Then the base voltage on the low-tension side is 13.8 kV.

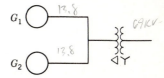

Figure 10.5 One-line diagram for Example 10.1.

Generator 1:

$$X_d'' = 0.25 \frac{75{,}000}{50{,}000} = 0.375 \text{ per unit}$$

$$E_{g1} = \frac{66}{69} = 0.957 \text{ per unit}$$

Generator 2:

$$X_d'' = 0.25 \frac{75{,}000}{25{,}000} = 0.750 \text{ per unit}$$

$$E_{g2} = \frac{66}{69} = 0.957 \text{ per unit}$$

Transformer:

$$X = 0.10 \text{ per unit}$$

Figure 10.6 shows the reactance diagram before the fault. A three-phase fault at P is simulated by closing switch S. The internal voltages of the two machines may be considered to be in parallel since they must be identical in magnitude and phase if no circulating current flows between them. The equivalent parallel subtransient reactance is

$$\frac{0.375 \times 0.75}{0.375 + 0.75} = 0.25 \text{ per unit}$$

Therefore, as a phasor with E_g as reference, the subtransient current in the short circuit is

$$I'' = \frac{0.957}{j0.25 + j0.10} = -j2.735 \text{ per unit}$$

The voltage on the delta side of the transformer is

$$(-j2.735)(j0.10) = 0.2735 \text{ per unit}$$

and in generators 1 and 2

$$I_1'' = \frac{0.957 - 0.274}{j0.375} = -j1.823 \text{ per unit}$$

$$I_2'' = \frac{0.957 - 0.274}{j0.75} = -j0.912 \text{ per unit}$$

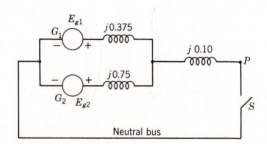

Figure 10.6 Reactance diagram for Example 10.1.

To find the current in amperes, the per-unit values are multiplied by the base current of the circuit:

$$|I_1''| = 1.823 \frac{75,000}{\sqrt{3} \times 13.8} = 5720 \text{ A}$$

$$|I_2''| = 0.912 \frac{75,000}{\sqrt{3} \times 13.8} = 2860 \text{ A}$$

Although machine reactances are not true constants of the machine and depend on the degree of saturation of the magnetic circuit, their values usually lie within certain limits and can be predicted for various types of machines. Table A.4 gives typical values of machine reactances that are needed in making fault calculations and in stability studies. In general, subtransient reactances of generators and motors are used to determine the initial current flowing on the occurrence of a short circuit. For determining the interrupting capacity of circuit breakers, except those which open instantaneously, subtransient reactance is used for generators and transient reactance is used for synchronous motors. In stability studies where the problem is to determine whether a fault will cause a machine to lose synchronism with the rest of the system if the fault is removed after a certain time interval, transient reactances apply.

10.3 INTERNAL VOLTAGES OF LOADED MACHINES UNDER TRANSIENT CONDITIONS

All the preceding discussion pertains to a synchronous generator that carries no current at the time a three-phase fault occurs at the terminals of the machine. Now consider a generator that is loaded when the fault occurs. Figure 10.7a is the equivalent circuit of a generator that has a balanced three-phase load. External impedance is shown between the generator terminals and the point P where the fault occurs. The current flowing before the fault occurs at point P is I_L, the voltage at the fault is V_f, and the terminal voltage of the generator is V_t. As discussed in Chap. 6, the equivalent circuit of the synchronous generator is its no-load voltage E_g in series with its synchronous reactance X_s. If a three-phase fault occurs at P in the system, we see that a short circuit from P to neutral in the equivalent circuit does not satisfy the conditions for calculating subtransient current since the reactance of the generator must be X_d'' if we are calculating subtransient current I'' or X_d' if we are calculating transient current I'.

The circuit shown in Fig. 10.7b gives us the desired result. Here a voltage E_g'' in series with X_d'' supplies the steady-state current I_L when switch S is open and supplies the current to the short circuit through X_d'' and Z_{ext} when switch S is closed. If we can determine E_g'', this current through X_d'' will be I''. With switch S open we see that

$$E_g'' = V_t + jI_L X_d'' \tag{10.6}$$

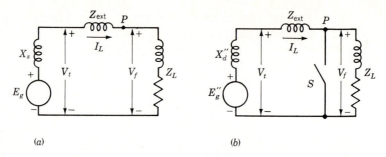

Figure 10.7 Equivalent circuits for a generator supplying a balanced three-phase load. Application of a three-phase fault at P is simulated by closing switch S. (a) Usual steady-state generator equivalent circuit with load. (b) Circuit for calculation of I''.

and this equation defines E_g'', which is called the *subtransient internal voltage.* Similarly when calculating transient current I' which must be supplied through the transient reactance X_d' the driving voltage is the *transient internal voltage* E_g', where

$$E_g' = V_t + jI_L X_d' \qquad (10.7)$$

Voltages E_g'' and E_g' are determined by I_L and are both equal to the no-load voltage E_g only when I_L is zero, at which time E_g is equal to V_t.

At this point it is important to note that E_g'' in series with X_d'' represents the generator before the fault occurs and immediately after the fault only if the prefault current in the generator is I_L. On the other hand, E_g in series with the synchronous reactance X_s is the equivalent circuit of the machine under steady-state conditions for any load. For a different value of I_L in the circuit of Fig. 10.7 E_g would remain the same but a new value of E_g'' would be required.

Synchronous motors have reactances of the same type as generators. When a motor is short-circuited, it no longer receives electric energy from the power line, but its field remains energized and the inertia of its rotor and connected load keeps it rotating for an indefinite period. The internal voltage of a synchronous motor causes it to contribute current to the system, for it is then acting like a generator. By comparison with the corresponding formulas for a generator, the subtransient internal voltage and transient internal voltage for a synchronous motor are given by

$$E_m'' = V_t - jI_L X_d'' \qquad (10.8)$$

$$E_m' = V_t - jI_L X_d' \qquad (10.9)$$

Systems that contain generators and motors under load may be solved either by Thévenin's theorem or by the use of transient or subtransient internal voltages, as is illustrated in the following examples.

Example 10.2 A synchronous generator and motor are rated 30,000 kVA, 13.2 kV, and both have subtransient reactances of 20%. The line connecting them has a reactance of 10% on the base of the machine ratings. The motor is drawing 20,000 kW at 0.8 power factor leading and a terminal voltage of 12.8 kV when a symmetrical three-phase fault occurs at the motor terminals. Find the subtransient current in the generator, motor, and fault by using the internal voltage of the machines.

SOLUTION Choose as base 30,000 kVA, 13.2 kV.

Figure 10.8a shows the equivalent circuit of the system described. We see that Fig. 10.8a is similar to Fig. 10.7b and that before the fault, E_g'' and E_m'' could be replaced by E_g and E_m provided we replaced subtransient reactances by synchronous reactances. However, to find subtransient current we need the representation of Fig. 10.8a.

If we use the voltage at the fault V_f as the reference phasor,

$$V_f = \frac{12.8}{13.2} = 0.97 \underline{/0°} \text{ per unit}$$

$$\text{Base current} = \frac{30,000}{\sqrt{3} \times 13.2} = 1312 \text{ A}$$

$$I_L = \frac{20,000 \underline{/36.9°}}{0.8 \times \sqrt{3} \times 12.8} = 1128 \underline{/36.9°} \text{ A}$$

$$= \frac{1128 \underline{/36.9°}}{1312} = 0.86 \underline{/36.9°} \text{ per unit}$$

$$= 0.86(0.8 + j0.6) = 0.69 + j0.52 \text{ per unit}$$

For the generator,

$$V_t = 0.970 + j0.1(0.69 + j0.52) = 0.918 + j0.069 \text{ per unit}$$

$$E_g'' = 0.918 + j0.069 + j0.2(0.69 + j0.52) = 0.814 + j0.207 \text{ per unit}$$

$$I_g'' = \frac{0.814 + j0.207}{j0.3} = 0.69 - j2.71 \text{ per unit}$$

$$= 1312(0.69 - j2.71) = 905 - j3550 \text{ A}$$

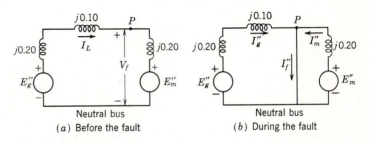

Neutral bus
(a) Before the fault

Neutral bus
(b) During the fault

Figure 10.8 Equivalent circuits for Example 10.2.

For the motor,

$$V_t = V_f = 0.97\underline{/0°} \text{ per unit}$$

$$E''_m = 0.97 + j0 - j0.2(0.69 + j0.52) = 0.97 - j0.138 + 0.104$$

$$= 1.074 - j0.138 \text{ per unit}$$

$$I''_m = \frac{1.074 - j0.138}{j0.2} = -0.69 - j5.37 \text{ per unit}$$

$$= 1312(-0.69 - j5.37) = -905 - j7050 \text{ A}$$

In the fault,

$$I''_f = I''_g + I''_m = 0.69 - j2.71 - 0.69 - j5.37 = -j8.08 \text{ per unit}$$

$$= -j8.08 \times 1312 = -j10,600 \text{ A}$$

Figure 10.8b shows the paths of I''_g, I''_m, and I''_f.

The subtransient current in the fault can be found by Thévenin's theorem, which is applicable to linear, bilateral circuits. When constant values are used for the reactances of synchronous machines, linearity is assumed. When the theorem is applied to the circuit of Fig. 10.7b, the equivalent circuit is a single generator and a single impedance terminating at the point of application of the fault. The new generator has an internal voltage equal to V_f, the voltage at the fault point before the fault occurs. The impedance is that measured at the point of application of the fault looking back into the circuit with all the generated voltages short-circuited. Subtransient reactances should be used if the initial current is desired. Figure 10.9 is the Thévenin equivalent of Fig. 10.7b. The impedance Z_{th} is equal to $(Z_{ext} + jX''_d)Z_L/(Z_L + Z_{ext} + jX''_d)$. Upon the occurrence of a three-phase short circuit at P, simulated by closing S, the subtransient current in the fault is

$$I'' = \frac{V_f}{Z_{th}} = \frac{V_f(Z_L + Z_{ext} + jX''_d)}{Z_L(Z_{ext} + jX''_d)} \quad (10.10)$$

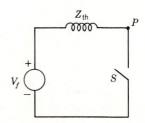

Figure 10.9 Thévenin equivalent of the circuit of Fig. 10.7b.

Example 10.3 Solve Example 10.2 by the use of Thévenin's theorem.

SOLUTION

$$Z_{th} = \frac{j0.3 \times j0.2}{j0.3 + j0.2} = j0.12 \text{ per unit}$$

$$V_f = 0.97\underline{/0°} \text{ per unit}$$

In the fault,

$$I_f'' = \frac{0.97 + j0}{j0.12} = -j8.08 \text{ per unit}$$

The above current, found by applying Thévenin's theorem, is that which flows out of the circuit at the fault because of the reduction of the voltage to zero at that point. If this current caused by the fault is divided between the parallel circuits of the machines inversely as their impedances, the resulting values are the currents from each machine due only to the *change* in voltage at the fault point. To the fault currents thus attributed to the two machines must be added the current flowing in each before the fault occurred to find the total current in the machines after the fault. The superposition theorem supplies the reason for adding the current flowing before the fault to the current computed by Thévenin's theorem. Figure 10.10*a* shows a generator having a voltage V_f connected at the fault and equal to the voltage at the fault before the fault occurs. This generator has no effect on the current flowing before the fault occurs, and the circuit corresponds to that of Fig. 10.8*a*. Adding in series with V_f another generator having an emf of equal magnitude but 180° out of phase with V_f gives the circuit of Fig. 10.10*b*, which corresponds to that of Fig. 10.8*b*. The principle of superposition, applied by first shorting E_g'', E_m'', and V_f, gives the currents found by distributing the fault current between the two generators inversely as the impedances of their circuits. Then shorting the remaining generator $-V_f$ with E_g'', E_m'', and V_f in the circuit gives the current flowing before the fault. Adding the two values of current in each branch gives the current in

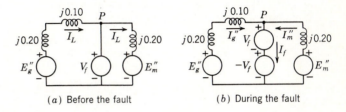

(*a*) Before the fault   (*b*) During the fault

Figure 10.10 Circuits illustrating the application of the superposition theorem to determine the proportion of the fault current in each branch of the system.

the branch after the fault. Applying the above principle to the present example gives

$$\text{Fault current from generator} = -j8.08 \times \frac{j0.2}{j0.5} = -j3.23 \text{ per unit}$$

$$\text{Fault current from motor} = -j8.08 \times \frac{j0.3}{j0.5} = -j4.85 \text{ per unit}$$

To these currents must be added the prefault current I_L to obtain the total subtransient currents in the machines:

$$I''_g = 0.69 + j0.52 - j3.23 = 0.69 - j2.71 \text{ per unit}$$

$$I''_m = -0.69 - j0.52 - j4.85 = -0.69 - j5.37 \text{ per unit}$$

Note that I_L is in the same direction as I''_g but opposite to I''_m. The per-unit values found for I''_f, I''_g, and I''_m are the same as in Example 10.2, and so the ampere values will also be the same.

Usually load current is omitted in determining the current in each line upon the occurrence of a fault. In the Thévenin method neglect of load current means that the prefault current in each line is not added to the component of current flowing toward the fault in the line. The method of Example 10.2 neglects load current if the subtransient internal voltages of all machines are assumed equal to the voltage V_f at the fault before the fault occurs, for such is the case if no current flows anywhere in the network prior to the fault.

Neglecting load current in Example 10.3 gives

$$\text{Fault current from generator} = 3.23 \times 1312 = 4240 \text{ A}$$

$$\text{Fault current from motor} = 4.85 \times 1312 = 6360 \text{ A}$$

$$\text{Current in fault} = 8.08 \times 1312 = 10{,}600 \text{ A}$$

The current in the fault is the same whether or not load current is considered, but the contributions from the lines differ. When load current is included, we find from Example 10.2

$$\text{Fault current from generator} = |905 - j3550| = 3660 \text{ A}$$

$$\text{Fault current from motor} = |-905 - j7050| = 7200 \text{ A}$$

The arithmetic sum of the generator and motor current magnitudes does not equal the fault current because the currents from the generator and motor are not in phase.

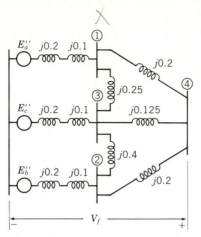

Figure 10.11 Reactance diagram obtained from Fig. 7.3 by substituting subtransient for synchronous reactances of the machines and subtransient internal voltages for no-load generated voltages. Reactance values are marked in per unit.

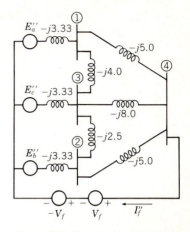

Figure 10.12 Circuit of Fig. 10.11 with admittances marked in per unit and a three-phase fault on bus 4 of the system simulated by V_f and $-V_f$ in series.

10.4 THE BUS IMPEDANCE MATRIX IN FAULT CALCULATIONS

Our discussion of fault calculations has been confined to simple circuits, but now we shall extend our study to general networks. However, let us proceed to the general equations by starting with a specific network with which we are already familiar. If we change the reactances in series with the generated voltages of the circuit shown in Fig. 7.3 to subtransient reactances, and if the generated voltages become subtransient internal voltages, we have the network shown in Fig. 10.11. If this network is the single-phase equivalent of a three-phase system and we choose to study a fault at bus 4, we can follow the procedure of Sec. 10.3 and call V_f the voltage at bus 4 before the fault occurs.

A three-phase fault at bus 4 is simulated by the network of Fig. 10.12 where the impedance values of Fig. 10.11 have been changed to admittances. The generated voltages V_f and $-V_f$ in series constitute the short circuit. Generated voltage V_f alone in this branch would cause no current in the branch. With V_f and $-V_f$ in series the branch is a short circuit, and the branch current is I_f''. Admittances rather than impedances have been marked in per unit on this diagram. If E_a'', E_b'', E_c'', and V_f are short-circuited, the voltages and currents are those due only to $-V_f$. Then the only current entering a node from a source is that from $-V_f$ and is $-I_f''$ *into* node 4 (I_f'' *from* node 4) since there is no current in this branch until the insertion of $-V_f$. The node equations in matrix form for the network with $-V_f$ the only source are

$$
\begin{bmatrix} 0 \\ 0 \\ 0 \\ -I_f'' \end{bmatrix} = j \begin{bmatrix} -12.33 & 0.0 & 4.0 & 5.0 \\ 0.0 & -10.83 & 2.5 & 5.0 \\ 4.0 & 2.5 & -17.83 & 8.0 \\ 5.0 & 5.0 & 8.0 & -18.0 \end{bmatrix} \begin{bmatrix} V_1^\Delta \\ V_2^\Delta \\ V_3^\Delta \\ -V_f \end{bmatrix} \qquad (10.11)
$$

when the superscript Δ indicates that the voltages are due only to $-V_f$. The Δ sign is chosen to indicate the change in voltage due to the fault.

By inverting the bus admittance matrix of the network of Fig. 10.12 we obtain the bus impedance matrix. The bus voltages due to $-V_f$ are given by

$$
\begin{bmatrix} V_1^\Delta \\ V_2^\Delta \\ V_3^\Delta \\ -V_f \end{bmatrix} = \mathbf{Z}_{\text{bus}} \begin{bmatrix} 0 \\ 0 \\ 0 \\ -I_f'' \end{bmatrix}
\tag{10.12}
$$

and so

$$
I_f'' = \frac{V_f}{Z_{44}}
\tag{10.13}
$$

and

$$
V_1^\Delta = -I_f'' Z_{14} = -\frac{Z_{14}}{Z_{44}} V_f
$$

$$
V_2^\Delta = -\frac{Z_{24}}{Z_{44}} V_f \qquad V_3^\Delta = -\frac{Z_{34}}{Z_{44}} V_f
\tag{10.14}
$$

When the generator voltage $-V_f$ is short-circuited in the network of Fig. 10.12 and E_a'', E_b'', E_c'', and V_f are in the circuit, the currents and voltages everywhere in the network are those existing before the fault. By the principle of superposition these prefault voltages added to those given by Eqs. (10.14) yield the voltages existing after the fault occurs. Usually the faulted network is assumed to have been without loads before the fault. In such a case no current is flowing before the fault, and all voltages throughout the network are the same and equal to V_f. This assumption simplifies our work considerably, and applying the principle of superposition gives

$$
V_1 = V_f + V_1^\Delta = V_f - I_f'' Z_{14}
$$

$$
V_2 = V_f + V_2^\Delta = V_f - I_f'' Z_{24}
$$

$$
V_3 = V_f + V_3^\Delta = V_f - I_f'' Z_{34}
$$

$$
V_4 = V_f - V_f = 0
$$

$$
\tag{10.15}
$$

These voltages exist when subtransient current flows and $\mathbf{Z}_{\text{bus}}$ has been formed for a network having subtransient values for generator reactances.

In general terms for a fault on bus k, and neglecting prefault currents,

$$
I_f = \frac{V_f}{Z_{kk}}
\tag{10.16}
$$

and the postfault voltage at bus n is

$$
V_n = V_f - \frac{Z_{nk}}{Z_{kk}} V_f
\tag{10.17}
$$

Using the numerical values of Eq. (10.11), we invert the square matrix $\mathbf{Y}_{bus}$ of that equation and find

$$\mathbf{Z}_{bus} = j \begin{bmatrix} 0.1488 & 0.0651 & 0.0864 & 0.0978 \\ 0.0651 & 0.1554 & 0.0799 & 0.0967 \\ 0.0864 & 0.0798 & 0.1341 & 0.1058 \\ 0.0978 & 0.0967 & 0.1058 & 0.1566 \end{bmatrix} \qquad (10.18)$$

Usually V_f is assumed to be $1.0\underline{/0^\circ}$ per unit, and with this assumption for our faulted network

$$I''_f = \frac{1}{j0.1566} = -j6.386 \text{ per unit}$$

$$V_1 = 1 - \frac{j0.0978}{j0.1566} = 0.3755 \text{ per unit}$$

$$V_2 = 1 - \frac{j0.0967}{j0.1566} = 0.3825 \text{ per unit}$$

$$V_3 = 1 - \frac{j0.1058}{j0.1566} = 0.3244 \text{ per unit}$$

Currents in any part of the network can be found from the voltages and impedances. For instance, the fault current in the branch connecting nodes 1 and 3 flowing toward node 3 is

$$I''_{13} = \frac{V_1 - V_3}{j0.25} = \frac{0.3755 - 0.3244}{j0.25}$$

$$= -j0.2044 \text{ per unit}$$

From the generator connected to node 1 the current is

$$I''_a = \frac{E''_a - V_1}{j0.3} = \frac{1 - 0.3755}{j0.3}$$

$$= -j2.0817 \text{ per unit}$$

Other currents can be found in a similar manner, and voltages and currents with the fault on any other bus are calculated just as easily from the impedance matrix.

Equation (10.16) is simply an application of Thévenin's theorem, and we recognize that the quantities on the principal diagonal of the bus impedance matrix are the Thévenin impedances of the network for calculating fault current at the various buses. Power companies furnish data to a customer who must determine the fault current to specify circuit breakers for an industrial plant or distribution system connected to the utility system at any point. Usually the data supplied lists the short-circuit megavoltamperes, where

$$\text{Short-circuit MVA} = \sqrt{3} \times (\text{nominal kV}) \times I_{sc} \times 10^{-3} \qquad (10.19)$$

With resistance and shunt capacitance neglected, the single-phase Thévenin equivalent circuit which represents the system is an emf equal to the nominal line voltage divided by $\sqrt{3}$ in series with an inductive reactance of

$$X_{th} = \frac{(\text{nominal kV}/\sqrt{3}) \times 1000}{I_{sc}} \ \Omega \tag{10.20}$$

Solving Eq. (10.19) for I_{sc} and substituting in Eq. (10.20) yield

$$X_{th} = \frac{(\text{nominal kV})^2}{\text{short-circuit MVA}} \ \Omega \tag{10.21}$$

If base kilovolts is equal to nominal kilovolts, converting to per unit yields

$$X_{th} = \frac{\text{base MVA}}{\text{short-circuit MVA}} \ \text{per unit} \tag{10.22}$$

$$X_{th} = \frac{I_{\text{base}}}{I_{sc}} \ \text{per unit} \tag{10.23}$$

10.5 A BUS IMPEDANCE MATRIX EQUIVALENT NETWORK

Although we cannot devise a physically realizable circuit employing the impedances of the bus impedance network, we can draw a circuit with transfer impedances *indicated* between branches. Such a diagram will be helpful in understanding the significance of the equations developed in Sec. 10.4.

In Fig. 10.13 brackets have been drawn between branch 4 and the other three branches of a network which has four nodes in addition to the reference node.† Associated with these brackets are the symbols Z_{14}, Z_{24}, and Z_{34}, which

† This equivalent network is drawn in the manner adopted in J. R. Neuenswander, *Modern Power Systems*, Intext Educational Publishers, New York, 1971, which refers to the bus impedance matrix equivalent network as the *rake equivalent*.

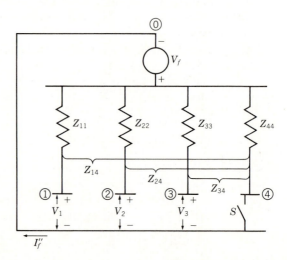

Figure 10.13 Bus impedance matrix equivalent network with four independent nodes. Closing switch S simulates a fault on node 4. Only the transfer admittances for node 4 are shown.

identify transfer impedances of node 4 of the bus impedance matrix. The driving-point impedances of the bus impedance matrix are Z_{11}, Z_{22}, Z_{33}, and Z_{44}. No current can flow in any branches when switch S is open. When S is closed, current flows in the circuit only toward node 4. This current is V_f/Z_{44}, which, according to Eq. (10.13), is I_f'' for a fault on node 4. We shall interpret the brackets to mean that the current I_f'' toward node 4 in the network induces voltage drops of $I_f''Z_{14}$, $I_f''Z_{24}$, and $I_f''Z_{34}$ in the branches connected to nodes 1, 2, and 3, respectively. These voltage drops are in the direction toward the respective nodes.

If switch S in the circuit of Fig. 10.13 is open, all nodes will be at the voltage of V_f, as in Fig. 10.11 if E_a'', E_b'', and E_c'' equal V_f. If S is closed, examination of the circuit shows that the voltages at all four nodes with respect to the reference node 0 will be the values specified by Eqs. (10.15). Therefore, if we interpret the indicated transfer impedances of this circuit as described above, the circuit is the equivalent of that shown in Fig. 10.11 with S open and Fig. 10.12 with switch S closed, where we are still neglecting prefault current.

Of course, we can simulate short circuits at the other buses in a similar manner and extend this approach to a general network having any number of nodes. We could indicate the other transfer impedances of the equivalent circuit by additional brackets and have not done so only because it becomes confusing to have so many brackets to indicate the transfer impedances. In fact we shall usually omit the brackets when drawing this network equivalent for the bus impedance matrix, but we have to realize that the transfer impedances do exist and must be considered in interpreting the network.

Example 10.4 Determine the bus impedance matrix for the network of Example 8.1, for which the results of a load-flow study are shown in Fig. 8.2. Generators at buses 1 and 3 are rated 270 and 225 MVA, respectively. The generator subtransient reactances plus the reactances of the transformers connecting them to the buses are each 0.30 per unit on the generator rating as base. The turns ratios of the transformers are such that the voltage base in each generator circuit is equal to the voltage rating of the generator. Include the generator and transformer reactances in the matrix. Find the subtransient current in a three-phase fault at bus 4 and the current coming to the faulted bus over each line. Prefault current is to be neglected and all voltages are assumed to be 1.0 per unit before the fault occurs. System base is 100 MVA. Neglect all resistances.

SOLUTION Converted to the 100-MVA base, the combined generator and transformer reactances are

Generator at bus 1: $\quad X = 0.30 \times \dfrac{100}{270} = 0.1111$ per unit

Generator at bus 3: $\quad X = 0.30 \times \dfrac{100}{225} = 0.1333$ per unit

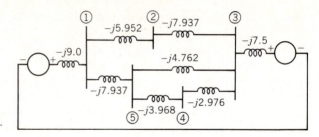

Figure 10.14 Admittance diagram for Example 10.4.

The network with admittances marked in per unit is shown in Fig. 10.14 from which the node admittance matrix is

$$\mathbf{Y}_{bus} = j \begin{bmatrix} -22.889 & 5.952 & 0.0 & 0.0 & 7.937 \\ 5.952 & -13.889 & 7.937 & 0.0 & 0.0 \\ 0.0 & 7.937 & -23.175 & 2.976 & 4.762 \\ 0.0 & 0.0 & 2.976 & -6.944 & 3.968 \\ 7.937 & 0.0 & 4.762 & 3.968 & -16.667 \end{bmatrix}$$

This 5×5 bus is inverted on a digital computer to yield the short-circuit matrix

$$\mathbf{Z}_{bus} = j \begin{bmatrix} 0.0793 & 0.0558 & 0.0382 & 0.0511 & 0.0608 \\ 0.0558 & 0.1338 & 0.0664 & 0.0630 & 0.0605 \\ 0.0382 & 0.0664 & 0.0875 & 0.0720 & 0.0603 \\ 0.0511 & 0.0630 & 0.0720 & 0.2321 & 0.1002 \\ 0.0608 & 0.0605 & 0.0603 & 0.1002 & 0.1301 \end{bmatrix}$$

Visualizing a network like that of Fig. 10.13 will help in finding the desired currents and voltages.

The subtransient current in a three-phase fault on bus 4 is

$$I'' = \frac{1.0}{j0.2321} = -j4.308 \text{ per unit}$$

At buses 3 and 5 the voltages are

$$V_3 = 1.0 - (-j4.308)(j0.0720) = 0.6898 \text{ per unit}$$
$$V_5 = 1.0 - (-j4.308)(j0.1002) = 0.5683 \text{ per unit}$$

Currents to the fault are

From bus 3: $0.6898(-j2.976) = -j2.053$

From bus 5: $0.5683(-j3.968) = \underline{-j2.255}$

$$-j4.308 \text{ per unit}$$

From the same short-circuit matrix we can find similar information for faults on any of the other buses.

10.6 THE SELECTION OF CIRCUIT BREAKERS

Much study has been given to circuit-breaker ratings and applications and our discussion here will give some introduction to the subject. For additional guidance necessary to specify the breakers the reader should consult the ANSI publications listed in the footnotes which accompany this section.

The subtransient current to which we have devoted most of our attention to this point is the initial symmetrical current and does not include the dc component. As we have seen, inclusion of the dc component results in a rms value of current immediately after the fault, which is higher than the subtransient current. For oil circuit breakers above 5 kV the subtransient current multiplied by 1.6 is considered to be the rms value of the current whose disruptive forces the breaker must withstand during the first half cycle after the fault occurs. This current is called the *momentary current*, and for many years circuit breakers were rated in terms of their momentary current as well as other criteria.†

The interrupting rating of a circuit breaker was specified in kilovoltamperes or megavoltamperes. The interrupting kilovoltamperes equal $\sqrt{3}$ times the kilovolts of the bus to which the breaker is connected times the current which the breaker must be capable of interrupting when its contacts part. This current is, of course, lower than the momentary current and depends on the speed of the breaker, such as 8, 5, 3, or $1\frac{1}{2}$ cycles, which is a measure of the time from the occurrence of the fault to the extinction of the arc.

The current which a breaker must interrupt is usually asymmetrical since it still contains some of the decaying dc component. A schedule of preferred ratings for ac high-voltage oil circuit breakers specifies the interrupting current ratings of breakers in terms of the component of the asymmetrical current which is symmetrical about the zero axis. This current is properly called the *required symmetrical interrupting capability* or simply the *rated symmetrical short-circuit current*. Often the adjective symmetrical is omitted. Selection of circuit breakers may also be made on the basis of total current (dc component included).‡ We shall limit our discussion to a brief treatment of the symmetrical basis of breaker selection.

Breakers are identified by nominal-voltage class, such as 69 kV. Among other factors specified are rated continuous current, rated maximum voltage, voltage range factor K, and rated short-circuit current at rated maximum kilovolts. K determines the range of voltage over which rated short-circuit current times operating voltage is constant. For a 69-kV breaker having a maximum

---

† See G. N. Lester, "High Voltage Circuit Breaker Standards in the USA: Past, Present, and Future," *IEEE Trans. Power Appar. Syst.*, vol. 93, 1974, pp. 590–600.

‡ See Schedules of Preferred Ratings and Related Required Capabilities for AC High-Voltage Circuit Breakers Rated on a Symmetrical Current Basis, ANSI C37.06-1971 and Guide for Calculation of Fault Currents for Application of AC High-Voltage Circuit Breakers Rated on a Total Current Basis, ANSI C37.5-1979, American National Standards Institute, New York.

rated voltage of 72.5 kV, a voltage range factor K of 1.21, and a continuous current rating of 1200 A, the rated short-circuit current at the maximum voltage (symmetrical current which can be interrupted at 72.5 kV) is 19,000 A. This means that the product 72.5 × 19,000 is the constant value of rated short-circuit current times operating voltage in the range 72.5 to 60 kV since 72.5/1.21 equals 60. The rated short-circuit current at 60 kV is 19,000 × 1.21, or 23,000 A. At lower operating voltages this short-circuit current cannot be exceeded. At 69 kV the rated short-circuit current is

$$\frac{72.5}{69} \times 19,000 = 20,000 \text{ A}$$

Breakers of the 115-kV class and higher have a K of 1.0.

A simplified procedure for calculating the symmetrical short-circuit current, called the E/X method,† disregards all resistance, all static loads, and all prefault current. Subtransient reactance is used for generators in the E/X method, and for synchronous motors the recommended reactance is the X_d'' of the motor times 1.5, which is the approximate value of the transient reactance of the motor. Induction motors below 50 hp are neglected, and various multiplying factors are applied to the X_d'' of larger induction motors depending on their size. If no motors are present, symmetrical short-circuit current equals subtransient current.

The impedance by which the voltage V_f at the fault is divided to find short-circuit current must be examined when the E/X method is used. In specifying a breaker for bus k this impedance is Z_{kk} of the bus impedance matrix with the proper machine reactances since the short-circuit current is expressed by Eq. (10.16). If the ratio of X/R of this impedance is 15 or less, a breaker of the correct voltage and kilovoltamperes may be used if its interrupting current rating is equal to or exceeds the calculated current. If the X/R ratio is unknown, the calculated current should be no more than 80% of the allowed value for the breaker at the existing bus voltage. The ANSI application guide specifies a corrected method to account for ac and dc time constants for the decay of the current amplitude if the X/R ratio exceeds 15. The corrected method also considers breaker speed.

This discussion of the selection of circuit breakers is presented not as a study of breaker applications but to indicate the importance of understanding fault calculations. The following example should clarify the principle.

Example 10.5 A 25,000-kVA 13.8-kV generator with $X_d'' = 15\%$ is connected through a transformer to a bus which supplies four identical motors, as shown in Fig. 10.15. The subtransient reactance X_d'' of each motor is 20% on a base of 5000 kVA, 6.9 kV. The three-phase rating of the transformer is

† See Application Guide for AC High-Voltage Circuit Breakers Rated on a Symmetrical Current Basis, ANSI C37.010-1972, American National Standards Institute, New York. This publication is also IEEE Std 320-1972.

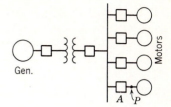

Figure 10.15 One-line diagram for Example 10.5.

25,000 kVA, 13.8/6.9 kV, with a leakage reactance of 10%. The bus voltage at the motors is 6.9 kV when a three-phase fault occurs at the point P. For the fault specified, determine (a) the subtransient current in the fault, (b) the subtransient current in breaker A, and (c) the symmetrical short-circuit interrupting current (as defined for circuit-breaker applications) in the fault and in breaker A.

SOLUTION (a) For a base of 25,000 kVA, 13.8 kV in the generator circuit, the base for the motors is 25,000 kVA, 6.9 kV. The subtransient reactance of each motor is

$$X''_d = 0.20 \frac{25,000}{5000} = 1.0 \text{ per unit}$$

Figure 10.16 is the diagram with subtransient values of reactance marked. For a fault at P,

$$V_f = 1.0 \text{ per unit}$$

$$Z_{th} = j0.125 \text{ per unit}$$

$$I''_f = \frac{1.0}{j0.125} = -j8.0 \text{ per unit}$$

The base current in the 6.9-kV circuit is

$$\frac{25,000}{\sqrt{3} \times 6.9} = 2090 \text{ A}$$

$$I''_f = 8 \times 2090 = 16,720 \text{ A}$$

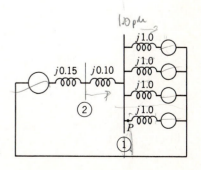

Figure 10.16 Reactance diagram for Example 10.5.

(b) Through breaker A comes the contribution from the generator and three of the four motors.

The generator contributes a current of

$$-j8.0 \times \frac{0.25}{0.50} = -j4.0 \text{ per unit}$$

Each motor contributes 25% of the remaining fault current, or $-j1.0$ A per unit.

Through breaker A,

$$I'' = -j4.0 + 3(-j1.0) = -j7.0 \text{ per unit} = 7 \times 2090 = 14,630 \text{ A}$$

(c) To compute the current through breaker A to be interrupted, replace the subtransient reactance of $j1.0$ by the transient reactance of $j1.5$ in the motor circuits of Fig. 10.16. Then

$$Z_{th} = j\frac{0.375 \times 0.25}{0.375 + 0.25} = j0.15 \text{ per unit}$$

The generator contributes a current of

$$\frac{1.0}{j0.15} \times \frac{0.375}{0.625} = -j4.0 \text{ per unit}$$

Each motor contributes a current of

$$\frac{1}{4} \times \frac{1.0}{j0.15} \times \frac{0.25}{0.625} = -j0.67 \text{ per unit}$$

The symmetrical short-circuit current to be interrupted is

$$(4.0 + 3 \times 0.67) \times 2090 = 12,560 \text{ A}$$

The usual procedure is to rate all the breakers connected to a bus on the basis of the current into a fault on the bus. In that case the short-circuit current interrupting rating of the breakers connected to the 6.9-kV bus must be at least

$$4 + 4 \times 0.67 = 6.67 \text{ per unit}$$

or

$$6.67 \times 2090 = 13,940 \text{ A}$$

A 14.4-kV circuit breaker has a rated maximum voltage of 15.5 kV and a K of 2.67. At 15.5 kV its rated short-circuit interrupting current is 8900 A. This breaker is rated for a symmetrical short-circuit interrupting current of $2.67 \times 8900 = 23,760$ A, at a voltage of $15.5/2.67 = 5.8$ kV. This current is the maximum that can be interrupted even though the breaker may be in a circuit of lower voltage. The short-circuit interrupting current rating at 6.9 kV is

$$\frac{15.5}{6.9} \times 8900 = 20,000 \text{ A}$$

The required capability of 13,940 A is well below 80% of 20,000 A, and the breaker is suitable with respect to short-circuit current.

The short-circuit current could have been found by using the bus impedance matrix. For this purpose two nodes have been identified in Fig. 10.16. Node 1 is the bus on the low-tension side of the transformer, and node 2 is on the high-tension side. For motor reactance of 1.5 per unit

$$Y_{11} = -j10 + \frac{1}{j1.5/4} = -j12.67$$

$$Y_{12} = j10$$

$$Y_{22} = -j10 - j6.67 = -j16.67$$

The node admittance matrix is

$$\mathbf{Y}_{bus} = j \begin{bmatrix} -12.67 & 10.0 \\ 10.0 & -16.67 \end{bmatrix}$$

and its inverse is

$$\mathbf{Z}_{bus} = j \begin{bmatrix} 0.150 & 0.090 \\ 0.090 & 0.114 \end{bmatrix}$$

Figure 10.17 is the network corresponding to the bus impedance matrix. Closing S_1 with S_2 open represents a fault on bus 1.

The symmetrical short-circuit interrupting current in a three-phase fault at node 1 is

$$I_{sc} = \frac{1.0}{j0.15} = -j6.67 \text{ per unit}$$

which agrees with our previous calculations. The bus impedance matrix also gives us the voltage at bus 2 with the fault on bus 1.

$$V_2 = 1.0 - I_{sc}Z_{21} = 1.0 - (-j6.67)(j0.09) = 0.4$$

and since the admittance between nodes 1 and 2 is $-j10$, the current into the

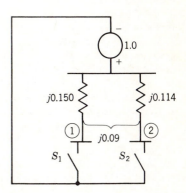

Figure 10.17 Bus impedance equivalent network for the bus impedance matrix of Example 10.5.

fault from the transformer is

$$(0.4 - 0.0)(-j10) = -j4.0 \text{ per unit}$$

which also agrees with our previous result.

We also know immediately the short-circuit current in a three-phase fault at node 2 which, by referring to Fig. 10.17 with S_1 open and S_2 closed is

$$I_{sc} = \frac{1.0}{j0.114} = -j8.77 \text{ per unit}$$

Even this simple example illustrates the value of the bus impedance matrix where the effects of a fault at a number of buses are to be studied. Matrix inversion is not necessary, for, as we have seen in Sec. 7.7, $\mathbf{Z}_{bus}$ can be generated directly by the computer.

PROBLEMS

10.1 A 60-Hz alternating voltage having a rms value of 100 V is applied to a series RL circuit by closing a switch. The resistance is 15 Ω, and the inductance is 0.12 H.

(a) Find the value of the dc component of current upon closing the switch if the instantaneous value of the voltage is 50 V at that time.

(b) What is the instantaneous value of the voltage which will produce the maximum dc component of current upon closing the switch?

(c) What is the instantaneous value of the voltage which will result in the absence of any dc component of current upon closing the switch?

(d) If the switch is closed when the instantaneous voltage is zero, find the instantaneous current 0.5, 1.5, and 5.5 cycles later.

10.2 A generator connected through a 5-cycle circuit breaker to a transformer is rated 100 MVA, 18 kV, with reactances of $X_d'' = 19\%$, $X_d' = 26\%$, and $X_d = 130\%$. It is operating at no load and rated voltage when a three-phase short circuit occurs between the breaker and the transformer. Find (a) the sustained short-circuit current in the breaker, (b) the initial symmetrical rms current in the breaker, and (c) the maximum possible dc component of the short-circuit current in the breaker.

10.3 The three-phase transformer connected to the generator described in Prob. 10.2 is rated 100 MVA, 240Y/18Δ kV, $X = 10\%$. If a three-phase short circuit occurs on the high-tension side of the transformer at rated voltage and no load, find (a) the initial symmetrical rms current in the transformer windings on the high-tension side and (b) the initial symmetrical rms current in the line on the low-tension side.

10.4 A 60-Hz generator is rated 500 MVA, 20 kV, with $X_d'' = 0.20$ per unit. It supplies a purely resistive load of 400 MW at 20 kV. The load is connected directly across the terminals of the generator. If all three phases of the load are short-circuited simultaneously, find the initial symmetrical rms current in the generator in per unit on a base of 500 MVA, 20 kV.

10.5 A generator is connected through a transformer to a synchronous motor. Reduced to the same base, the per-unit subtransient reactances of the generator and motor are 0.15 and 0.35, respectively, and the leakage reactance of the transformer is 0.10 per unit. A three-phase fault occurs at the terminals of the motor when the terminal voltage of the generator is 0.9 per unit and the output current of the generator is 1.0 per unit at 0.8 power factor leading. Find the subtransient current in per unit in the fault, in the generator, and in the motor. Use the terminal voltage of the generator as the reference phasor, and obtain the solution (a) by computing the voltages behind subtransient reactance in the generator and motor and (b) by using Thévenin's theorem.

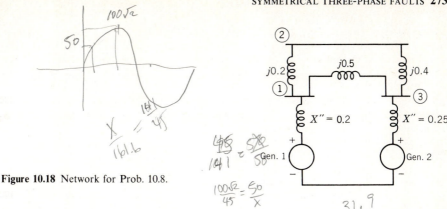

Figure 10.18 Network for Prob. 10.8.

10.6 Two synchronous motors having subtransient reactances of 0.80 and 0.25 per unit, respectively, on a base of 480 V, 2000 kVA are connected to a bus. This motor bus is connected by a line having a reactance of 0.023 Ω to a bus of a power system. At the power-system bus the short-circuit megavolt-amperes of the power system are 9.6 MVA for a nominal voltage of 480 V. When the voltage at the motor bus is 440 V, neglect load current and find the initial symmetrical rms current in a three-phase fault at the motor bus.

10.7 The bus impedance matrix of a four-bus network with values in per unit is

$$\mathbf{Z}_{bus} = j \begin{bmatrix} 0.15 & 0.08 & 0.04 & 0.07 \\ 0.08 & 0.15 & 0.06 & 0.09 \\ 0.04 & 0.06 & 0.13 & 0.05 \\ 0.07 & 0.09 & 0.05 & 0.12 \end{bmatrix}$$

Generators are connected to buses 1 and 2, and their subtransient reactances were included when finding $\mathbf{Z}_{bus}$. If prefault current is neglected, find the subtransient current in per unit in the fault for a three-phase fault on bus 4. Assume the voltage at the fault is 1.0 per unit before the fault occurs. Find also the per-unit current from generator 2 whose subtransient reactance is 0.2 per unit.

10.8 For the network shown in Fig. 10.18 find the subtransient current in per unit from generator 1 and in line 1–2 and the voltages at buses 1 and 3 for a three-phase fault on bus 2. Assume that no current is flowing prior to the fault and that the prefault voltage at bus 2 is 1.0 per unit. Use the bus impedance matrix in the calculations.

10.9 If a three-phase fault occurs at bus 1 of the network of Fig. 10.11 when there is no load (all node voltages 1.0 per unit), find the subtransient current in the fault, the voltages at buses 2, 3, and 4, and the current from the generator connected to bus 2.

10.10 Find the subtransient current in per unit in a three-phase fault on bus 5 of the network of Example 8.1. Neglect prefault current, assume all bus voltages are 1.0 per unit before the fault occurs, and make use of calculations already made in Example 10.4. Find the current in lines 1-5 and 3-5 also.

10.11 A 625-kVA 2.4-kV generator with $X''_d = 0.20$ per unit is connected to a bus through a circuit breaker as shown in Fig. 10.19. Connected through circuit breakers to the same bus are three

Figure 10.19 One-line diagram for Prob. 10.11.

synchronous motors rated 250 hp, 2.4 kV, 1.0 power factor, 90% efficiency, with $X''_d = 0.20$ per unit. The motors are operating at full load, unity power factor, and rated voltage, with the load equally divided between the machines.

(a) Draw the impedance diagram with the impedances marked in per unit on a base of 625 kVA, 2.4 kV.

(b) Find the symmetrical short-circuit current in amperes which must be interrupted by breakers A and B for a three-phase fault at point P. Simplify the calculations by neglecting the prefault current.

(c) Repeat part (b) for a three-phase fault at point Q.

(d) Repeat part (b) for a three-phase fault at point R.

10.12 A circuit breaker having a nominal rating of 34.5 kV and a continuous current rating of 1500 A has a voltage range factor K of 1.65. Rated maximum voltage is 38 kV, and the rated short-circuit current at that voltage is 22 kA. Find (a) the voltage below which rated short-circuit current does not increase as operating voltage decreases and the value of that current and (b) rated short-circuit current at 34.5 kV.

SYMMETRICAL COMPONENTS

In 1918 one of the most powerful tools for dealing with unbalanced polyphase circuits was discussed by C. L. Fortescue at a meeting of the American Institute of Electrical Engineers.† Since that time the method of symmetrical components has become of great importance and has been the subject of many articles and experimental investigations. Unsymmetrical faults on transmission systems, which may consist of short circuits, impedance between lines, impedance from one or two lines to ground, or open conductors, are studied by the method of symmetrical components.

11.1 SYNTHESIS OF UNSYMMETRICAL PHASORS FROM THEIR SYMMETRICAL COMPONENTS

Fortescue's work proves that an unbalanced system of *n* related phasors can be resolved into *n* systems of balanced phasors called the *symmetrical components* of the original phasors. The *n* phasors of each set of components are equal in length, and the angles between adjacent phasors of the set are equal. Although the method is applicable to any unbalanced polyphase system, we shall confine our discussion to three-phase systems.

---

† C. L. Fortescue, "Method of Symmetrical Coordinates Applied to the Solution of Polyphase Networks," *Trans. AIEE*, vol. 37, pp. 1027–1140, 1918.

According to Fortescue's theorem, three unbalanced phasors of a three-phase system can be resolved into three balanced systems of phasors. The balanced sets of components are:

1. Positive-sequence components consisting of three phasors equal in magnitude, displaced from each other by 120° in phase, and having the same phase sequence as the original phasors
2. Negative-sequence components consisting of three phasors equal in magnitude, displaced from each other by 120° in phase, and having the phase sequence opposite to that of the original phasors
3. Zero-sequence components consisting of three phasors equal in magnitude and with zero phase displacement from each other

It is customary, when solving a problem by symmetrical components, to designate the three phases of the system as a, b, and c in such a manner that the phase sequence of the voltages and currents in the system is abc. Thus the phase sequence of the positive-sequence components of the unbalanced phasors is abc, and the phase sequence of the negative-sequence components is acb. If the original phasors are voltages, they may be designated V_a, V_b, and V_c. The three sets of symmetrical components are designated by the additional subscript 1 for the positive-sequence components, 2 for the negative-sequence components, and 0 for the zero-sequence components. The positive-sequence components of V_a, V_b, and V_c are V_{a1}, V_{b1}, and V_{c1}. Similarly, the negative-sequence components are V_{a2}, V_{b2}, and V_{c2}, and the zero-sequence components are V_{a0}, V_{b0}, and V_{c0}. Figure 11.1 shows three such sets of symmetrical components. Phasors representing currents will be designated by I with subscripts as for voltages.

Since each of the original unbalanced phasors is the sum of its components, the original phasors expressed in terms of their components are

$$V_a = V_{a1} + V_{a2} + V_{a0} \tag{11.1}$$

$$V_b = V_{b1} + V_{b2} + V_{b0} \tag{11.2}$$

$$V_c = V_{c1} + V_{c2} + V_{c0} \tag{11.3}$$

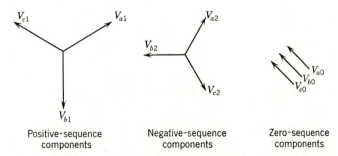

| Positive-sequence components | Negative-sequence components | Zero-sequence components |

Figure 11.1 Three sets of balanced phasors which are the symmetrical components of three unbalanced phasors.

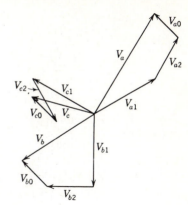

Figure 11.2 Graphical addition of the components shown in Fig. 11.1 to obtain three unbalanced phasors.

The synthesis of a set of three unbalanced phasors from the three sets of symmetrical components of Fig. 11.1 is shown in Fig. 11.2.

The many advantages of analysis of power systems by the method of symmetrical components will become apparent gradually as we apply the method to the study of unsymmetrical faults on otherwise symmetrical systems. It is sufficient to say here that the method consists in finding the symmetrical components of current at the fault. Then the values of current and voltage at various points in the system can be found. The method is simple and leads to accurate predictions of system behavior.

11.2 OPERATORS

Because of the phase displacement of the symmetrical components of the voltages and currents in a three-phase system, it is convenient to have a shorthand method of indicating the rotation of a phasor through 120°. The result of the multiplication of two complex numbers is the product of their magnitudes and the sum of their angles. If the complex number expressing a phasor is multiplied by a complex number of unit magnitude and angle θ, the resulting complex number represents a phasor equal to the original phasor displaced by the angle θ.

The complex number of unit magnitude and associated angle θ is an operator that rotates the phasor on which it operates through the angle θ.

We are already familiar with the operator j, which causes rotation through 90°, and the operator -1, which causes rotation through 180°. Two successive applications of the operator j cause rotation through $90° + 90°$, which leads us to the conclusion that $j \times j$ causes rotation through 180°, and thus we recognize that j^2 is equal to -1. Other powers of the operator j are found by similar analysis.

The letter a is commonly used to designate the operator that causes a rotation of 120° in the counterclockwise direction. Such an operator is a complex number of unit magnitude with an angle of 120° and is defined by

$$a = 1\,\underline{/120°} = 1\varepsilon^{j2\pi/3} = -0.5 + j0.866$$

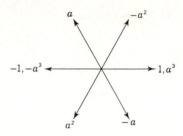

Figure 11.3 Phasor diagram of the various powers of the operator a.

If the operator a is applied to a phasor twice in succession, the phasor is rotated through $240°$. Three successive applications of a rotate the phasor through $360°$. Thus,

$$a^2 = 1\underline{/240°} = -0.5 - j0.866$$

and

$$a^3 = 1\underline{/360°} = 1\underline{/0°} = 1$$

Figure 11.3 shows phasors representing various powers of a.

11.3 THE SYMMETRICAL COMPONENTS OF UNSYMMETRICAL PHASORS

We have seen (Fig. 11.2) the synthesis of three unsymmetrical phasors from three sets of symmetrical phasors. The synthesis was made in accordance with Eqs. (11.1) to (11.3). Now let us examine these same equations to determine how to resolve three unsymmetrical phasors into their symmetrical components.

First, we note that the number of unknown quantities can be reduced by expressing each component of V_b and V_c as the product of some function of the operator a and a component of V_a. Reference to Fig. 11.1 verifies the following relations:

$$V_{b1} = a^2 V_{a1} \qquad V_{c1} = a V_{a1}$$
$$V_{b2} = a V_{a2} \qquad V_{c2} = a^2 V_{a2} \qquad (11.4)$$
$$V_{b0} = V_{a0} \qquad V_{c0} = V_{a0}$$

Repeating Eq. (11.1) and substituting Eqs. (11.4) in Eqs. (11.2) and (11.3) yield

$$V_a = V_{a1} + V_{a2} + V_{a0} \qquad (11.5)$$
$$V_b = a^2 V_{a1} + a V_{a2} + V_{a0} \qquad (11.6)$$
$$V_c = a V_{a1} + a^2 V_{a2} + V_{a0} \qquad (11.7)$$

or in matrix form

$$\begin{bmatrix} V_a \\ V_b \\ V_c \end{bmatrix} = \begin{bmatrix} 1 & 1 & 1 \\ 1 & a^2 & a \\ 1 & a & a^2 \end{bmatrix} \begin{bmatrix} V_{a0} \\ V_{a1} \\ V_{a2} \end{bmatrix} \qquad (11.8)$$

For convenience we let

$$\mathbf{A} = \begin{bmatrix} 1 & 1 & 1 \\ 1 & a^2 & a \\ 1 & a & a^2 \end{bmatrix} \tag{11.9}$$

Then, as may be verified easily,

$$\mathbf{A}^{-1} = \frac{1}{3} \begin{bmatrix} 1 & 1 & 1 \\ 1 & a & a^2 \\ 1 & a^2 & a \end{bmatrix} \tag{11.10}$$

and premultiplying both sides of Eq. (11.8) by $\mathbf{A}^{-1}$ yields

$$\begin{bmatrix} V_{a0} \\ V_{a1} \\ V_{a2} \end{bmatrix} = \frac{1}{3} \begin{bmatrix} 1 & 1 & 1 \\ 1 & a & a^2 \\ 1 & a^2 & a \end{bmatrix} \begin{bmatrix} V_a \\ V_b \\ V_c \end{bmatrix} \tag{11.11}$$

which shows us how to resolve three unsymmetrical phasors into their symmetrical components. These relations are so important that we shall write the separate equations in ordinary fashion. From Eq. (11.11)

$$V_{a0} = \tfrac{1}{3}(V_a + V_b + V_c) \tag{11.12}$$

$$V_{a1} = \tfrac{1}{3}(V_a + aV_b + a^2V_c) \tag{11.13}$$

$$V_{a2} = \tfrac{1}{3}(V_a + a^2V_b + aV_c) \tag{11.14}$$

If required, the components V_{b0}, V_{b1}, V_{b2}, V_{c0}, V_{c1}, and V_{c2} can be found by Eqs. (11.4).

Equation (11.12) shows that no zero-sequence components exist if the sum of the unbalanced phasors is zero. Since the sum of the line-to-line voltage phasors in a three-phase system is always zero, zero-sequence components are never present in the line voltages, regardless of the amount of unbalance. The sum of the three line-to-neutral voltage phasors is not necessarily zero, and voltages to neutral may contain zero-sequence components.

The preceding equations could have been written for any set of related phasors, and we might have written them for currents instead of for voltages. They may be solved either analytically or graphically. Because some of the preceding equations are so fundamental, they are summarized for currents:

$$I_a = I_{a1} + I_{a2} + I_{a0} \tag{11.15}$$

$$I_b = a^2I_{a1} + aI_{a2} + I_{a0} \tag{11.16}$$

$$I_c = aI_{a1} + a^2I_{a2} + I_{a0} \tag{11.17}$$

$$I_{a0} = \tfrac{1}{3}(I_a + I_b + I_c) \tag{11.18}$$

$$I_{a1} = \tfrac{1}{3}(I_a + aI_b + a^2I_c) \tag{11.19}$$

$$I_{a2} = \tfrac{1}{3}(I_a + a^2I_b + aI_c) \tag{11.20}$$

In a three-phase system the sum of the line currents is equal to the current I_n in the return path through the neutral. Thus,

$$I_a + I_b + I_c = I_n \qquad (11.21)$$

Comparing Eqs. (11.18) and (11.21) gives

$$I_n = 3I_{a0} \qquad (11.22)$$

In the absence of a path through the neutral of a three-phase system, I_n is zero, and the line currents contain no zero-sequence components. A Δ-connected load provides no path to neutral, and the line currents flowing to a Δ-connected load can contain no zero-sequence components.

Example 11.1 One conductor of a three-phase line is open. The current flowing to the Δ-connected load through line a is 10 A. With the current in line a as reference and assuming that line c is open, find the symmetrical components of the line currents.

SOLUTION Figure 11.4 is a diagram of the circuit. The line currents are

$$I_a = 10\underline{/0°} \text{ A} \qquad I_b = 10\underline{/180°} \text{ A} \qquad I_c = 0 \text{ A}$$

From Eqs. (11.18) to (11.20)

$$I_{a0} = \tfrac{1}{3}(10\underline{/0°} + 10\underline{/180°} + 0) = 0$$

$$I_{a1} = \tfrac{1}{3}(10\underline{/0°} + 10\underline{/180° + 120°} + 0)$$

$$= 5 - j2.89 = 5.78 \underline{/-30°} \text{ A}$$

$$I_{a2} = \tfrac{1}{3}(10\underline{/0°} + 10\underline{/180° + 240°} + 0)$$

$$= 5 + j2.89 = 5.78\underline{/30°} \text{ A}$$

From Eqs. (11.4)

$$I_{b1} = 5.78 \underline{/-150°} \text{ A} \qquad I_{c1} = 5.78\underline{/90°} \text{ A}$$

$$I_{b2} = 5.78\underline{/150°} \text{ A} \qquad I_{c2} = 5.78 \underline{/-90°} \text{ A}$$

$$I_{b0} = 0 \qquad I_{c0} = 0$$

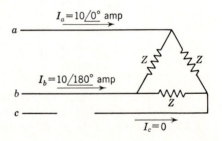

Figure 11.4 Circuit for Example 11.1.

We note that components I_{c1} and I_{c2} have definite values although line c is open and can carry no net current. As is expected, therefore, the sum of the components in line c is zero. Of course, the sum of the components in line a is $10\underline{/0°}$ A, and the sum of the components in line b is $10\underline{/180°}$ A.

11.4 PHASE SHIFT OF SYMMETRICAL COMPONENTS IN Y-Δ TRANSFORMER BANKS

In discussing symmetrical components for three-phase transformers, we need to examine the standard method of marking transformer terminals. In Sec. 6.5 we discussed the placing of dots at one end of each winding on the same iron core of a transformer to indicate that current flowing from the dotted terminal to the unmarked terminal of each winding produced a magnetomotive force acting in the same direction in the magnetic circuit. We noted that, if the small effect of magnetizing current is neglected, two currents I_1 and I_2 flowing in the only two windings on a common transformer core would be in phase if we chose the current to be positive when entering the dotted terminal of one winding and when leaving the dotted terminal of the other winding.

The standard marking of single-phase, two-winding transformers substitutes H_1 and X_1 for the dots on the high- and low-tension windings, respectively. The other ends of the windings are marked H_2 and X_2. Figure 11.5 shows both dots and standard markings, and I_p and I_s must be in phase. In Sec. 6.5 we noted that the dots on the windings of a single-phase transformer indicated that the voltage drops from dotted to unmarked terminals of the two windings are in phase. So in the single-phase transformer the terminals H_1 and X_1 are positive at the same time with respect to H_2 and X_2. If the direction of the arrow marked I_s in Fig. 11.5 were reversed while the direction of the arrow marked I_p remained the same, I_s and I_p would be 180° out of phase. Therefore, the primary and secondary currents are either in phase or 180° out of phase, depending on the direction assumed to be positive for the flow of current. Similarly, primary and secondary voltages may be in phase or 180° out of phase, depending on which terminal is assumed to be positive for specifying the voltage drop of each.

The high-tension terminals of three-phase transformers are marked H_1, H_2, and H_3, and the low-tension terminals are marked X_1, X_2, and X_3. In Y-Y or Δ-Δ transformers the markings are such that voltages to neutral from terminals

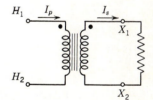

Figure 11.5 Schematic diagram of single-phase transformer windings showing standard markings and the directions assumed positive for primary and secondary current.

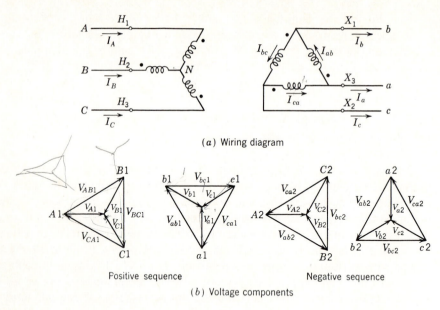

(*a*) Wiring diagram

Positive sequence   Negative sequence

(*b*) Voltage components

Figure 11.6 Wiring diagram and voltage phasors for a three-phase transformer connected Y-Δ where the Y side is the high-tension side.

H_1, H_2, and H_3 are in phase with the voltages to neutral from terminals X_1, X_2, and X_3, respectively.

Figure 11.6*a* is the wiring diagram of a Y-Δ transformer. The high-tension terminals H_1, H_2, and H_3 are connected to phases A, B, and C, respectively, and the phase sequence is ABC. The arrangement and notation of the diagram conform to a convention that we shall follow in all our computations. Windings that are drawn in parallel directions are those linked magnetically by being wound on the same core. When capital letters are assigned to phases on one side of the transformer, lowercase letters will be assigned to the phases on the other side. It is customary to use uppercase letters on the high-tension side of the transformer and lowercase letters on the low-tension side. In Fig. 11.6*a* winding AN is the phase on the Y-connected side which is linked magnetically with the phase winding bc on the Δ-connected side. The location of the dots on the windings shows that V_{AN} is *in phase* with V_{bc}. We shall examine later the case where the Y-connected side is the low-tension winding. If H_1 is the terminal to which line A is connected, it is customary to connect phase B to H_2 and phase C to H_3.

The American standard for designating terminals H_1 and X_1 on Y-Δ transformers requires that the positive-sequence voltage drop from H_1 to neutral lead the positive-sequence voltage drop from X_1 to neutral by 30°, regardless of whether the Y or the Δ winding is on the high-tension side. Similarly, the voltage from H_2 to neutral leads the voltage from X_2 to neutral by 30°, and the voltage from H_3 to neutral leads the voltage from X_3 to neutral by 30°. The phasor

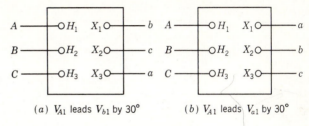

(a) V_{A1} leads V_{b1} by 30° (b) V_{A1} leads V_{a1} by 30°

Figure 11.7 Labeling of lines connected to a three-phase Y-Δ transformer.

diagrams for the sequence components of voltage are shown in Fig. 11.6b. We designate the positive-sequence voltage V_{AN1} as V_{A1} and other voltages to neutral in a similar manner and see that V_{A1} leads V_{b1} by 30°, which enables us to determine that the terminal to which phase b is connected should be labeled X_1.

Figure 11.7a shows the connections of the phases to the transformer terminals so that the positive-sequence voltage to neutral V_{A1} leads the positive-sequence voltage to neutral V_{b1} by 30°. It is not necessary, however, to label the lines attached to the transformer terminals as we have done since no standards have been adopted for such labeling. Very frequently the designation of lines will be as shown in Fig. 11.7b. We shall follow the scheme of Fig. 11.7a which conforms to the wiring and phasor diagrams of Fig. 11.6 since such nomenclature is most convenient for computations. If the scheme of Fig. 11.7b is preferred, it is necessary only to exchange a for b, b for c, and c for a in the work which follows.

Inspection of the positive- and negative-sequence phasor diagrams of Fig. 11.6 shows that V_{a1} leads V_{A1} by 90° and that V_{a2} lags V_{A2} by 90°. The diagrams show V_{A1} and V_{A2} in phase, which is not necessarily true, but phase shift between V_{A1} and V_{A2} does not alter the 90° relation between V_{A1} and V_{a1} or between V_{A2} and V_{a2}.

Since the direction specified for I_A in Fig. 11.6a is away from the dot in the transformer winding and the direction of I_{bc} is also away from the dot in its winding, these currents are 180° out of phase. Therefore, the phase relation between the Y and Δ currents is as shown in Fig. 11.8. We note that I_{a1} leads I_{A1}

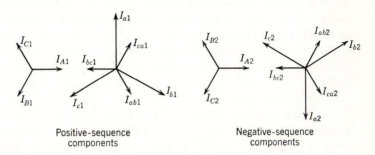

Positive-sequence components

Negative-sequence components

Figure 11.8 Current phasors of a three-phase transformer connected Y-Δ where the Y side is the high-tension side.

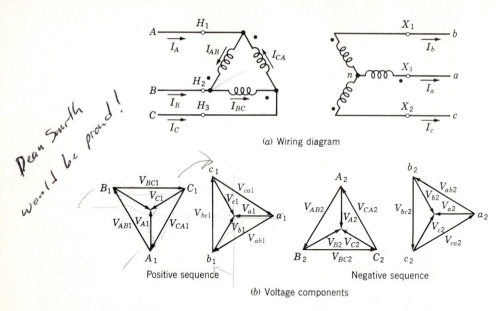

(a) Wiring diagram

Positive sequence   Negative sequence

(b) Voltage components

Figure 11.9 Wiring diagram and voltage phasors for a three-phase transformer connected Y-Δ where the Δ side is the high-tension side.

by 90° and I_{a2} lags I_{A2} by 90°. Summarizing the relations between the symmetrical components of the line currents on the two sides of the transformer gives

$$V_{a1} = +jV_{A1} \qquad I_{a1} = +jI_{A1}$$
$$V_{a2} = -jV_{A2} \qquad I_{a2} = -jI_{A2}$$

$$(11.23)$$

where each voltage and current is expressed in per unit. Transformer impedance and magnetizing current are neglected, which explains why the *per-unit* magnitudes of voltage and current are exactly the same on both sides of the transformer (for instance, $|V_{a1}|$ equals $|V_{A1}|$).

Up to this point our discussion of the Y-Δ transformer has been confined to the case where the high-tension windings are connected in Y. Figure 11.9 shows the Δ-connected windings on the high-tension side of the transformer. The figure shows that in order to have the positive-sequence voltage from H_1 to neutral lead the positive-sequence voltage from X_1 to neutral by 30° V_{BC1} and V_{a1} must be 180° out of phase, and the currents I_{BC1} and I_{a1} must be 180° out of phase as shown in Fig. 11.10. The phasor diagrams for the voltages and currents show that Eqs. (11.23) are still valid.

We have assumed the power flow from the high-tension to the low-tension windings by showing I_A, I_B, and I_C toward the transformer and I_a, I_b, and I_c away from the transformer. If we were assuming power flow in the opposite direction, voltage relationships would remain the same but all line currents would be shown in the opposite direction. However, this would cause no change

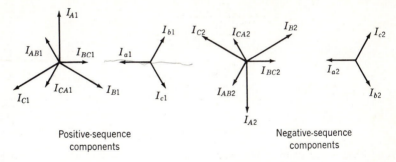

Positive-sequence
components

Negative-sequence
components

Figure 11.10 Current phasors of a three-phase transformer connected Y-Δ where the Δ side is the high-tension side.

in the phase angles of primary and secondary line currents with respect to each other. Therefore Eqs. (11.23) are valid for both voltages and currents regardless of which windings are the primary.

Example 11.2 Three identical resistors are Y-connected to the low-tension Y side of a Δ-Y transformer. The voltages at the resistor load are

$$|V_{ab}| = 0.8 \text{ per unit} \qquad |V_{bc}| = 1.2 \text{ per unit} \qquad |V_{ca}| = 1.0 \text{ per unit}$$

Assume that the neutral of the load is not connected to the neutral of the transformer secondary. Find the line voltages and currents in per unit on the Δ side of the transformer.

SOLUTION Assuming an angle of 180° for V_{ca} and using the law of cosines to find the angles of the other line voltages, we have

$$V_{ab} = 0.8\underline{/82.8°} \text{ per unit}$$

$$V_{bc} = 1.2 \underline{/-41.4°} \text{ per unit}$$

$$V_{ca} = 1.0\underline{/180°} \text{ per unit}$$

The symmetrical components of the line voltages are

$$V_{ab1} = \tfrac{1}{3}(0.8\underline{/82.8°} + 1.2\underline{/120° - 41.4°} + 1.0\underline{/240° + 180°})$$

$$= \tfrac{1}{3}(0.1 + j0.794 + 0.237 + j1.177 + 0.5 + j0.866)$$

$$= 0.279 + j0.946 = 0.985\underline{/73.6°} \text{ per unit (line-line voltage base)}$$

$$V_{ab2} = \tfrac{1}{3}(0.8\underline{/82.8°} + 1.2\underline{/240° - 41.4°} + 1.0\underline{/120° + 180°})$$

$$= \tfrac{1}{3}(1.0 + j0.794 - 1.138 - j0.383 + 0.5 - j0.866)$$

$$= -0.179 - j0.152 = 0.235\underline{/220.3°} \text{ per unit (line-line voltage base)}$$

To determine the positive- and negative-sequence voltages to neutral we need to examine the phase difference between line and phase voltages of

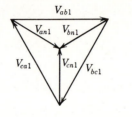

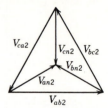

Positive-sequence components   Negative-sequence components

Figure 11.11 Positive- and negative-sequence components of line-to-line and line-to-neutral voltages of a three-phase system.

balanced-Y loads for positive and negative sequences. Consider Fig. 11-11, where V_{ab1} and V_{ab2} are taken as reference arbitrarily. The choice of the reference has no effect on the results. We see that

$$V_{an1} = \frac{1}{\sqrt{3}} V_{ab1} \underline{/-30°} \tag{11.24}$$

and

$$V_{an2} = \frac{1}{\sqrt{3}} V_{ab2} \underline{/30°} \tag{11.25}$$

We find V_{an} as the sum of its components:

$$V_{an} = V_{an1} + V_{an2} \tag{11.26}$$

The other voltages to neutral are found by obtaining their components from V_{an1} and V_{an2} by Eqs. (11.4). If the voltages to neutral are in per unit referred to the base voltage to neutral and the line voltages are in per unit referred to the base voltage from line to line, the $1/\sqrt{3}$ factor must be omitted in Eqs. (11.24) and (11.25). If both voltages are referred to the same base, the equations are correct as given.

The absence of a neutral connection means that zero-sequence currents are not present. Therefore, the phase voltages at the load contain positive- and negative-sequence components only. The phase voltages are found from Eqs. (11.24) and (11.25) with the factor $1/\sqrt{3}$ omitted, since the line voltages are expressed in terms of the base voltage from line to line and the phase voltages are desired in per unit of the base voltage to neutral. Thus

$$V_{an1} = 0.985\underline{/73.6° - 30°}$$

$$= 0.985 \underline{/43.6°} \text{ per unit (line-neutral voltage base)}$$

$$V_{an2} = 0.235\underline{/220.3° + 30°}$$

$$= 0.235\underline{/250.3°} \text{ per unit (line-neutral voltage base)}$$

Since each resistor has an impedance of $1.0\underline{/0^\circ}$ per unit,

$$I_{a1} = \frac{V_{a1}}{1.0\underline{/0^\circ}} = 0.985\underline{/43.6^\circ} \text{ per unit}$$

$$I_{a2} = \frac{V_{a2}}{1.0\underline{/0^\circ}} = 0.235\underline{/250.3^\circ} \text{ per unit}$$

The direction assumed to be positive for the currents is from the supply toward the Δ primary of the transformer and away from the Y side toward the load.

Multiplying both sides of Eqs. (11.23) by j, we obtain for the high-tension side of the transformer

$$V_{A1} = -jV_{a1} = 0.985 \underline{/-46.4^\circ} = 0.680 - j0.713$$

$$V_{A2} = jV_{a2} = 0.235 \underline{/-19.7^\circ} = 0.221 - j0.079$$

$$V_A = V_{A1} + V_{A2} = 0.901 - j0.792$$

$$= 1.20 \underline{/-41.3^\circ} \text{ per unit}$$

$$V_{B1} = a^2 V_{A1} = 0.985\underline{/193.6^\circ} = -0.958 - j0.232$$

$$V_{B2} = aV_{A2} = 0.235\underline{/100.3^\circ} = -0.042 + j0.232$$

$$V_B = V_{B1} + V_{B2} = -1.0$$

$$= 1.0\underline{/180^\circ} \text{ per unit}$$

$$V_{C1} = aV_{A1} = 0.985\underline{/73.6^\circ} = 0.278 + j0.944$$

$$V_{C2} = a^2 V_{A2} = 0.235\underline{/220.3^\circ} = -0.179 - j0.152$$

$$V_C = V_{C1} + V_{C2} = 0.099 + j0.792$$

$$= 0.8\underline{/82.9^\circ} \text{ per unit}$$

$$V_{AB} = V_A - V_B = 0.901 - j0.792 + 1.0 = 1.901 - j0.792$$

$$= 2.06 \underline{/-22.6^\circ} \text{ per unit (line-neutral voltage base)}$$

$$= \frac{2.06}{\sqrt{3}} \underline{/-22.6} = 1.19 \underline{/-22.6^\circ} \text{ per unit (line-line voltage base)}$$

$$V_{BC} = V_B - V_C = -1.0 - 0.099 - j0.792 = -1.099 - j0.792$$

$$= 1.355\underline{/215.8^\circ} \text{ per unit (line-neutral voltage base)}$$

$$= \frac{1.355}{\sqrt{3}} = 0.782\underline{/215.8^\circ} \text{ per unit (line-line voltage base)}$$

$$V_{CA} = V_C - V_A = 0.099 + j0.792 - 0.901 + j0.792 = -0.802 + j1.584$$

$$= 1.78\underline{/116.9^\circ} \text{ per unit (line-neutral voltage base)}$$

$$= \frac{1.78}{\sqrt{3}} \underline{/116.9^\circ} = 1.028\underline{/116.9^\circ} \text{ per unit (line-line voltage base)}$$

Since the load impedance in each phase is resistance of $1.0\underline{/0°}$ per unit, I_{a1} and V_{a1} are found to have identical per-unit values in this problem. Likewise, I_{a2} and V_{a2} are identical in per unit. Therefore, I_A must be identical to V_A expressed in per unit. Thus

$$I_A = 1.20 \underline{/-41.3°} \text{ per unit}$$

$$I_B = 1.0\underline{/180°} \text{ per unit}$$

$$I_C = 0.80\underline{/82.9°} \text{ per unit}$$

When problems involving unsymmetrical faults are solved, positive- and negative-sequence components are found separately and phase shift is taken into account, if necessary, by applying Eq. (11.23). Digital-computer programs can be written to incorporate the effects of phase shift.

11.5 POWER IN TERMS OF SYMMETRICAL COMPONENTS

If the symmetrical components of current and voltage are known, the power expended in a three-phase circuit can be computed directly from the components. Demonstration of this statement is a good example of the matrix manipulation of symmetrical components.

The total complex power flowing into a three-phase circuit through three lines a, b, and c is

$$S = P + jQ = V_a I_a^* + V_b I_b^* + V_c I_c^* \qquad (11.27)$$

where V_a, V_b, and V_c are voltages to neutral at the terminals and I_a, I_b, and I_c are the currents flowing into the circuit in the three lines. A neutral connection may or may not be present. In matrix notation

$$\mathbf{S} = \begin{bmatrix} V_a & V_b & V_c \end{bmatrix} \begin{bmatrix} I_a \\ I_b \\ I_c \end{bmatrix}^* = \begin{bmatrix} V_a \\ V_b \\ V_c \end{bmatrix}^T \begin{bmatrix} I_a \\ I_b \\ I_c \end{bmatrix}^* \qquad (11.28)$$

where the conjugate of a matrix is understood to be composed of elements that are the conjugates of the corresponding elements of the original matrix.

To introduce the symmetrical components of the voltages and currents we make use of Eqs. (11.8) and (11.9) to obtain

$$\mathbf{S} = [\mathbf{AV}]^T[\mathbf{AI}]^* \qquad (11.29)$$

where

$$\mathbf{V} = \begin{bmatrix} V_{a0} \\ V_{a1} \\ V_{a2} \end{bmatrix} \quad \text{and} \quad \mathbf{I} = \begin{bmatrix} I_{a0} \\ I_{a1} \\ I_{a2} \end{bmatrix} \qquad (11.30)$$

The *reversal rule* of matrix algebra states that the transpose of the product of two matrices is equal to the product of the transposes of the matrices in reverse order. According to this rule

$$[\mathbf{AV}]^T = \mathbf{V}^T\mathbf{A}^T \tag{11.31}$$

and so

$$\mathbf{S} = \mathbf{V}^T\mathbf{A}^T[\mathbf{AI}]^* = \mathbf{V}^T\mathbf{A}^T\mathbf{A}^*\mathbf{I}^* \tag{11.32}$$

Noting that $\mathbf{A}^T = \mathbf{A}$ and that a and a^2 are conjugates, we obtain

$$\mathbf{S} = \begin{bmatrix} V_{a0} & V_{a1} & V_{a2} \end{bmatrix} \begin{bmatrix} 1 & 1 & 1 \\ 1 & a^2 & a \\ 1 & a & a^2 \end{bmatrix} \begin{bmatrix} 1 & 1 & 1 \\ 1 & a & a^2 \\ 1 & a^2 & a \end{bmatrix} \begin{bmatrix} I_{a0} \\ I_{a1} \\ I_{a2} \end{bmatrix}^* \tag{11.33}$$

or, since $\mathbf{A}^T\mathbf{A}^*$ is equal to

$$3\begin{bmatrix} 1 & 0 & 0 \\ 0 & 1 & 0 \\ 0 & 0 & 1 \end{bmatrix}$$

$$\mathbf{S} = 3\begin{bmatrix} V_{a0} & V_{a1} & V_{a2} \end{bmatrix} \begin{bmatrix} I_{a0} \\ I_{a1} \\ I_{a2} \end{bmatrix}^* \tag{11.34}$$

So complex power is

$$V_a I_a^* + V_b I_b^* + V_c I_c^* = 3V_0 I_0^* + 3V_1 I_1^* + 3V_2 I_2^* \tag{11.35}$$

which shows how complex power can be computed from the symmetrical components of the voltages and currents of an unbalanced three-phase circuit.

11.6 UNSYMMETRICAL SERIES IMPEDANCES

We shall be concerned particularly with systems that are normally balanced and become unbalanced only upon the occurrence of an unsymmetrical fault. Let us look, however, at the equations of a three-phase circuit when the series impedances are unequal. We shall reach a conclusion that is important in analysis by symmetrical components. Figure 11.12 shows the unsymmetrical part of a system with three unequal series impedances Z_a, Z_b, and Z_c. If we assume no

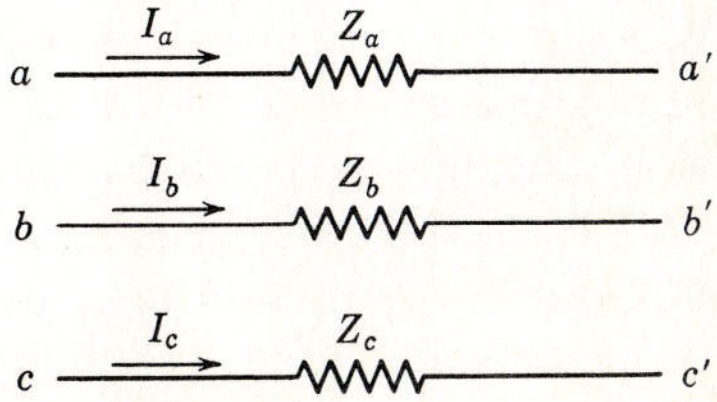

Figure 11.12 Portion of a three-phase system showing three unequal series impedances.

mutual inductance (no coupling) between the three impedances, the voltage drops across the part of the system shown are given by the matrix equation

$$\begin{bmatrix} V_{aa'} \\ V_{bb'} \\ V_{cc'} \end{bmatrix} = \begin{bmatrix} Z_a & 0 & 0 \\ 0 & Z_b & 0 \\ 0 & 0 & Z_c \end{bmatrix} \begin{bmatrix} I_a \\ I_b \\ I_c \end{bmatrix} \tag{11.36}$$

and in terms of the symmetrical components of voltage and current

$$\mathbf{A}\begin{bmatrix} V_{aa'0} \\ V_{aa'1} \\ V_{aa'2} \end{bmatrix} = \begin{bmatrix} Z_a & 0 & 0 \\ 0 & Z_b & 0 \\ 0 & 0 & Z_c \end{bmatrix} \mathbf{A}\begin{bmatrix} I_{a0} \\ I_{a1} \\ I_{a2} \end{bmatrix} \tag{11.37}$$

where $\mathbf{A}$ is the matrix defined by Eq. (11.9). Premultiplying both sides of the equation by $\mathbf{A}^{-1}$ yields the matrix equation from which we obtain

$$V_{aa'1} = \tfrac{1}{3}I_{a1}(Z_a + Z_b + Z_c) + \tfrac{1}{3}I_{a2}(Z_a + a^2 Z_b + a Z_c)$$
$$+ \tfrac{1}{3}I_{a0}(Z_a + a Z_b + a^2 Z_c)$$

$$V_{aa'2} = \tfrac{1}{3}I_{a1}(Z_a + a Z_b + a^2 Z_c) + \tfrac{1}{3}I_{a2}(Z_a + Z_b + Z_c) \tag{11.38}$$
$$+ \tfrac{1}{3}I_{a0}(Z_a + a^2 Z_b + a Z_c)$$

$$V_{aa'0} = \tfrac{1}{3}I_{a1}(Z_a + a^2 Z_b + a Z_c) + \tfrac{1}{3}I_{a2}(Z_a + a Z_b + a^2 Z_c)$$
$$+ \tfrac{1}{3}I_{a0}(Z_a + Z_b + Z_c)$$

If the impedances are made equal (that is, if $Z_a = Z_b = Z_c$), Eqs. (11.38) reduce to

$$V_{aa'1} = I_{a1} Z_a \qquad V_{aa'2} = I_{a2} Z_a \qquad V_{aa'0} = I_{a0} Z_a \tag{11.39}$$

Thus, we conclude that the symmetrical components of unbalanced currents flowing in a balanced-Y load or in balanced series impedances produce voltage drops of like sequence only, provided no coupling exists between phases. If the impedances are unequal, however, Eqs. (11.38) show that the voltage drop of any one sequence is dependent on the currents of all three sequences. If coupling such as mutual inductance existed between the three impedances of Fig. 11.13, the square matrix of Eqs. (11.36) and (11.37) would contain off-diagonal elements and Eqs. (11.38) would have additional terms.

Although current in any conductor of a three-phase transmission line induces a voltage in the other phases, the way in which reactance is calculated eliminates consideration of coupling. The self-inductance calculated on the basis of complete transposition includes the effect of mutual reactance. The assumption of transposition yields equal series impedances. Thus the component currents of any one sequence produce only voltage drops of like sequence in a transmission line; that is, positive-sequence currents produce positive-sequence voltage drops only. Likewise negative-sequence currents produce negative-sequence voltage drops only, and zero-sequence currents produce zero-sequence voltage drops only. Equations (11.38) apply to unbalanced-Y loads because points a', b', and c' may be connected to form a neutral. We could study variations of these equations for special cases such as single-phase loads where $Z_b = Z_c = 0$, but we shall confine our discussion to systems that are balanced before a fault occurs.

11.7 SEQUENCE IMPEDANCES AND SEQUENCE NETWORKS

In any part of a circuit, the voltage drop caused by current of a certain sequence depends on the impedance of that part of the circuit to current of that sequence. The impedance of any section of a balanced network to current of one sequence may be different from impedance to current of another sequence.

The impedance of a circuit when positive-sequence currents alone are flowing is called the *impedance to positive-sequence current*. Similarly, when only negative-sequence currents are present, the impedance is called the *impedance to negative-sequence current*. When only zero-sequence currents are present, the impedance is called the *impedance to zero-sequence current*. These names of the impedances of a circuit to currents of the different sequences are usually shortened to the less descriptive terms *positive-sequence impedance*, *negative-sequence impedance*, and *zero-sequence impedance*.

The analysis of an unsymmetrical fault on a symmetrical system consists in finding the symmetrical components of the unbalanced currents that are flowing. Since the component currents of one phase sequence cause voltage drops of like sequence only and are independent of currents of other sequences, in a balanced system, currents of any one sequence may be considered to flow in an independent network composed of the impedances to the current of that sequence only. The single-phase equivalent circuit composed of the impedances to current of any one sequence only is called the *sequence network* for that particular sequence. The sequence network includes any generated emfs of like sequence. Sequence networks carrying the currents I_{a1}, I_{a2}, and I_{a0} are interconnected to represent various unbalanced fault conditions. Therefore, to calculate the effect of a fault by the method of symmetrical components, it is essential to determine the sequence impedances and to combine them to form the sequence networks.

11.8 SEQUENCE NETWORKS OF UNLOADED GENERATORS

An unloaded generator, grounded through a reactor, is shown in Fig. 11.13. When a fault (not indicated in the figure) occurs at the terminals of the generator, currents I_a, I_b, and I_c flow in the lines. If the fault involves ground, the current flowing into the neutral of the generator is designated I_n. One or two of the line currents may be zero, but the currents can be resolved into their symmetrical components regardless of how unbalanced they may be.

Drawing the sequence networks is simple. The generated voltages are of positive sequence only, since the generator is designed to supply balanced three-phase voltages. Therefore, the positive-sequence network is composed of an emf

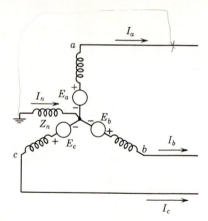

Figure 11.13 Circuit diagram of an unloaded generator grounded through a reactance. The emfs of each phase are E_a, E_b, and E_c.

in series with the positive-sequence impedance of the generator. The negative- and zero-sequence networks contain no emfs but include the impedances of the generator to negative- and zero-sequence currents, respectively. The sequence components of current are shown in Fig. 11.14. They are flowing through impedances of their own sequence only, as indicated by the appropriate subscripts on the impedances shown in the figure. The sequence networks shown in Fig. 11.14 are the single-phase equivalent circuits of the balanced three-phase circuits through which the symmetrical components of the unbalanced currents are considered to flow. The generated emf in the positive-sequence network is the no-load terminal voltage to neutral, which is also equal to the transient and subtransient internal voltages since the generator is not loaded. The reactance in the positive-sequence network is the subtransient, transient, or synchronous reactance, depending on whether subtransient, transient, or steady-state conditions are being studied.

The reference bus for the positive- and negative-sequence networks is the neutral of the generator. So far as positive- and negative-sequence components are concerned, the neutral of the generator is at ground potential if there is a connection between neutral and ground having a finite or zero impedance since the connection will carry no positive- or negative-sequence current.

The current flowing in the impedance Z_n between neutral and ground is $3I_{a0}$. By referring to Fig. 11.14e, we see that the voltage *drop* of zero sequence *from point a to ground* is $-3I_{a0}Z_n - I_{a0}Z_{g0}$, where Z_{g0} is the zero-sequence impedance per phase of the generator. The zero-sequence network, which is a single-phase circuit assumed to carry only the zero-sequence current of one phase, must therefore have an impedance of $3Z_n + Z_{g0}$, as shown in Fig. 11.14f. The total zero-sequence impedance through which I_{a0} flows is

$$Z_0 = 3Z_n + Z_{g0} \tag{11.40}$$

Usually the components of current and voltage for phase a are found from equations determined by the sequence networks. The equations for the components of voltage drop from point a of phase a to the reference bus (or ground) are,

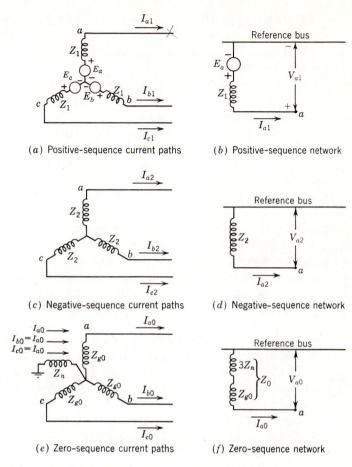

(a) Positive-sequence current paths (b) Positive-sequence network

(c) Negative-sequence current paths (d) Negative-sequence network

(e) Zero-sequence current paths (f) Zero-sequence network

Figure 11.14 Paths for current of each sequence in a generator and the corresponding sequence networks.

as may be deduced from Fig. 11.14,

$$V_{a1} = E_a - I_{a1} Z_1 \tag{11.41}$$

$$V_{a2} = -I_{a2} Z_2 \tag{11.42}$$

$$V_{a0} = -I_{a0} Z_0 \tag{11.43}$$

where E_a is the positive-sequence no-load voltage to neutral, Z_1 and Z_2 are the positive- and negative-sequence impedances of the generator, and Z_0 is defined by Eq. (11.40). The above equations, which apply to any generator carrying unbalanced currents, are the starting points for the derivation of equations for the components of current for different types of faults. They apply to the case of a loaded generator under steady-state conditions. When computing transient or subtransient conditions the equations apply to a loaded generator if E_g' or E_g'' is substituted for E_a.

11.9 SEQUENCE IMPEDANCES OF CIRCUIT ELEMENTS

The positive- and negative-sequence impedances of linear, symmetrical, static circuits are identical because the impedance of such circuits is independent of phase order provided the applied voltages are balanced. The impedance of a transmission line to zero-sequence currents differs from the impedance to positive- and negative-sequence currents.

The impedances of rotating machines to currents of the three sequences will generally be different for each sequence. The mmf produced by negative-sequence armature current rotates in the direction opposite to that of the rotor on which is the dc field winding. Unlike the flux produced by positive-sequence current, which is stationary with respect to the rotor, the flux produced by the negative-sequence current is sweeping rapidly over the face of the rotor. The currents induced in the field and damper windings by the rotating armature flux keep the flux from penetrating the rotor. This condition is similar to the rapidly changing flux immediately upon the occurrence of a short circuit at the terminals of a machine. The flux path is the same as that encountered in evaluating subtransient reactance. So, in a cylindrical-rotor machine subtransient and negative-sequence reactances are equal. Values given in Table A.4 confirm this statement.

When only zero-sequence current flows in the armature winding of a three-phase machine, the current and mmf of one phase are a maximum at the same time as the current and mmf of each of the other phases. The windings are so distributed around the circumference of the armature that the point of maximum mmf produced by one phase is displaced 120 electrical degrees in space from the point of maximum mmf of each of the other phases. If the mmf produced by the current of each phase had a perfectly sinusoidal distribution in space, a plot of mmf around the armature would result in three sinusoidal curves whose sum would be zero at every point. No flux would be produced across the air gap, and the only reactance of any phase winding would be that due to leakage and end turns. In an actual machine, the winding is not distributed to produce perfectly sinusoidal mmf. The flux resulting from the sum of the mmfs is very small but makes the zero-sequence reactance somewhat higher than in the ideal case where there is no air-gap flux due to zero-sequence current.†

In deriving the equations for inductance and capacitance of transposed transmission lines, we assumed balanced three-phase currents and did not specify phase order. Therefore, the resulting equations are valid for both positive- and negative-sequence impedances. When only zero-sequence current flows in a transmission line, the current in each phase is identical. The current

---

† The reader who wishes to study machine impedances should consult such books as Central Station Engineers of Westinghouse Electric Corporation, *Electrical Transmission and Distribution Reference Book*, 4th ed., chap. 6, pp. 145–194, East Pittsburgh, Pa., 1964; or A. E. Fitzgerald, C. Kingsley, Jr., and A. Kusko, *Electric Machinery*, 3d ed., chaps. 6 and 9, McGraw-Hill Book Company, New York, 1971.

returns through the ground, through overhead ground wires, or through both. Because zero-sequence current is identical in each phase conductor (rather than equal only in magnitude and displaced in phase by 120° from other phase currents), the magnetic field due to zero-sequence current is very different from the magnetic field caused by either positive- or negative-sequence current. The difference in magnetic field results in the zero-sequence inductive reactance of overhead transmission lines being 2 to 3.5 times as large as the positive-sequence reactance. The ratio is toward the higher portion of the specified range for double-circuit lines and lines without ground wires.

A transformer in a three-phase circuit may consist of three individual single-phase units, or it may be a three-phase transformer. Although the zero-sequence series impedances of three-phase units may differ slightly from the positive- and negative-sequence values, it is customary to assume that series impedances of all sequences are equal regardless of the type of transformer. Table A.5 lists transformer reactances. Reactance and impedance are almost equal for transformers of 1000 kVA or larger. For simplicity in our calculations we shall omit shunt admittance, which accounts for exciting current.

The zero-sequence impedance of balanced Y- and Δ-connected loads equals the positive- and negative-sequence impedance. The zero-sequence network for such loads is discussed in Sec. 11.11.

11.10 POSITIVE- AND NEGATIVE-SEQUENCE NETWORKS

The object of obtaining the values of the sequence impedances of a power system is to enable us to construct the sequence networks for the complete system. The network of a particular sequence shows all the paths for the flow of current of that sequence in the system.

We discussed the construction of some rather complex positive-sequence networks in Chap. 6. The transition from a positive-sequence network to a negative-sequence network is simple. Three-phase synchronous generators and motors have internal voltages of positive sequence only, since they are designed to generate balanced voltages. Since the positive- and negative-sequence impedances are the same in a static symmetrical system, conversion of a positive-sequence network to a negative-sequence network is accomplished by changing, if necessary, only the impedances that represent rotating machinery and by omitting the emfs. Electromotive forces are omitted on the assumption of balanced generated voltages and the absence of negative-sequence voltages induced from outside sources.

Since all the neutral points of a symmetrical three-phase system are at the same potential when balanced three-phase currents are flowing, all the neutral points must be at the same potential for either positive- or negative-sequence currents. Therefore the neutral of a symmetrical three-phase system is the logical reference potential for specifying positive- and negative-sequence voltage drops

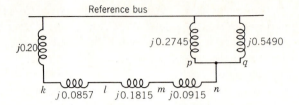

Figure 11.15 Negative-sequence network for Example 11.3.

and is the reference bus of the positive- and negative-sequence networks. Impedance connected between the neutral of a machine and ground is not a part of either the positive- or negative-sequence network because neither positive- nor negative-sequence current can flow in an impedance so connected.

Negative-sequence networks, like the positive-sequence networks of Chap. 6, may contain the exact equivalent circuits of parts of the system or may be simplified by omitting series resistance and shunt admittance.

Example 11.3 Draw the negative-sequence network for the system described in Example 6.10. Assume that the negative-sequence reactance of each machine is equal to its subtransient reactance. Omit resistance.

SOLUTION Since all the negative-sequence reactances of the system are equal to the positive-sequence reactances, the negative-sequence network is identical to the positive-sequence network of Fig. 6.30 except for the omission of emfs from the negative-sequence network. The required network is drawn in Fig. 11.15.

11.11 ZERO-SEQUENCE NETWORKS

A three-phase system operates single phase insofar as the zero-sequence currents are concerned, for the zero-sequence currents are the same in magnitude and phase at any point in all the phases of the system. Therefore zero-sequence currents will flow only if a return path exists through which a completed circuit is provided. The reference for zero-sequence voltages is the potential of the ground at the point in the system at which any particular voltage is specified. Since zero-sequence currents may be flowing in the ground, the ground is not necessarily at the same potential at all points and the reference bus of the zero-sequence network does not represent a ground of uniform potential. The impedance of the ground and ground wires is included in the zero-sequence impedance of the transmission line, and the return circuit of the zero-sequence network is a conductor of zero impedance, which is the reference bus of the system. It is because the impedance of the ground is included in the zero-sequence impedance that voltages measured to the reference bus of the zero-sequence network give the correct voltage to ground.

If a circuit is Y-connected, with no connection from the neutral to ground or to another neutral point in the circuit, the sum of the currents flowing into the neutral in the three phases is zero. Since currents whose sum is zero have no zero-sequence components, the impedance to zero-sequence current is infinite beyond the neutral point; this fact is indicated by an open circuit in the zero-sequence network between the neutral of the Y-connected circuit and the reference bus, as shown in Fig. 11.16a.

If the neutral of a Y-connected circuit is grounded through zero impedance, a zero-impedance connection is inserted to connect the neutral point and the reference bus of the zero-sequence network, as shown in Fig. 11.16b.

If the impedance Z_n is inserted between the neutral and ground of a Y-connected circuit, an impedance of $3Z_n$ must be placed between the neutral and reference bus of the zero-sequence network, as shown in Fig. 11.16c. As explained in Sec. 11.8, the zero-sequence voltage drop caused in the zero-sequence network by I_{a0} flowing through $3Z_n$ is the same as in the actual system where

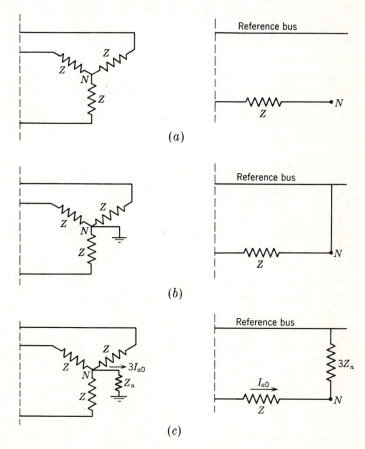

Figure 11.16 Zero-sequence networks for Y-connected loads.

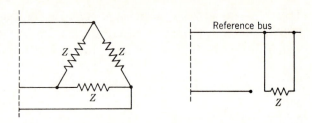

Figure 11.17 Δ-connected load and its zero-sequence network.

$3I_{a0}$ flows through Z_n. Impedance consisting of a resistor or reactor is usually connected between the neutral of a generator and ground to limit the zero-sequence current during a fault. The impedance of such a current-limiting resistor or reactor is represented in the zero-sequence network in the manner described.

A Δ-connected circuit, since it can provide no return path, offers infinite impedance to zero-sequence line currents. The zero-sequence network is open at the Δ-connected circuit. Zero-sequence currents may circulate inside the Δ circuit since the Δ is a closed series circuit for circulating single-phase currents. Such currents would have to be produced in the Δ, however, by induction from an outside source or by zero-sequence generated voltages. A Δ circuit and its zero-sequence network are shown in Fig. 11.17. Even when zero-sequence voltages are generated in the phases of the Δ, no zero-sequence voltage exists between the Δ terminals, for the rise in voltage in each phase of the generator is matched by the voltage drop in the zero-sequence impedance of each phase.

The zero-sequence equivalent circuits of three-phase transformers deserve special attention. The various possible combinations of the primary and secondary windings in Y and Δ alter the zero-sequence network. Transformer theory enables us to construct the equivalent circuit for the zero-sequence network. We remember that no current flows in the primary of a transformer unless current flows in the secondary, if we neglect the relatively small magnetizing current. We know, also, that the primary current is determined by the secondary current and the turns ratio of the windings, again with magnetizing current neglected. These principles guide us in the analysis of individual cases. Five possible connections of two-winding transformers will be discussed. These connections are shown in Fig. 11.18. The arrows on the connection diagrams show the possible paths for the flow of zero-sequence current. Absence of an arrow indicates that the transformer connection is such that zero-sequence current cannot flow. The zero-sequence approximately equivalent circuit, with resistance and a path for magnetizing current omitted, is shown in Fig. 11.18 for each connection. The letters P and Q identify corresponding points on the connection diagram and equivalent circuit. The reasoning to justify the equivalent circuit for each connection follows.

CASE 1: *Y-Y Bank, One Neutral Grounded* If either one of the neutrals of a Y-Y bank is ungrounded, zero-sequence current cannot flow in either winding.

The absence of a path through one winding prevents current in the other. An open circuit exists for zero-sequence current between the two parts of the system connected by the transformer.

CASE 2: *Y-Y Bank, Both Neutrals Grounded* Where both neutrals of a Y-Y bank are grounded, a path through the transformer exists for zero-sequence currents in both windings. Provided the zero-sequence current can follow a complete circuit outside the transformer on both sides, it can flow in both windings of the transformer. In the zero-sequence network, points on the two sides of the transformer are connected by the zero-sequence impedance of the transformer in the same manner as in the positive- and negative-sequence networks.

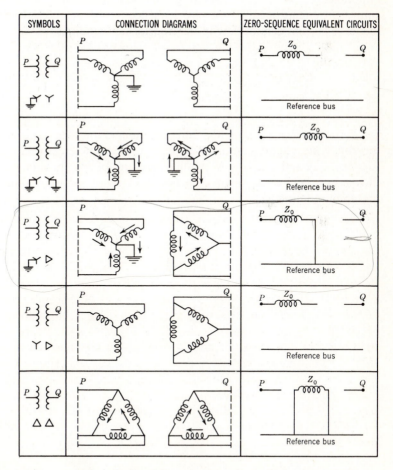

Figure 11.18 Zero-sequence equivalent circuits of three-phase transformer banks, together with diagrams of connections and the symbols for one-line diagrams.

CASE 3: *Y-Δ Bank, Grounded Y* If the neutral of a Y-Δ bank is grounded, zero-sequence currents have a path to ground through the Y because corresponding induced currents can circulate in the Δ. The zero-sequence current circulating in the Δ to balance the zero-sequence current in the Y cannot flow in the lines connected to the Δ. The equivalent circuit must provide for a path from the line on the Y side through the equivalent resistance and leakage reactance of the transformer to the reference bus. An open circuit must exist between the line and the reference bus on the Δ side. If the connection from neutral to ground contains an impedance Z_n, the zero-sequence equivalent circuit must have an impedance of $3Z_n$ in series with the equivalent resistance and leakage reactance of the transformer to connect the line on the Y side to ground.

CASE 4: *Y-Δ Bank, Ungrounded Y* An ungrounded Y is a case where the impedance Z_n between neutral and ground is infinite. The impedance $3Z_n$ in the equivalent circuit of case 3 for zero-sequence impedance becomes infinite. Zero-sequence current cannot flow in the transformer windings.

CASE 5: *Δ-Δ Bank* Since a Δ circuit provides no return path for zero-sequence current, no zero-sequence current can flow into a Δ-Δ bank, although it can circulate within the Δ windings.

Zero-sequence equivalent circuits determined for various parts of the system separately are readily combined to form the complete zero-sequence network. Figures 11.19 and 11.20 show one-line diagrams of two small power systems and their corresponding zero-sequence networks simplified by omitting resistances and shunt admittances.

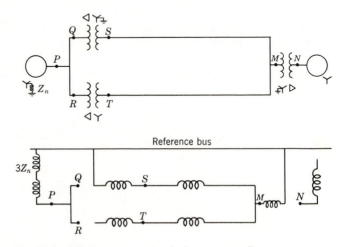

Figure 11.19 One-line diagram of a small power system and the corresponding zero-sequence network.

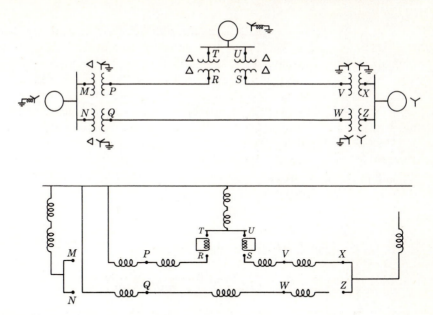

Figure 11.20 One-line diagram of a small power system and the corresponding zero-sequence network.

Example 11.4 Draw the zero-sequence network for the system described in Example 6.10. Assume zero-sequence reactances for the generators and motors of 0.05 per unit. Current limiting reactors of 0.4 Ω each are in the neutral of the generator and the larger motor. The zero-sequence reactance of the transmission line is 1.5 Ω/km.

SOLUTION The zero-sequence leakage reactance of transformers is equal to the positive-sequence reactance. So, for the transformers, $X_0 = 0.0857$ per unit and 0.0915 per unit, as in Example 6.10.

Zero-sequence reactances of the generator and motors are:

Generator: $X_0 = 0.05$ per unit

Motor 1: $X_0 = 0.05 \dfrac{300}{200} \left(\dfrac{13.2}{13.8}\right)^2 = 0.0686$ per unit

Motor 2: $X_0 = 0.05 \dfrac{300}{100} \left(\dfrac{13.2}{13.8}\right)^2 = 0.1372$ per unit

In the generator circuit

$$\text{Base } Z = \frac{(20)^2}{300} = 1.333 \ \Omega$$

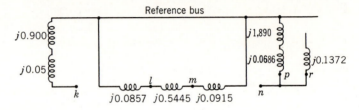

Figure 11.21 Zero-sequence network for Example 11.4.

and in the motor circuit

$$\text{Base } Z = \frac{(13.8)^2}{300} = 0.635 \ \Omega$$

In the impedance network for the generator

$$3Z_n = 3 \ \frac{0.4}{1.333} = 0.900 \text{ per unit}$$

and for the motor

$$3Z_n = 3 \ \frac{0.4}{0.635} = 1.890 \text{ per unit}$$

For the transmission line

$$X_0 = \frac{1.5 \times 64}{176.3} = 0.5445 \text{ per unit}$$

The zero-sequence network is shown in Fig. 11.21.

11.12 CONCLUSIONS

Unbalanced voltages and currents can be resolved into their symmetrical components. Problems are solved by treating each set of components separately and superimposing the results.

In balanced networks having no coupling between phases the currents of one phase sequence induce voltage drops of like sequence only. Impedances of circuit elements to currents of different sequences are not necessarily equal.

A knowledge of the positive-sequence network is necessary for load studies on power systems, for fault calculations, and for stability studies. If the fault calculations or stability studies involve unsymmetrical faults on otherwise symmetrical systems, the negative- and zero-sequence networks are needed also. Synthesis of the zero-sequence network requires particular care, because the zero-sequence network may differ from the others considerably.

PROBLEMS

$I_{ab}=$

11.1 Evaluate the following expressions in polar form:

(a) $a - 1$
(b) $1 - a^2 + a$
(c) $a^2 + a + j$
(d) $ja + a^2$

11.2 If $V_{an1} = 50\underline{/0°}$, $V_{an2} = 20\underline{/90°}$, and $V_{an0} = 10\underline{/180°}$ V, determine analytically the voltages to neutral V_{an}, V_{bn} and V_{cn}, and also show graphically the sum of the given symmetrical components which determine the line-to-neutral voltages.

11.3 When a generator has terminal a open and the other two terminals are connected to each other with a short circuit from this connection to ground, typical values for the symmetrical components of current in phase a are $I_{a1} = 600\underline{/-90°}$ A, $I_{a2} = 250\underline{/90°}$ A, and $I_{a0} = 350\underline{/90°}$ A. Find the current into the ground and the current in each phase of the generator.

11.4 Determine the symmetrical components of the three currents $I_a = 10\underline{/0°}$, $I_b = 10\underline{/230°}$, and $I_c = 10\underline{/130°}$ A.

11.5 The currents flowing in the lines toward a balanced load connected in Δ are $I_a = 100\underline{/0°}$ $I_b = 141.4\underline{/225°}$, and $I_c = 100\underline{/90°}$ A. Determine a general expression for the relation between the symmetrical components of the line currents flowing toward a Δ-connected load and the phase currents in the load, that is, between I_{a1} and I_{ab1} and between I_{a2} and I_{ab2}. Start by drawing phasor diagrams of the positive- and negative-sequence line and phase currents. Find I_{ab} in amperes from the symmetrical components of the given line currents.

11.6 The voltages at the terminals of a balanced load consisting of three 10-Ω resistors connected in Y are $V_{ab} = 100\underline{/0°}$, $V_{bc} = 80.8\underline{/-121.44°}$, and $V_{ca} = 90\underline{/130°}$ V. Determine a general expression for the relation between the symmetrical components of the line and phase voltages, that is, between V_{ab1} and V_{an1} and between V_{ab2} and V_{an2}. Assume that there is no connection to the neutral of the load. Find the line currents from the symmetrical components of the given line voltages.

11.7 Find the power expended in the three 10-Ω resistors of Prob. 11.6 from the symmetrical components of currents and voltages. Check the answer.

11.8 Three single-phase transformers are connected as shown in Fig. 11.22 to form a Y-Δ transformer. The high-tension windings are Y-connected with polarity marks as indicated. Magnetically coupled windings are drawn in parallel directions. Determine the correct placement of polarity marks on the low-tension windings. Identify the numbered terminals on the low-tension side (a) with the letters a, b, and c where I_{A1} leads I_{a1} by 30° and (b) with the letters a', b', and c' so that $I_{a'1}$ is 90° out of phase with I_{A1}.

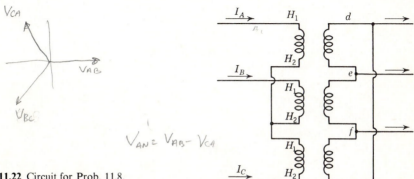

Figure 11.22 Circuit for Prob. 11.8.

11.9 Assume that the currents specified in Prob. 11.5 are flowing toward a load from lines connected to the Y side of a Δ-Y transformer rated 10 MVA, 13.2Δ/66Y kV. Determine the currents flowing in the lines on the Δ side by converting the symmetrical components of the currents to per unit on the base of the transformer rating and by shifting the components according to Eq. (11.23). Check the results by computing the currents in each phase of the Δ windings in amperes directly from the currents on the Y side by multiplying by the turns ratio of the windings. Complete the check by computing the line currents from the phase currents on the Δ side.

11.10 Balanced three-phase voltages of 100 V line-to-line are applied to a Y-connected load consisting of three resistors. The neutral of the load is not grounded. The resistance in phase a is 10 Ω, in phase b is 20 Ω, and in phase c is 30 Ω. Select voltage to neutral of the three-phase line as reference and determine the current in phase a and the voltage V_{an}.

11.11 Draw the negative- and zero-sequence impedance networks for the power system of Prob. 6.15. Mark the values of all reactances in per unit on a base of 50 MVA, 13.8 kV in the circuit of generator 1. Letter the networks to correspond to the one-line diagram. The neutrals of generators 1 and 3 are connected to ground through current-limiting reactors having a reactance of 5%, each on the base of the machine to which it is connected. Each generator has negative- and zero-sequence reactances of 20 and 5%, respectively, on its own rating as base. The zero-sequence reactance of the transmission line is 210 Ω from B to C and 250 Ω from C to E.

11.12 Draw the negative- and zero-sequence impedance networks for the power system of Prob. 6.16. Choose a base of 50 MVA, 138 kV in the 40-Ω transmission line, and mark all reactances in per unit. The negative-sequence reactance of each synchronous machine is equal to its subtransient reactance. The zero-sequence reactance of each machine is 8% based on its own rating. The neutrals of the machines are connected to ground through current-limiting reactors having a reactance of 5%, each on the base of the machine to which it is connected. Assume that the zero-sequence reactances of the transmission lines are 300% of their positive-sequence reactances.

TWELVE

UNSYMMETRICAL FAULTS

Most of the faults that occur on power systems are unsymmetrical faults, which may consist of unsymmetrical short circuits, unsymmetrical faults through impedances, or open conductors. Unsymmetrical faults occur as single line-to-ground faults, line-to-line faults, or double line-to-ground faults. The path of the fault current from line to line or line to ground may or may not contain impedance. One or two open conductors result in unsymmetrical faults, either through the breaking of one or two conductors or through the action of fuses and other devices that may not open the three phases simultaneously.

Since any unsymmetrical fault causes unbalanced currents to flow in the system, the method of symmetrical components is very useful in an analysis to determine the currents and voltages in all parts of the system after the occurrence of the fault. First, we shall discuss faults at the terminals of an unloaded generator. Then we shall consider faults on a power system by applying Thévenin's theorem, which allows us to find the current in the fault by replacing the entire system by a single generator and series impedance. Finally, we shall investigate the bus impedance matrix as applied to the analysis of unsymmetrical faults.

Regardless of the type of fault which occurs at the terminals of a generator we can apply Eqs. (11.41) to (11.43), derived in Sec. 11.8. In matrix form these equations become

$$
\begin{bmatrix} V_{a0} \\ V_{a1} \\ V_{a2} \end{bmatrix} = \begin{bmatrix} 0 \\ E_a \\ 0 \end{bmatrix} - \begin{bmatrix} Z_0 & 0 & 0 \\ 0 & Z_1 & 0 \\ 0 & 0 & Z_2 \end{bmatrix} \begin{bmatrix} I_{a0} \\ I_{a1} \\ I_{a2} \end{bmatrix} \tag{12.1}
$$

For each type of fault we shall use Eq. (12.1), together with equations that describe conditions at the fault, to derive I_{a1} in terms of E_a, Z_1, Z_2, and Z_0.

12.1 SINGLE LINE-TO-GROUND FAULT ON AN UNLOADED GENERATOR

The circuit diagram for a single line-to-ground fault on an unloaded Y-connected generator with its neutral grounded through a reactance is shown in Fig. 12.1, where phase a is the one on which the fault occurs. The relations to be developed for this type of fault will apply only when the fault is on phase a, but this should cause no difficulty since the phases are labeled arbitrarily and any phase can be designated as phase a. The conditions at the fault are expressed by the following equations:

$$I_b = 0 \qquad I_c = 0 \qquad V_a = 0$$

With $I_b = 0$ and $I_c = 0$ the symmetrical components of current are given by

$$\begin{bmatrix} I_{a0} \\ I_{a1} \\ I_{a2} \end{bmatrix} = \frac{1}{3} \begin{bmatrix} 1 & 1 & 1 \\ 1 & a & a^2 \\ 1 & a^2 & a \end{bmatrix} \begin{bmatrix} I_a \\ 0 \\ 0 \end{bmatrix}$$

so that I_{a0}, I_{a1}, and I_{a2} each equal $I_a/3$ and

$$I_{a1} = I_{a2} = I_{a0} \tag{12.2}$$

Substituting I_{a1} for I_{a2} and I_{a0} in Eq. (12.1), we obtain

$$\begin{bmatrix} V_{a0} \\ V_{a1} \\ V_{a2} \end{bmatrix} = \begin{bmatrix} 0 \\ E_a \\ 0 \end{bmatrix} - \begin{bmatrix} Z_0 & 0 & 0 \\ 0 & Z_1 & 0 \\ 0 & 0 & Z_2 \end{bmatrix} \begin{bmatrix} I_{a1} \\ I_{a1} \\ I_{a1} \end{bmatrix} \tag{12.3}$$

Performing the indicated matrix multiplication and subtraction yields an equality of two column matrices. Premultiplying both column matrices by the row matrix $[1 \quad 1 \quad 1]$ gives

$$V_{a0} + V_{a1} + V_{a2} = -I_{a1} Z_0 + E_a - I_{a1} Z_1 - I_{a1} Z_2 \tag{12.4}$$

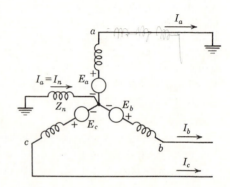

Figure 12.1 Circuit diagram for a single line-to-ground fault on phase a at the terminals of an unloaded generator whose neutral is grounded through a reactance.

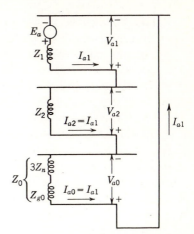

Figure 12.2 Connection of the sequence networks of an unloaded generator for a single line-to-ground fault on phase a at the terminals of the generator.

Since $V_a = V_{a0} + V_{a1} + V_{a2} = 0$, we solve Eq. (12.4) for I_{a1} and obtain

$$I_{a1} = \frac{E_a}{Z_1 + Z_2 + Z_0} \qquad (12.5)$$

Equations (12.2) and (12.5) are the special equations for a single line-to-ground fault. They are used with Eq. (12.1) and the symmetrical-component relations to determine all the voltages and currents at the fault. If the three sequence networks of the generator are connected in series, as shown in Fig. 12.2, we see that the currents and voltages resulting therefrom satisfy the equations above, for the three sequence impedances are then in series with the voltage E_a. With the sequence networks so connected, the voltage across each sequence network is the symmetrical component of V_a of that sequence. The connection of the sequence networks as shown in Fig. 12.2 is a convenient means of remembering the equations for the solution of the single line-to-ground fault, for all the necessary equations can be determined from the sequence-network connection.

If the neutral of the generator is not grounded, the zero-sequence network is open-circuited and Z_0 is infinite. Since Eq. (12.5) shows that I_{a1} is zero when Z_0 is infinite, I_{a2} and I_{a0} must be zero. Thus no current flows in line a since I_a is the sum of its components, all of which are zero. The same result can be seen without the use of symmetrical components since inspection of the circuit shows that no path exists for the flow of current in the fault unless there is a ground at the generator neutral.

Example 12.1 A salient-pole generator without dampers is rated 20 MVA, 13.8 kV and has a direct-axis subtransient reactance of 0.25 per unit. The negative- and zero-sequence reactances are, respectively, 0.35 and 0.10 per unit. The neutral of the generator is solidly grounded. Determine the subtransient current in the generator and the line-to-line voltages for subtransient conditions when a single line-to-ground fault occurs at the generator terminals with the generator operating unloaded at rated voltage. Neglect resistance.

SOLUTION On a base of 20 MVA, 13.8 kV, $E_a = 1.0$ per unit since the internal voltage is equal to the terminal voltage at no load. Then, in per unit,

$$I_{a1} = \frac{E_a}{Z_1 + Z_2 + Z_0} = \frac{1.0 + j0}{j0.25 + j0.35 + j0.10} = -j1.43 \text{ per unit}$$

$$I_a = 3I_{a1} = -j4.29 \text{ per unit}$$

$$\text{Base current} = \frac{20,000}{\sqrt{3} \times 13.8} = 837 \text{ A}$$

Subtransient current in line a is

$$I_a = -j4.29 \times 837 = -j3,590 \text{ A}$$

The symmetrical components of the voltage from point a to ground are

$$V_{a1} = E_a - I_{a1} Z_1 = 1.0 - (-j1.43)(j0.25)$$
$$= 1.0 - 0.357 = 0.643 \text{ per unit}$$

$$V_{a2} = -I_{a2} Z_2 = -(-j1.43)(j0.35) = -0.50 \text{ per unit}$$

$$V_{a0} = -I_{a0} Z_0 = -(-j1.43)(j0.10) = -0.143 \text{ per unit}$$

Line-to-ground voltages are

$$V_a = V_{a1} + V_{a2} + V_{a0} = 0.643 - 0.50 - 0.143 = 0$$

$$V_b = a^2 V_{a1} + a V_{a2} + V_{a0}$$
$$= 0.643(-0.5 - j0.866) - 0.50(-0.5 + j0.866) - 0.143$$
$$= -0.322 - j0.557 + 0.25 - j0.433 - 0.143$$
$$= -0.215 - j0.990 \text{ per unit}$$

$$V_c = a V_{a1} + a^2 V_{a2} + V_{a0}$$
$$= 0.643(-0.5 + j0.866) - 0.50(-0.5 - j0.866) - 0.143$$
$$= -0.322 + j0.557 + 0.25 + j0.433 - 0.143$$
$$= -0.215 + j0.990 \text{ per unit}$$

Line-to-line voltages are

$$V_{ab} = V_a - V_b = 0.215 + j0.990 = 1.01 \underline{/77.7°} \text{ per unit}$$

$$V_{bc} = V_b - V_c = 0 - j1.980 = 1.980 \underline{/270°} \text{ per unit}$$

$$V_{ca} = V_c - V_a = -0.215 + j0.990 = 1.01 \underline{/102.3°} \text{ per unit}$$

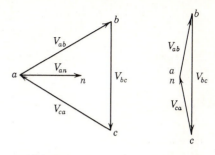

Figure 12.3 Phasor diagrams of the line voltages of Example 12.1 before and after the fault.

(*a*) Prefault   (*b*) Postfault

Since the generated voltage-to-neutral E_a was taken as 1.0 per unit, the above line-to-line voltages are expressed in per unit of the base voltage to neutral. Expressed in volts, the postfault line voltages are

$$V_{ab} = 1.01 \times \frac{13.8}{\sqrt{3}} \underline{/77.7°} = 8.05 \underline{/77.7°} \text{ kV}$$

$$V_{bc} = 1.980 \times \frac{13.8}{\sqrt{3}} \underline{/270°} = 15.78 \underline{/270°} \text{ kV}$$

$$V_{ca} = 1.01 \times \frac{13.8}{\sqrt{3}} \underline{/102.3°} = 8.05 \underline{/102.3°} \text{ kV}$$

Before the fault the line voltages were balanced and equal to 13.8 kV. For comparison with the line voltages after the fault occurs, the prefault voltages, with $V_{an} = E_a$ as reference, are given as

$$V_{ab} = 13.8 \underline{/30°} \text{ kV} \qquad V_{bc} = 13.8 \underline{/270°} \text{ kV} \qquad V_{ca} = 13.8 \underline{/150°} \text{ kV}$$

The phasor diagrams of prefault and postfault voltages are shown in Fig. 12.3.

12.2 LINE-TO-LINE FAULT ON AN UNLOADED GENERATOR

The circuit diagram for a line-to-line fault on an unloaded, Y-connected generator is shown in Fig. 12.4 with the fault on phases *b* and *c*. The conditions at the fault are expressed by the following equations:

$$V_b = V_c \qquad I_a = 0 \qquad I_b = -I_c$$

With $V_b = V_c$ the symmetrical components of voltage are given by

$$\begin{bmatrix} V_{a0} \\ V_{a1} \\ V_{a2} \end{bmatrix} = \frac{1}{3} \begin{bmatrix} 1 & 1 & 1 \\ 1 & a & a^2 \\ 1 & a^2 & a \end{bmatrix} \begin{bmatrix} V_a \\ V_b \\ V_b \end{bmatrix}$$

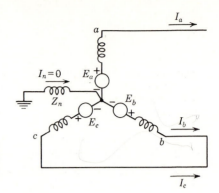

Figure 12.4 Circuit diagram for a line-to-line fault between phases b and c at the terminals of an unloaded generator whose neutral is grounded through a reactance.

from which we find

$$V_{a1} = V_{a2} \tag{12.6}$$

Since $I_b = -I_c$ and $I_a = 0$, the symmetrical components of current are given by

$$
\begin{bmatrix} I_{a0} \\ I_{a1} \\ I_{a2} \end{bmatrix} = \frac{1}{3} \begin{bmatrix} 1 & 1 & 1 \\ 1 & a & a^2 \\ 1 & a^2 & a \end{bmatrix} \begin{bmatrix} 0 \\ -I_c \\ I_c \end{bmatrix}
$$

and therefore

$$I_{a0} = 0 \tag{12.7}$$

$$I_{a2} = -I_{a1} \tag{12.8}$$

With a connection from the generator neutral to ground, Z_0 is finite, and so

$$V_{a0} = 0 \tag{12.9}$$

since I_{a0} is zero by Eq. (12.7).

Equation (12.1), with the substitutions allowed by Eqs. (12.6) to (12.9), becomes

$$
\begin{bmatrix} 0 \\ V_{a1} \\ V_{a1} \end{bmatrix} = \begin{bmatrix} 0 \\ E_a \\ 0 \end{bmatrix} - \begin{bmatrix} Z_0 & 0 & 0 \\ 0 & Z_1 & 0 \\ 0 & 0 & Z_2 \end{bmatrix} \begin{bmatrix} 0 \\ I_{a1} \\ -I_{a1} \end{bmatrix} \tag{12.10}
$$

Performing the indicated matrix operations and premultiplying the resulting matrix equation by the row matrix $\begin{bmatrix} 1 & 1 & -1 \end{bmatrix}$ gives

$$0 = E_a - I_{a1} Z_1 - I_{a1} Z_2 \tag{12.11}$$

and solving for I_{a1} yields

$$I_{a1} = \frac{E_a}{Z_1 + Z_2} \tag{12.12}$$

Equations (12.6) to (12.8) and (12.12) are the special equations for a line-to-line fault. They are used with Eq. (12.1) and the symmetrical-component relations to determine all the voltages and currents at the fault. The special

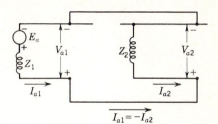

Figure 12.5 Connection of the sequence networks of an unloaded generator for a line-to-line fault between phases b and c at the terminals of the generator.

equations indicate how the sequence networks are connected to represent the fault. Since Z_0 does not enter into the equations, the zero-sequence network is not used. The positive- and negative-sequence networks must be in parallel since $V_{a1} = V_{a2}$. The parallel connection of the positive- and negative-sequence networks without the zero-sequence network makes $I_{a1} = -I_{a2}$, as specified by Eq. (12.8). The connection of the sequence networks for a line-to-line fault is shown in Fig. 12.5. The currents and voltages in the sequence networks, when so connected, satisfy all the equations derived for the line-to-line fault.

Since there is no ground at the fault, there is only one ground in the circuit (at the generator neutral) and no current can flow in the ground. In the derivation of the relations for the line-to-line fault we found that $I_{a0} = 0$. This is consistent with the fact that no ground current can flow, since the ground current I_n is equal to $3I_{a0}$. The presence or absence of a grounded neutral at the generator does not affect the fault current. If the generator neutral is not grounded, Z_0 is infinite and V_{a0} is indeterminate, but line-to-line voltages may be found since they contain no zero-sequence components.

Example 12.2 Find the subtransient currents and the line-to-line voltages at the fault under subtransient conditions when a line-to-line fault occurs at the terminals of the generator described in Example 12.1. Assume that the generator is unloaded and operating at rated terminal voltage when the fault occurs. Neglect resistance.

SOLUTION

$$I_{a1} = \frac{1.0 + j0}{j0.25 + j0.35} = -j1.667 \text{ per unit}$$

$$I_{a2} = -I_{a1} = j1.667 \text{ per unit}$$

$$I_{a0} = 0$$

$$I_a = I_{a1} + I_{a2} + I_{a0} = -j1.667 + j1.667 = 0$$

$$I_b = a^2 I_{a1} + a I_{a2} + I_{a0}$$

$$= -j1.667(-0.5 - j0.866) + j1.667(-0.5 + j0.866)$$

$$= j0.833 - 1.443 - j0.833 - 1.443 = -2.886 + j0 \text{ per unit}$$

$$I_c = -I_b = 2.886 + j0 \text{ per unit}$$

As in Example 12.1, base current is 837 A, and so

$$I_a = 0$$

$$I_b = -2.886 \times 837 = 2416\underline{/180°} \text{ A}$$

$$I_c = 2.886 \times 837 = 2416\underline{/0°} \text{ A}$$

The symmetrical components of the voltage from a to ground are

$$V_{a1} = V_{a2} = 1 - (-j1.667)(j0.25) = 1 - 0.417 = 0.583 \text{ per unit}$$

$$V_{a0} = 0 \text{ (neutral of generator grounded)}$$

Line-to-ground voltages are

$$V_a = V_{a1} + V_{a2} + V_{a0} = 0.583 + 0.583 = 1.166\underline{/0°} \text{ per unit}$$

$$V_b = a^2 V_{a1} + a V_{a2} + V_{a0}$$

$$V_c = V_b = 0.583(-0.5 - j0.866) + 0.583(-0.5 + j0.866)$$

$$= -0.583 \text{ per unit}$$

Line-to-line voltages are

$$V_{ab} = V_a - V_b = 1.166 + 0.583 = 1.749\underline{/0°} \text{ per unit}$$

$$V_{bc} = V_b - V_c = -0.583 + 0.583 = 0 \text{ per unit}$$

$$V_{ca} = V_c - V_a = -0.583 - 1.166 = 1.749\underline{/180°} \text{ per unit}$$

Expressed in volts, the line-to-line voltages are

$$V_{ab} = 1.749 \times \frac{13.8}{\sqrt{3}} = 13.94\underline{/0°} \text{ kV}$$

$$V_{bc} = 0 \text{ kV}$$

$$V_{ca} = -1.749 \times \frac{13.8}{\sqrt{3}} = 13.94\underline{/180°} \text{ kV}$$

12.3 DOUBLE LINE-TO-GROUND FAULT ON AN UNLOADED GENERATOR

The circuit diagram for a double line-to-ground fault on an unloaded, Y-connected generator having a grounded neutral is shown in Fig. 12.6. The faulted phases are b and c. The conditions at the fault are expressed by the following equations:

$$V_b = 0 \qquad V_c = 0 \qquad I_a = 0$$

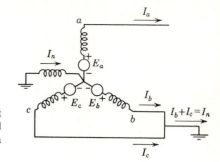

Figure 12.6 Circuit for a double line-to-ground fault on phases b and c at the terminals of an unloaded generator whose neutral is grounded through a reactance.

With $V_b = 0$ and $V_c = 0$, the symmetrical components of voltage are given by

$$\begin{bmatrix} V_{a0} \\ V_{a1} \\ V_{a2} \end{bmatrix} = \frac{1}{3} \begin{bmatrix} 1 & 1 & 1 \\ 1 & a & a^2 \\ 1 & a^2 & a \end{bmatrix} \begin{bmatrix} V_a \\ 0 \\ 0 \end{bmatrix}$$

Therefore V_{a0}, V_{a1}, and V_{a2} equal $V_a/3$, and

$$V_{a1} = V_{a2} = V_{a0} \tag{12.13}$$

Substituting $E_a - I_{a1} Z_1$ for V_{a1}, V_{a2}, and V_{a0} in Eq. (12.1) and premultiplying both sides by $\mathbf{Z}^{-1}$, where

$$\mathbf{Z}^{-1} = \begin{bmatrix} Z_0 & 0 & 0 \\ 0 & Z_1 & 0 \\ 0 & 0 & Z_2 \end{bmatrix}^{-1} = \begin{bmatrix} \dfrac{1}{Z_0} & 0 & 0 \\ 0 & \dfrac{1}{Z_1} & 0 \\ 0 & 0 & \dfrac{1}{Z_2} \end{bmatrix}$$

give

$$\begin{bmatrix} \dfrac{1}{Z_0} & 0 & 0 \\ 0 & \dfrac{1}{Z_1} & 0 \\ 0 & 0 & \dfrac{1}{Z_2} \end{bmatrix} \begin{bmatrix} E_a - I_{a1} Z_1 \\ E_a - I_{a1} Z_1 \\ E_a - I_{a1} Z_1 \end{bmatrix} = \begin{bmatrix} \dfrac{1}{Z_0} & 0 & 0 \\ 0 & \dfrac{1}{Z_1} & 0 \\ 0 & 0 & \dfrac{1}{Z_2} \end{bmatrix} \begin{bmatrix} 0 \\ E_a \\ 0 \end{bmatrix} - \begin{bmatrix} I_{a0} \\ I_{a1} \\ I_{a2} \end{bmatrix} \tag{12.14}$$

Premultiplying both sides of Eq. (12.14) by the row matrix $[1 \quad 1 \quad 1]$ and recognizing that $I_{a1} + I_{a2} + I_{a0} = I_a = 0$, we have

$$\frac{E_a}{Z_0} - I_{a1} \frac{Z_1}{Z_0} + \frac{E_a}{Z_1} - I_{a1} + \frac{E_a}{Z_2} - I_{a1} \frac{Z_1}{Z_2} = \frac{E_a}{Z_1} \tag{12.15}$$

and upon collecting terms we obtain

$$I_{a1} \left(1 + \frac{Z_1}{Z_0} + \frac{Z_1}{Z_2} \right) = \frac{E_a(Z_2 + Z_0)}{Z_2 Z_0} \tag{12.16}$$

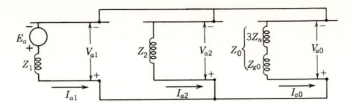

Figure 12.7 Connection of the sequence networks of an unloaded generator for a double line-to-ground fault on phases b and c at the terminals of the generator.

and

$$I_{a1} = \frac{E_a(Z_2 + Z_0)}{Z_1 Z_2 + Z_1 Z_0 + Z_2 Z_0} = \frac{E_a}{Z_1 + Z_2 Z_0/(Z_2 + Z_0)} \qquad (12.17)$$

Equations (12.13) and (12.17) are the special equations for a double line-to-ground fault. They are used with Eq. (12.1) and the symmetrical-component relations to determine all the voltages and currents at the fault. Equation (12.13) indicates that the sequence networks should be connected in parallel, as shown in Fig. 12.7, since the positive-, negative-, and zero-sequence voltages are equal at the fault. Examination of Fig. 12.7 shows that all the conditions derived above for the double line-to-ground fault are satisfied by this connection. The diagram of network connections shows that the positive-sequence current I_{a1} is determined by the voltage E_a impressed on Z_1 in series with the parallel combination of Z_2 and Z_0. The same relation is given by Eq. (12.17).

In the absence of a ground connection at the generator no current can flow into the ground at the fault. In this case Z_0 would be infinite and I_{a0} would be zero. Insofar as current is concerned, the result would be the same as in a line-to-line fault. Equation (12.17) for a double line-to-ground fault approaches Eq. (12.12) for a line-to-line fault as Z_0 approaches infinity, as may be seen by dividing the numerator and denominator of the second term in the denominator of Eq. (12.17) by Z_0 and letting Z_0 be infinitely large.

Example 12.3 Find the subtransient currents and the line-to-line voltages at the fault under subtransient conditions when a double line-to-ground fault occurs at the terminals of the generator described in Example 12.1. Assume that the generator is unloaded and operating at rated voltage when the fault occurs. Neglect resistance.

SOLUTION

$$I_{a1} = \frac{E_a}{Z_1 + Z_2 Z_0/(Z_2 + Z_0)} = \frac{1.0 + j0}{j0.25 + (j0.35 \times j0.10)/(j0.35 + j0.10)}$$

$$= \frac{1.0}{j0.25 + j0.0778} = \frac{1.0}{j0.3278} = -j3.05 \text{ per unit}$$

$$V_{a1} = V_{a2} = V_{a0} = E_a - I_{a1}Z_1 = 1 - (-j3.05)(j0.25)$$

$$= 1.0 - 0.763 = 0.237 \text{ per unit}$$

$$I_{a2} = -\frac{V_{a2}}{Z_2} = -\frac{0.237}{j0.35} = j0.68 \text{ per unit}$$

$$I_{a0} = -\frac{V_{a0}}{Z_0} = -\frac{0.237}{j0.10} = j2.37 \text{ per unit}$$

$$I_a = I_{a1} + I_{a2} + I_{a0} = -j3.05 + j0.68 + j2.37 = 0$$

$$I_b = a^2 I_{a1} + a I_{a2} + I_{a0}$$

$$= (-0.5 - j0.866)(-j3.05) + (-0.5 + j0.866)(j0.68) + j2.37$$

$$= j1.525 - 2.641 - j0.34 - 0.589 + j2.37$$

$$= -3.230 + j3.555 = 4.80\underline{/132.3°} \text{ per unit}$$

$$I_c = a I_{a1} + a^2 I_{a2} + I_{a0}$$

$$= (-0.5 + j0.866)(-j3.05) + (-0.5 - j0.866)(j0.68) + j2.37$$

$$= j1.525 + 2.641 - j0.34 + 0.589 + j2.37 = 3.230 + j3.555$$

$$= 4.80\underline{/47.7°} \text{ per unit}$$

$$I_n = 3I_{a0} = 3 \times j2.37 = j7.11 \text{ per unit}$$

$$I_n = I_b + I_c = -3.230 + j3.555 + 3.230 + j3.555 = j7.11 \text{ per unit}$$

$$V_a = V_{a1} + V_{a2} + V_{a0} = 3V_{a1} = 3 \times 0.237 = 0.711 \text{ per unit}$$

$$V_b = V_c = 0$$

$$V_{ab} = V_a - V_b = 0.711 \text{ per unit}$$

$$V_{bc} = 0$$

$$V_{ca} = V_c - V_a = -0.711 \text{ per unit}$$

Expressed in amperes and volts,

$$I_a = 0$$

$$I_b = 837 \times 4.80 \underline{/132.3°} = 4017\underline{/132.3°} \text{ A}$$

$$I_c = 837 \times 4.80 \underline{/47.7°} = 4017\underline{/47.7°} \text{ A}$$

$$I_n = 837 \times 7.11\underline{/90°} = 5951\underline{/90°} \text{ A}$$

$$V_{ab} = 0.711 \times \frac{13.8}{\sqrt{3}} = 5.66\underline{/0°} \text{ kV}$$

$$V_{bc} = 0$$

$$V_{ca} = -0.711 \times \frac{13.8}{\sqrt{3}} = 5.66\underline{/180°} \text{ kV}$$

12.4 UNSYMMETRICAL FAULTS ON POWER SYSTEMS

In the derivation of equations for the symmetrical components of currents and voltages in a general network during a fault, we shall designate as I_a, I_b, and I_c the currents flowing out of the original balanced system at the fault from phases a, b, and c, respectively. We can visualize the currents I_a, I_b, and I_c by referring to Fig. 12.8, which shows the three lines of the three-phase system at the part of the network where the fault occurs. The flow of current from each line into the fault is indicated by arrows shown on the diagram beside hypothetical stubs connected to each line at the fault location. Appropriate connections of the stubs represent various types of faults. For instance, connecting stubs b and c produces a line-to-line fault through zero impedance. The current in stub a is then zero, and I_b is equal to $-I_c$.

The line-to-ground voltages at the fault will be designated V_a, V_b, and V_c. Before the fault occurs, the line-to-neutral voltage of phase a at the fault will be called V_f, which is a positive-sequence voltage since the system is assumed to be balanced. We met the prefault voltage V_f previously in Chap. 10 in calculations to determine the currents in a power system when a symmetrical three-phase fault occurred.

A single-line diagram of a power system containing three synchronous machines is shown in Fig. 12.9. Such a system is sufficiently general for equations derived therefrom to be applicable to any balanced system regardless of the complexity. Figure 12.9 also shows the sequence networks of the system. The point where a fault is assumed to occur is marked P on the single-line diagram and on the sequence networks. As we saw in Chap. 10, the load current flowing in the positive-sequence network is the same, and the voltages to ground external to the machines are the same, regardless of whether the machines are represented by their subtransient internal voltages and their subtransient reactances, by their transient internal voltages and their transient reactances, or by their no load voltages and their synchronous reactances.

Since linearity is assumed in drawing the sequence networks, each of the networks can be replaced by its Thévenin equivalent between the two terminals composed of its reference bus and the point of application of the fault. The Thévenin equivalent circuit of each sequence network is shown adjacent to the diagram of the corresponding network in Fig. 12.9. The internal voltage of the single generator of the equivalent circuit for the positive-sequence network is V_f, the prefault voltage to neutral at the point of application of the fault. The

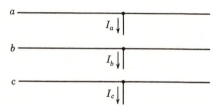

Figure 12.8 Three conductors of a three-phase system. The stubs carrying currents I_a, I_b, and I_c may be interconnected to represent different types of faults.

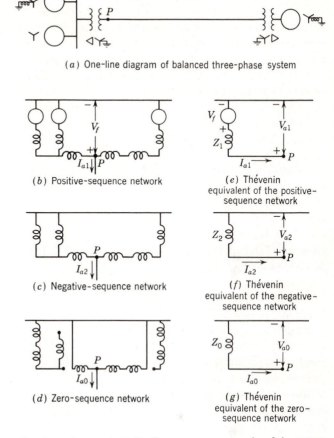

(a) One-line diagram of balanced three-phase system

(b) Positive-sequence network

(e) Thévenin equivalent of the positive-sequence network

(c) Negative-sequence network

(f) Thévenin equivalent of the negative-sequence network

(d) Zero-sequence network

(g) Thévenin equivalent of the zero-sequence network

Figure 12.9 One-line diagram of a three-phase system, the three sequence networks of the system, and the Thévenin equivalent of each network for a fault at P.

impedance Z_1 of the equivalent circuit is the impedance measured between point P and the reference bus of the positive-sequence network with all the internal emfs short-circuited. The value of Z_1 is dependent on the reactances used in the network. We recall, for instance, that subtransient reactances of generators and 1.5 times the subtransient reactances of synchronous motors or the transient reactances of the motors are the values used to calculate the symmetrical current to be interrupted.

Since no negative- or zero-sequence currents are flowing before the fault occurs, the prefault voltage between point P and the reference bus is zero in the negative- and zero-sequence networks. Therefore, no emfs appear in the equivalent circuits of the negative- and zero-sequence networks. The impedances Z_2 and Z_0 are measured between point P and the reference bus in their respective networks and depend on the location of the fault.

Since I_a is the current flowing from the system into the fault, its components I_{a1}, I_{a2}, and I_{a0} flow out of their respective sequence networks and out of the equivalent circuits of the networks at P, as shown in Fig. 12.9. The Thévenin equivalents of the positive-, negative-, and zero-sequence networks of the system are the same as the networks of a single generator. The matrix equations for the symmetrical components of voltages at the fault, therefore, must be the same as Eq. (12.1) except that V_f replaces E_a; that is,

$$\begin{bmatrix} V_{a0} \\ V_{a1} \\ V_{a2} \end{bmatrix} = \begin{bmatrix} 0 \\ V_f \\ 0 \end{bmatrix} - \begin{bmatrix} Z_0 & 0 & 0 \\ 0 & Z_1 & 0 \\ 0 & 0 & Z_2 \end{bmatrix} \begin{bmatrix} I_{a0} \\ I_{a1} \\ I_{a2} \end{bmatrix} \qquad (12.18)$$

Of course, we must evaluate the sequence impedances properly according to Thévenin's theorem and realize that the currents are the sequence components in the hypothetical stubs.

12.5 SINGLE LINE-TO-GROUND FAULT ON A POWER SYSTEM

For a single line-to-ground fault, the hypothetical stubs on the three lines are connected as shown in Fig. 12.10. The following relations exist at the fault:

$$I_b = 0 \qquad I_c = 0 \qquad V_a = 0$$

These three equations are the same as those which apply to a line-to-ground fault on a single generator. These equations with Eq. (12.18) and the relations of symmetrical components must have the same solutions as are found for similar equations in Sec. 12.1, except that V_f replaces E_a. Thus, for a line-to-ground fault,

$$I_{a1} = I_{a2} = I_{a0} \qquad (12.19)$$

and

$$I_{a1} = \frac{V_f}{Z_1 + Z_2 + Z_0} \qquad (12.20)$$

Equations (12.19) and (12.20) indicate that the three sequence networks should be connected in series through the fault point in order to simulate a single line-to-ground fault.

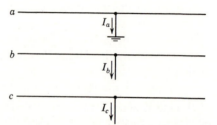

Figure 12.10 Connection diagram of the hypothetical stubs for a single line-to-ground fault.

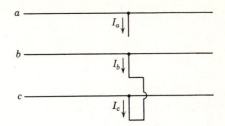

Figure 12.11 Connection diagram of the hypothetical stubs for a line-to-line fault.

12.6 LINE-TO-LINE FAULT ON A POWER SYSTEM

For a line-to-line fault, the hypothetical stubs on the three lines at the fault are connected as shown in Fig. 12.11. The following relations exist at the fault:

$$V_b = V_c \qquad I_a = 0 \qquad I_b = -I_c$$

The above equations are identical in form to those which apply to a line-to-line fault on an isolated generator. Their solution in the manner of Sec. 12.2, with Eq. (12.18) replacing Eq. (12.1), yields

$$V_{a1} = V_{a2} \tag{12.21}$$

$$I_{a1} = \frac{V_f}{Z_1 + Z_2} \tag{12.22}$$

Equations (12.21) and (12.22) indicate that the positive- and negative-sequence networks should be connected in parallel at the fault point in order to simulate a line-to-line fault.

12.7 DOUBLE LINE-TO-GROUND FAULT ON A POWER SYSTEM

For a double line-to-ground fault, the stubs are connected as shown in Fig. 12.12. The following relations exist at the fault:

$$V_b = V_c = 0$$

$$I_a = 0$$

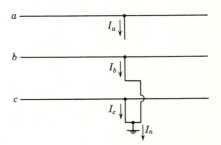

Figure 12.12 Connection diagram of the hypothetical stubs for a double line-to-ground fault.

By comparison with the derivation made in Sec. 12.3,

$$V_{a1} = V_{a2} = V_{a0} \tag{12.23}$$

$$I_{a1} = \frac{V_f}{Z_1 + Z_2 Z_0/(Z_2 + Z_0)} \tag{12.24}$$

Equations (12.23) and (12.24) indicate that the three sequence networks should be connected in parallel at the fault point in order to simulate a double line-to-ground fault.

12.8 INTERPRETATION OF THE INTERCONNECTED SEQUENCE NETWORKS

In the preceding sections we have seen that the sequence networks of a power system can be so interconnected that solving the resulting network yields the symmetrical components of current and voltage at the fault. The connections of the sequence networks to simulate various types of faults, including a symmetrical three-phase fault, are shown in Fig. 12.13. The sequence networks are indicated schematically by rectangles enclosing a heavy line to represent the reference bus of the network and a point marked P to represent the point in the network where the fault occurs. The positive-sequence network contains emfs that represent the internal voltages of the machines.

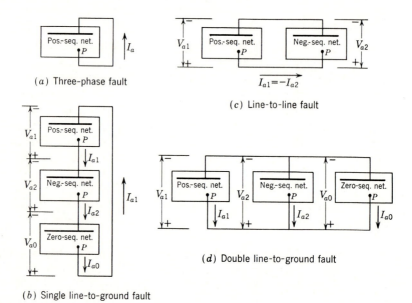

(a) Three-phase fault

$$I_{a1} = -I_{a2}$$

(c) Line-to-line fault

(b) Single line-to-ground fault

(d) Double line-to-ground fault

Figure 12.13 Connections of the sequence networks to simulate various types of faults. The sequence networks are indicated by rectangles. The point at which the fault occurs is P.

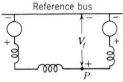

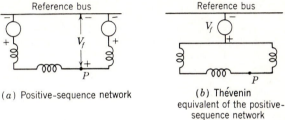

Figure 12.14 Positive-sequence network and its Thevenin equivalent.

(a) Positive-sequence network

(b) Thévenin equivalent of the positive-sequence network

If the emfs in a positive-sequence network like that shown in Fig. 12.14a are replaced by short circuits, the impedance between the fault point P and the reference bus is the positive-sequence impedance Z_1 in the equations developed for faults on a power system and is the series impedance of the Thévenin equivalent of the circuit between P and the reference bus. If the voltage V_f is connected in series with this modified positive-sequence network, the resulting circuit, shown in Fig. 12.14b, is the Thévenin equivalent of the original positive-sequence network. The circuits shown in Fig. 12.14 are equivalent only in their effect on any external connections made between P and the reference bus of the original networks. We can easily see that no current flows in the branches of the equivalent circuit in the absence of an external connection, but current will flow in the branches of the original positive-sequence network if any difference exists in the phase or magnitude of the two emfs in the network. In Fig. 12.14a the current flowing in the branches in the absence of an external connection is the prefault load current.

When the other sequence networks are interconnected with the positive-sequence network of Fig. 12.14a or its equivalent shown in Fig. 12.14b, the current flowing out of the network or its equivalent is I_{a1} and the voltage between P and the reference bus is V_{a1}. With such an external connection, the current in any branch of the original positive-sequence network is the positive-sequence current in phase a of that branch during the fault. The prefault component of this current is included. The current in any branch of the Thévenin equivalent of Fig. 12.14b, however, is only that portion of the actual positive-sequence current found by apportioning I_{a1} of the fault among the branches according to their impedances and does not include the prefault component.

An alternate method of studying unsymmetrical faults is by means of the bus impedance matrix. We shall discuss this method after we look at the following example to become more familiar with the sequence networks.

Example 12.4 A group of identical synchronous motors is connected through a transformer to a 4.16-kV bus at a location remote from the generating plants of a power system. The motors are rated 600 V and operate at 89.5 efficiency when carrying full load at unity power factor and rated voltage. The sum of their output ratings is 4476 kW (6000 hp). The reactances in per unit of each motor based on its own input kVA rating are

$X'' = 0.20$, $X_2 = 0.20$, and $X_0 = 0.04$ and each is grounded through a reactance of 0.02 per unit. The motors are connected to the 4.16-kV bus through a transformer bank composed of three single-phase units, each of which is rated 2400/600 V, 2500 kVA. The 600-V windings are connected in Δ to the motors and the 2400-V windings are connected in Y. The leakage reactance of each transformer is 10%.

The power system which supplies the 4.16 kV bus is represented by a Thévenin equivalent generator rated 7500 kVA, 4.16 kV with reactances of $X'' = X_2 = 0.10$ per unit, $X_0 = 0.05$ per unit, and X_n from neutral to ground equal to 0.05 per unit.

Each of the identical motors is supplying an equal share of a total load of 3730 kW (5000 hp) and is operating at rated voltage, 85% power-factor lag, and 88% efficiency when a single line-to-ground fault occurs on the low-tension side of the transformer bank. Treat the group of motors as a single equivalent motor. Draw the sequence networks showing values of the impedances. Determine the subtransient line currents in all parts of the system with prefault current neglected.

SOLUTION The one-line diagram of the system is shown in Fig. 12.15. The 600-V bus and the 4.16 kV bus are numbered 1 and 2, respectively. Choose the rating of the equivalent generator as base: 7500 kVA, 4.16 kV at the system bus.

Since

$$\sqrt{3} \times 2400 = 4160 \text{ V}$$

and

$$3 \times 2500 = 7500 \text{ kVA}$$

the three-phase rating of the transformer is 7500 kVA, 4160Y/600Δ V. So the base for the motor circuit is 7500 kVA, 600 V.

The input rating of the single equivalent motor is

$$\frac{6000 \times 0.746}{0.895} = 5000 \text{ kVA}$$

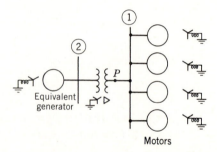

Figure 12.15 One-line diagram of the system of Example 12.4.

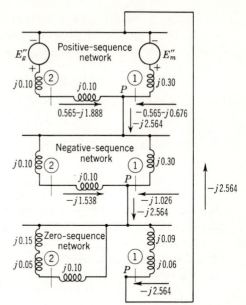

Figure 12.16 Connection of the sequence networks of Example 12.4. Subtransient currents are marked in per unit for a single line-to-ground fault at P. Prefault current is included.

and the reactances of the equivalent motor in percent are the same on the base of the combined rating as the reactances of the individual motors on the base of the rating of an individual motor. The reactances of the equivalent motor on the selected base are

$$X'' = 0.2 \frac{7500}{5000} = 0.3 \text{ per unit}$$

$$X_2 = 0.2 \frac{7500}{5000} = 0.3 \text{ per unit}$$

$$X_0 = 0.04 \frac{7500}{5000} = 0.06 \text{ per unit}$$

The reactance in the zero-sequence network to account for the reactance between neutral and ground of the equivalent motor is

$$3X_n = 3 \times 0.02 \frac{7500}{5000} = 0.09 \text{ per unit}$$

For the equivalent generator the reactance from neutral to ground in the zero-sequence network is

$$3X_n = 3 \times 0.05 = 0.15 \text{ per unit}$$

Figure 12.16 shows the connection of the sequence networks. Reactances are shown in per unit.

Since the motors are operating at rated voltage equal to the base voltage of the motor circuit, the prefault voltage of phase a at the fault is

$$V_f = 1.0 \text{ per unit}$$

Base current for the motor circuit is

$$\frac{7,500,000}{\sqrt{3} \times 600} = 7217 \text{ A}$$

and the actual motor current is

$$\frac{746 \times 5000}{0.88 \times \sqrt{3} \times 600 \times 0.85} = 4798 \text{ A}$$

The per-unit current drawn by the motor through line a before the fault occurs is

$$\frac{4798}{7217} \underline{/-\cos^{-1} 0.85} = 0.665 \underline{/-31.8°} = 0.565 - j0.350 \text{ per unit}$$

If prefault current is neglected, E_g'' and E_m'' are made equal to $1.0 \underline{/0°}$, or the positive-sequence network is replaced by its Thévenin equivalent circuit which is shown in Fig. 12.17. The computations follow.

$$Z_1 = \frac{(j0.1 + j0.1)(j0.3)}{j(0.1 + 0.1 + 0.3)} = j0.12 \text{ per unit}$$

$$Z_2 = \frac{(j0.1 + j0.1)(j0.3)}{j(0.1 + 0.1 + 0.3)} = j0.12 \text{ per unit}$$

$$Z_0 = j0.15 \text{ per unit}$$

$$I_{a1} = \frac{V_f}{Z_1 + Z_2 + Z_0} = \frac{1.0}{j0.12 + j0.12 + j0.15} = \frac{1.0}{j0.39} = -j2.564$$

$$I_{a2} = I_{a1} = -j2.564$$

$$I_{a0} = I_{a1} = -j2.564$$

Current in the fault $= 3I_{a0} = 3(-j2.564) = -j7.692$ per unit. The component of I_{a1} flowing toward P from the transformer is

$$\frac{-j2.564 \times j0.30}{j0.50} = -j1.538$$

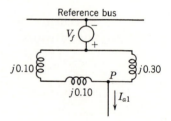

Figure 12.17 Thévenin equivalent of the positive-sequence network of Example 12.4.

and the component of I_{a1} flowing from the motor toward P is

$$\frac{-j2.564 \times j0.20}{j0.50} = -j1.026$$

Similarly the component of I_{a2} from the transformer is $-j1.538$, and the component of I_{a2} from the motor is $-j1.026$. All of I_{a0} flows toward P from the motor.

Currents in the lines at the fault are:

To P from the transformer in per unit:

$$\begin{bmatrix} I_a \\ I_b \\ I_c \end{bmatrix} = \begin{bmatrix} 1 & 1 & 1 \\ 1 & a^2 & a \\ 1 & a & a^2 \end{bmatrix} \begin{bmatrix} 0 \\ -j1.538 \\ -j1.538 \end{bmatrix} = \begin{bmatrix} -j3.076 \\ j1.538 \\ j1.538 \end{bmatrix}$$

To P from the motors in per unit:

$$\begin{bmatrix} I_a \\ I_b \\ I_c \end{bmatrix} = \begin{bmatrix} 1 & 1 & 1 \\ 1 & a^2 & a \\ 1 & a & a^2 \end{bmatrix} \begin{bmatrix} -j2.564 \\ -j1.026 \\ -j1.026 \end{bmatrix} = \begin{bmatrix} -j4.616 \\ -j1.538 \\ -j1.538 \end{bmatrix}$$

Our method of labeling the lines is such that currents I_{A1} and I_{A2} on the high-tension side of the transformer are related to the currents I_{a1} and I_{a2} on the low-tension side by

$$I_{a1} = jI_{A1} \qquad I_{a2} = -jI_{A2} \qquad\qquad (12.25)$$

So,

$$I_{A1} = -j(-j1.538) = -1.538$$

$$I_{A2} = j(-j1.538) = 1.538$$

and

$$I_{A0} = 0$$

since there are no zero-sequence currents on the high-voltage side of the transformer. Then

$$I_A = I_{A1} + I_{A2} = 0$$

$$I_{B1} = a^2 I_{A1} = (-0.5 - j0.866)(-1.538) = \quad 0.769 + j1.332$$

$$I_{B2} = aI_{A2} = (-0.5 + j0.866)(1.538) \quad = -0.769 + j1.332$$

$$\overline{\qquad\qquad I_B = I_{B1} + I_{B2} \quad = \quad 0 \quad + j2.664 \text{ per unit}}$$

$$I_{C1} = aI_{A1} = (-0.5 + j0.866)(-1.538) = \quad 0.769 - j1.332$$

$$I_{C2} = a^2 I_{A2} = (-0.5 - j0.866)(1.538) \quad = -0.769 - j1.332$$

$$\overline{\qquad\qquad I_C = I_{C1} + I_{C2} \quad = \quad 0 \quad - j2.664 \text{ per unit}}$$

If voltages throughout the system are to be found, their components at any point can be calculated from the currents and reactances of the sequence networks. Components of voltages on the high-voltage side of the transformer are found first without regard for phase shift. Then the effect of phase shift must be determined.

By evaluating the base currents on the two sides of the transformer we can convert the above per-unit currents to amperes. Base current for the motor circuit was found previously and equals 7217 A. Base current for the high-voltage circuit is

$$\frac{7,500,000}{\sqrt{3} \times 4160} = 1041 \text{ A}$$

Current in the fault is

$$7.692 \times 7217 = 55,500 \text{ A}$$

Currents in the lines between the transformer and the fault are:

In line a: $3.076 \times 7217 = 22,200$ A

In line b: $1.538 \times 7217 = 11,100$ A

In line c: $1.538 \times 7217 = 11,100$ A

Currents in the lines between the motor and the fault are:

In line a: $4.616 \times 7217 = 33,300$ A

In line b: $1.538 \times 7217 = 11,100$ A

In line c: $1.538 \times 7217 = 11,100$ A

Currents in the lines between the 4.16 kV bus and the transformer are:

In line A: 0

In line B: $2.664 \times 1041 = 2773$ A

In line C: $2.664 \times 1041 = 2773$ A

The currents we have calculated are those which would flow upon the occurrence of a fault when there is no load on the motors. These currents are correct only if the motors are drawing no current whatsoever. The statement of the problem specifies the load conditions at the time of the fault, however, and the load can be considered. To account for the load, we add the per-unit current drawn by the motor through line a before the fault occurs to the component of I_{a1} flowing toward P from the transformer and subtract the same current from the component of I_{a1} flowing from the motor to P. The

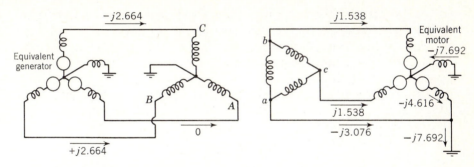

Figure 12.18 Per-unit values of subtransient line currents in all parts of the system of Example 12.4, prefault current neglected.

new value of positive-sequence current from the transformer to the fault in phase a is

$$0.565 - j0.350 - j1.538 = 0.565 - j1.888$$

and the new value of positive-sequence current from the motor to the fault in phase a is

$$-0.565 + j0.350 - j1.026 = -0.565 - j0.676$$

These values are shown in Fig. 12.16. The remainder of the calculation, using these new values, proceeds as in the example.

Figure 12.18 gives the per-unit values of subtransient line currents in all parts of the system when the fault occurs at no load. Figure 12.19 shows the values for the fault occurring on the system when the load specified in the example is considered. In a larger system where the fault current is much higher in comparison with the load current, the effect of neglecting the load current is less than is indicated by comparing Figs. 12.18 and 12.19. In the large system, however, the prefault currents determined by a load-flow study could simply be added to the fault current found with the load neglected.

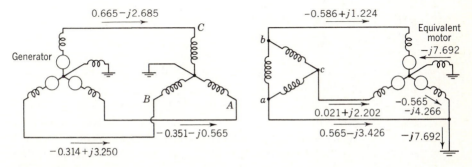

Figure 12.19 Per-unit values of subtransient line currents in all parts of the system of Example 12.4, prefault current considered.

12.9 ANALYSIS OF UNSYMMETRICAL FAULTS USING THE BUS IMPEDANCE MATRIX

In Chap. 10 we used the bus impedance matrix composed of positive-sequence impedances to determine currents and voltages upon the occurrence of a three-phase fault. The method can be extended easily to apply to unsymmetrical faults

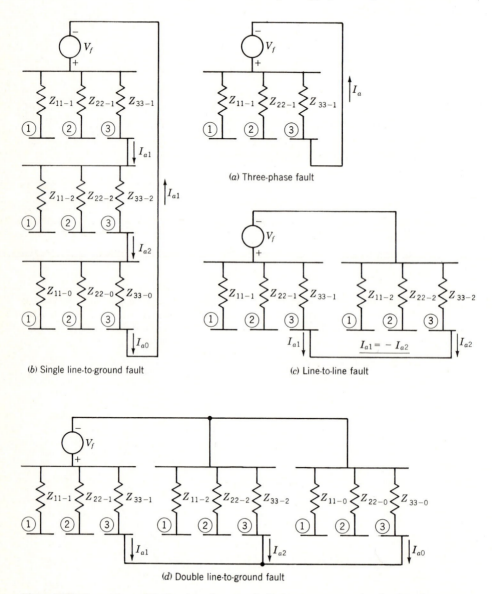

(a) Three-phase fault

(b) Single line-to-ground fault

(c) Line-to-line fault

(d) Double line-to-ground fault

Figure 12.20 Connections of the bus impedance equivalent sequence networks of a three-bus system to simulate various types of faults. Transfer impedances not shown.

by realizing that the negative- and zero-sequence networks can be represented by bus impedance equivalent networks just as the positive-sequence network was. Figure 12.20 corresponds to Fig. 12.13 and shows the interconnection of the bus impedance networks for a three-bus system with the fault on bus 3. The actual networks have merely been replaced by the bus impedance equivalent network. The additional subscripts 1, 2, and 0 have been attached to the impedances to identify the sequence networks to which they belong.

For the single line-to-ground fault on bus 3 examination of Fig. 12.20 shows

$$I_{a1} = \frac{V_f}{Z_{33-1} + Z_{33-2} + Z_{33-0}} \tag{12.26}$$

which should be compared with Eq. (12.20). Obviously Z_{33-1}, Z_{33-2}, and Z_{33-0} are equal to the values of Z_1, Z_2, and Z_0 of Eq. (12.20) if the fault is on bus 3. The positive-, negative-, and zero-sequence bus impedance matrices enable us to see immediately the values to be used for Z_1, Z_2, and Z_0 in Eqs. (12.20), (12.22), and (12.24). The transfer admittances (not shown in Fig. 12.20) enable us to calculate the voltages at the unfaulted buses, from which the currents in the lines are found.

Example 12.5 Solve for the subtransient current in a single line-to-ground fault first on bus 1 and then on bus 2 of the network of Example 12.4. Use the bus impedance matrices. Also find the voltages to neutral at bus 2 with the fault on bus 1.

SOLUTION We refer to Fig. 12.16 to find the elements of the node admittance matrices of the three sequence networks, as follows:

$$Y_{11-1} = Y_{11-2} = \frac{1}{j0.1} + \frac{1}{j0.3} = -j13.3$$

$$Y_{12-1} = Y_{12-2} = \frac{-1}{j0.1} = j10$$

$$Y_{22-1} = \frac{1}{j0.1} + \frac{1}{j0.1} = -j20$$

$$Y_{11-0} = \frac{1}{j0.15} = -j6.67 \qquad Y_{12-0} = 0$$

$$Y_{22-0} = \frac{1}{j0.2} + \frac{1}{j0.1} = -j15.0$$

$$\mathbf{Y}_{bus-1} = \mathbf{Y}_{bus-2} = j\begin{bmatrix} -13.3 & 10.0 \\ 10.0 & -20.0 \end{bmatrix}$$

$$\mathbf{Y}_{bus-0} = j\begin{bmatrix} -6.67 & 0.0 \\ 0.0 & -15.0 \end{bmatrix}$$

Inverting the three matrices above gives the three bus impedance matrices

$$\mathbf{Z}_{bus-1} = \mathbf{Z}_{bus-2} = j\begin{bmatrix} 0.12 & 0.06 \\ 0.06 & 0.08 \end{bmatrix}$$

$$\mathbf{Z}_{bus-0} = j\begin{bmatrix} 0.150 & 0.0 \\ 0.0 & 0.067 \end{bmatrix}$$

The current in the fault on bus 1 is

$$I_f'' = \frac{3 \times 1.0}{j0.12 + j0.12 + j0.15} = -j7.692 \text{ per unit}$$

which agrees with the value found in Example 12.4. If the fault is on bus 2,

$$I_f'' = \frac{3 \times 1.0}{j0.08 + j0.08 + j0.067} = -j13.216 \text{ per unit}$$

To find the voltages to neutral at bus 2 with the fault at bus 1 we observe first that

$$I_{a1} = I_{a2} = I_{a0} = \frac{-j7.692}{3} = -j2.564$$

By studying Fig. 12.20 and realizing that transfer impedances are present we see that at bus 2 with the fault on bus 1 and *neglecting phase shift* the sequence components of voltage are

$$V_{a1} = V_f - I_{a1}Z_{21-1} = 1 - (-j2.564)(j0.06) = 0.8462$$
$$V_{a2} = -I_{a2}Z_{21-2} = -(-j2.564)(j0.06) = -0.1538$$

and, since Z_{21-0} is zero,

$$V_{a0} = 0$$

Accounting for phase shift we have from Eqs. (11.23)

$$V_{A1} = -jV_{a1} = -j0.8462$$
$$V_{A2} = jV_{a2} = -j0.1538$$
$$V_A = V_{A1} + V_{A2} = -j0.8462 - j0.1538 = 0 - j1.000 \text{ per unit}$$
$$V_B = a^2V_{A1} + aV_{A2} = j0.4231 - 0.7328 + j0.0769 + 0.1332$$
$$= -0.600 + j0.500 \text{ per unit}$$
$$V_C = aV_{A1} + a^2V_{A2} = j0.4231 + 0.7328 + j0.0769 - 0.1332$$
$$= 0.600 + j0.500 \text{ per unit}$$

All per-unit values are on a line-to-neutral base. Examination of Fig. 12.18 will show why the magnitude V_A is 1.0.

Although the method of solution using the bus impedance matrix does not appear to have any great advantage over the method of Example 12.4 in this very simple network, it does give us the current for the fault on each of the buses. For a large network the method is well suited to the digital computer, which can build the bus impedance matrix directly and add or remove particular lines quite easily. Thus, with the bus impedance matrix for each sequence network all the features of digital-computer solutions for symmetrical three-phase faults can be extended to unsymmetrical faults.

12.10 FAULTS THROUGH IMPEDANCE

All the faults discussed in the preceding sections consisted of direct short circuits between lines and from one or two lines to ground. Although such direct short circuits result in the highest value of fault current and are therefore the most conservative values to use when determining the effects of anticipated faults, the fault impedance is seldom zero. Most faults are the result of insulator flashovers, where the impedance between the line and ground depends on the resistance of the arc, of the tower itself, and of the tower footing if ground wires are not used. Tower-footing resistances form the major part of the resistance between line and ground and depend on the soil conditions. The resistance of dry earth is 10 to 100 times the resistance of swampy ground. The effect of impedance in the fault is found by deriving equations similar to those for faults through zero impedance. Connections of the hypothetical stubs for faults through impedance are shown in Fig. 12.21.

A balanced system remains symmetrical after the occurrence of a *three-phase fault* having the same impedance between each line and a common point. Only

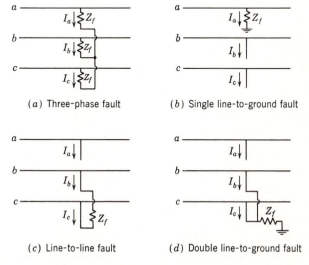

(*a*) Three-phase fault   (*b*) Single line-to-ground fault

(*c*) Line-to-line fault   (*d*) Double line-to-ground fault

Figure 12.21 Connection diagrams of the hypothetical stubs for various faults through impedance.

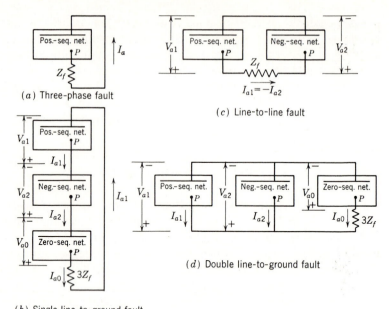

(a) Three-phase fault

(c) Line-to-line fault

(d) Double line-to-ground fault

(b) Single line-to-ground fault

Figure 12.22 Connections of the sequence networks to simulate various types of faults through impedance at point P.

positive-sequence currents flow. With the fault impedance Z_f equal in all phases, as shown in Fig. 12.21a, the voltage at the fault is

$$V_a = I_a Z_f$$

and since only positive-sequence currents flow,

$$V_{a1} = I_{a1} Z_f = V_f - I_{a1} Z_1$$

and

$$I_{a1} = \frac{V_f}{Z_1 + Z_f} \qquad (12.27)$$

The sequence-network connection is shown in Fig. 12.22a.

A formal derivation can be made for the *single line-to-ground* and *double line-to-ground* faults through impedance shown in Fig. 12.21b and d, but we shall arrive at the correct sequence-network connections by comparison with faults without impedance. Consider a generator with all terminals open and its neutral grounded. On such a generator a single or double line-to-ground fault through Z_f is no different with respect to the value of the fault current than the same type of fault without impedance but with Z_f placed in the connection between the generator neutral and ground. To account for an impedance Z_f in the neutral of a generator we add $3Z_f$ to the zero-sequence network. Thévenin's theorem enables us to apply the same reasoning to these types of faults on a power system, and so the sequence-network connections for a single line-to-ground fault and for a double line-to-ground fault are as shown in Fig. 12.22*b* and *d*. From these figures, for a single line-to-ground fault through Z_f

$$I_{a1} = I_{a2} = I_{a0}$$

$$I_{a1} = \frac{V_f}{Z_1 + Z_2 + Z_0 + 3Z_f} \tag{12.28}$$

and for a double line-to-ground fault through Z_f

$$V_{a1} = V_{a2}$$

$$I_{a1} = \frac{V_f}{Z_1 + Z_2(Z_0 + 3Z_f)/(Z_2 + Z_0 + 3Z_f)} \tag{12.29}$$

A *line-to-line* fault through impedance is shown in Fig. 12.21*c*. The conditions at the fault are

$$I_a = 0 \qquad I_b = -I_c \qquad V_c = V_b - I_b Z_f$$

I_a, I_b, and I_c bear the same relations to each other as in the line-to-line fault without impedance. Therefore,

$$I_{a1} = -I_{a2}$$

The sequence components of voltage are given by

$$\begin{bmatrix} V_{a0} \\ V_{a1} \\ V_{a2} \end{bmatrix} = \frac{1}{3} \begin{bmatrix} 1 & 1 & 1 \\ 1 & a & a^2 \\ 1 & a^2 & a \end{bmatrix} \begin{bmatrix} V_a \\ V_b \\ V_b - I_b Z_f \end{bmatrix} \tag{12.30}$$

or

$$3V_{a1} = V_a + (a + a^2)V_b - a^2 I_b Z_f \tag{12.31}$$

$$3V_{a2} = V_a + (a + a^2)V_b - a I_b Z_f \tag{12.32}$$

therefore

$$3(V_{a1} - V_{a2}) = (a - a^2)I_b Z_f = j\sqrt{3}\, I_b Z_f \tag{12.33}$$

Since $I_{a1} = -I_{a2}$,

$$I_b = a^2 I_{a1} + a I_{a2} = (a^2 - a)I_{a1} = -j\sqrt{3}\, I_{a1} \tag{12.34}$$

and, upon substituting I_b from Eq. (12.34) in Eq. (12.33), we obtain

$$V_{a1} - V_{a2} = I_{a1} Z_f \tag{12.35}$$

Equation (13.35) requires the insertion of Z_f between the fault points in the positive- and negative-sequence networks to fulfill the required conditions for the fault. The connections of the sequence networks for a line-to-line fault through impedance are shown in Fig. 12.22*c*. Of course the bus impedance matrix can be used to advantage to find Z_1, Z_2, and Z_0 of Eqs. (12.27), (12.28), (12.29), and (12.35).

Faults through impedance are similar to single-phase loads. The impedance Z_f of the single line-to-ground fault is equivalent to connecting a single-phase load Z_f from line a to neutral. The impedance Z_f of the line-to-line fault is equivalent to connecting a single-phase load Z_f from line b to line c.

12.11 COMPUTER CALCULATIONS OF FAULT CURRENTS

Modern fault-current programs for the digital computer are usually based on the bus impedance matrix. Three-phase and single line-to-ground faults are usually the only types of fault studied. Since circuit-breaker applications are made according to the symmetrical short-circuit current that must be interrupted, this current is calculated for the two types of fault. The printout includes the total fault current and the contributions from each line. The results also list those quantities when each line connected to the faulted bus is opened in turn while all others are in operation.

The program uses the data listed for the lines and their impedances as provided for the load-flow program and includes the appropriate reactance for each machine in forming the positive- and zero-sequence bus impedance matrices. As far as impedances are concerned the negative-sequence network is the same as the positive-sequence network. So, for a single line-to-ground fault at bus 1, I_{a1} is calculated in per unit as 1.0 divided by the sum of $2Z_{11-1}$ and Z_{11-0}.

The bus voltages are included in the computer printout, if called for, as well as the current in lines other than those connected to the faulted bus since this information can easily be found from the bus impedance matrix.

PROBLEMS

12.1 A 60-Hz turbogenerator is rated 500 MVA, 22 kV. It is Y-connected and solidly grounded and is operating at rated voltage at no load. It is disconnected from the rest of the system. Its reactances are $X'' = X_2 = 0.15$ and $X_0 = 0.05$ per unit. Find the ratio of the subtransient line current for a single line-to-ground fault to the subtransient line current for a symmetrical three-phase fault.

12.2 Find the ratio of the subtransient line current for a line-to-line fault to the subtransient current for a symmetrical three-phase fault on the generator of Prob. 12.1.

12.3 Determine the ohms of inductive reactance to be inserted in the neutral connection of the generator of Prob. 12.1 to limit the subtransient line current for a single line-to-ground fault to that for a three-phase fault.

12.4 With the inductive reactance found in Prob. 12.3 inserted in the neutral of the generator of Prob. 12.1, find the ratios of the subtransient line currents for the following faults to the subtransient line current for a three-phase fault: (a) single line-to-ground fault, (b) line-to-line fault, (c) double line-to-ground fault.

12.5 How many ohms of resistance in the neutral connection of the generator of Prob. 12.1 would limit the subtransient line current for a single line-to-ground fault to that for a three-phase fault?

12.6 A generator rated 100 MVA, 20 kV has $X'' = X_2 = 20\%$ and $X_0 = 5\%$. Its neutral is grounded through a reactor of 0.32 Ω. The generator is operating at rated voltage without load and is disconnected from the system when a single line-to-ground fault occurs at its terminals. Find the subtransient current in the faulted phase.

12.7 A 100-MVA 18-kV turbogenerator having $X'' = X_2 = 20\%$ and $X_0 = 5\%$ is about to be connected to a power system. The generator has a current-limiting reactor of 0.162 Ω in the neutral. Before the generator is connected to the system, its voltage is adjusted to 16 kV when a double line-to-ground fault develops at terminals b and c. Find the initial symmetrical rms current in the ground and in line b.

12.8 The reactances of a generator rated 100 MVA, 20 kV, are $X'' = X_2 = 20\%$ and $X_0 = 5\%$. The generator is connected to a Δ-Y transformer rated 100 MVA, 20Δ–230Y kV, with a reactance of 10%. The neutral of the transformer is solidly grounded. The terminal voltage of the generator is 20 kV when a single line-to-ground fault occurs on the open-circuited, high-tension side of the transformer. Find the initial symmetrical rms current in all phases of the generator.

12.9 A generator supplies a motor through a Y-Δ transformer. The generator is connected to the Y side of the transformer. A fault occurs between the motor terminals and the transformer. The symmetrical components of the subtransient current in the motor flowing toward the fault are $I_{a1} = -0.8 - j2.6$ per unit, $I_{a2} = -j2.0$ per unit, and $I_{a0} = -j3.0$ per unit. From the transformer toward the fault $I_{a1} = 0.8 - j0.4$ per unit, $I_{a2} = -j1.0$ per unit, and $I_{a0} = 0$. Assume $X_1'' = X_2$ for both the motor and the generator. Describe the type of fault. Find (a) the prefault current, if any, in line a, (b) the subtransient fault current in per unit, and (c) the subtransient current in each phase of the generator in per unit.

12.10 Calculate the subtransient currents in all parts of the system of Example 12.4 with prefault current neglected if the fault on the low-tension side of the transformer is a line-to-line fault.

12.11 Repeat Prob. 12.10 for a double line-to-ground fault.

12.12 The machines connected to the two high-tension buses shown in the one-line diagram of Fig. 12.23 are each rated 100 MVA, 20 kV with reactances of $X'' = X_2 = 20\%$ and $X_0 = 4\%$. Each three-phase transformer is rated 100 MVA, 345Y/20Δ kV, with leakage reactance of 8%. On a base of 100 MVA, 345 kV the reactances of the transmission line are $X_1 = X_2 = 15\%$ and $X_0 = 50\%$. Find the 2×2 bus impedance matrix for each of the three sequence networks. If no current is flowing in the network, find the subtransient current to ground for a double line-to-ground fault on lines B and C at bus 1. Repeat for a fault at bus 2. When the fault is at bus 2, determine the current in phase b of machine 2 if the lines are so named that V_{A1} and V_{a1} are 90° out of phase. If the phases are named so that I_{A1} leads I_{a1} by 30° what letter (a, b, or c) would identify the phase of machine 2 which would carry the current found for phase b above?

12.13 Two generators G_1 and G_2 are connected through transformers T_1 and T_2 to a high-tension bus which supplies a transmission line. The line is open at the far end at which point F a fault occurs. The prefault voltage at point F is 515 kV. Apparatus ratings and reactances are:

$$G_1 - 1000 \text{ MVA, 20 kV, } X_s = 100\% \qquad X'' = X_2 = 10\% \qquad X_0 = 5\%$$

$$G_2 - 800 \text{ MVA, 22 kV, } X_s = 120\% \qquad X'' = X_2 = 15\% \qquad X_0 = 8\%$$

$$T_1 - 1000 \text{ MVA, 500Y/20Δ kV, } X = 17.5\%$$

$$T_2 = 800 \text{ MVA, 500Y/22Y kV, } X = 16.0\%$$

$$\text{Line} - X_1 = 15\%, X_0 = 40\% \text{ on base of 1500 MVA, 500 kV}$$

Figure 12.23 Circuit for Prob. 12.12.

The neutral of G_1 is grounded through a reactance of 0.04 Ω. The neutral of G_2 is not grounded. Neutrals of all transformers are solidly grounded. Work on a base of 1000 MVA, 500 kV in the transmission line. Neglect prefault current and find subtransient current (a) in phase c of G_1 for a three-phase fault at F, (b) in phase B at F for a line-to-line fault on lines B and C, (c) in phase A at F for a line-to-ground fault on line A, and (d) in phase c of G_2 for a line-to-ground fault on line A. Assume V_{A1} leads V_{a1} by 90° in T_1.

12.14 For the network shown in Fig. 10.18, find the subtransient current in per unit (a) in a single line-to-ground fault on bus 2, and (b) in the faulted phase of line 1-2. Assume no current is flowing prior to the fault and that the prefault voltage at all buses is 1.0 per unit. Both generators are Y-connected. Transformers are at the ends of each transmission line in the system and are Y-Y with grounded neutrals except that the transformers connecting the lines to bus 3 are Y-Δ with the neutral of the Y solidly grounded. The Δ sides of the Y-Δ transformers are connected to bus 3. All line reactances shown in Fig. 10.18 between buses include the reactances of the transformers. Zero-sequence reactance values for these lines including transformers are 2.0 times those shown in Fig. 10.18. Zero-sequence reactances of generators connected to buses 1 and 3 are 0.04 and 0.08 per unit, respectively. The neutral of the generator at bus 1 is connected to ground through a reactor of 0.02 per unit, and the neutral of the generator at bus 3 is solidly grounded.

12.15 Find the subtransient current in per unit in a line-to-line fault on bus 2 of the network of Example 8.1. Neglect resistance and prefault current, assume all bus voltages are 1.0 before the fault occurs, and make use of calculations already made in Example 10.4. Find the current in lines 1-2 and 3-2 also. Assume that lines 1-2 and 3-2 are connected to bus 2 directly rather than through transformers and that the positive- and negative-sequence reactances are identical.

THIRTEEN

SYSTEM PROTECTION

In connection with our study of transmission-line transients in Chap. 5 we discussed briefly the protection of apparatus against surges resulting from lightning and switching. Failure of apparatus due to surges or other causes leads to faults on a power system. Now that we have studied balanced and unbalanced faults that can occur on a power system and know how to calculate the currents and voltages that exist during short circuits we are ready to study protection of a system by isolation of the faulted portion. Although occurrence of short circuits is somewhat of a rare event, it is of utmost importance that steps be taken to remove the short circuits from a power system as quickly as possible. In modern power systems this short circuit removal process is executed automatically, that is, without human intervention. The equipment that does this job is known collectively as the *protection system*. We shall stress protection of transmission lines and transformers in order to develop some important principles, but we shall also discuss a method that is used for the protection of generators, motors, and buses. An extensive treatment of system protection is beyond the scope of this book.

Strictly speaking a fault is any abnormal state of the system, so that faults in general consist of short circuits as well as open circuits. We will limit our discussion here to faults that are short circuits. Open circuit faults are much more unusual than short circuits, and often they are transformed into short circuits by subsequent events. In terms of the seriousness of consequences of a fault, short circuits are of far greater concern than open circuits, although some open circuits may present a potential hazard to personnel.

If short circuits are allowed to persist on a power system for an extended period, many or all of the following undesirable effects are likely to occur:

1. Reduced stability margins for the power system, a subject discussed in Chap. 14.
2. Damage to the equipment that is in the vicinity of the fault due to heavy currents, unbalanced currents, or low voltages produced by the short circuit.
3. Explosions which may occur in equipment containing insulating oil during a short circuit and which may cause fire resulting in a serious hazard to personnel and damage to other equipment.
4. Disruptions in the entire power system service area by a succession of protective actions taken by different protection systems, an occurrence known as cascading.

Which one of these effects will predominate in a given case depends upon the nature and operating conditions of the power system.

13.1 ATTRIBUTES OF PROTECTION SYSTEMS

Speedy elimination of a fault by the protection system requires correct operation of a number of subsystems of the protection system. The job of each of these subsystems can best be understood by describing the events that take place from the time of occurrence of a fault to its eventual elimination from the power system. Although complex sequential faults do occur occasionally on a power system, and some unusual protection-system operations may come into play from time to time, we will devote most of our attention to the simple case of the occurrence of a three-phase short circuit on a transmission line and the resulting operation of the appropriate protection system. Consider the system shown in Fig. 13.1. Buses 1 and 2 are at the two ends of a transmission line. At each end of the transmission line, two identical protection systems are shown enclosed by dotted lines. These constitute the protection system for transmission line 1-2. The protection system can be subdivided into three subsystems:

1. Circuit breakers (CB, or B)
2. Transducers (T)
3. Relays (R)

Circuit breakers have been mentioned briefly in Chap. 10. Relays are the devices which sense the fault and cause the circuit-breaker trip circuits to be energized and the breakers to open their contacts. Transducers provide the input to the relays. Each of these subsystems will be discussed further as we continue to develop the material of this chapter.

Usually we will use a double-numbering notation to identify circuit breakers and relays. Thus the line 1-2 in Fig. 13.1 has a circuit breaker B12 at the bus-1

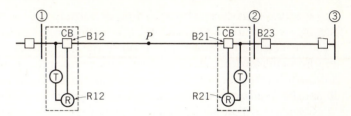

Figure 13.1 One-line diagram showing two transmission lines and elements of the protection system for line 1-2.

end of the line and a circuit breaker B21 at the bus-2 end. Relays at these points are labeled R12 and R21, respectively. A relay R23, although not shown here, is understood to be associated with B23. Sometimes, however, for the simple systems we will be examining it is more convenient to refer to circuit breakers by letters. For example, B12 and B21 could have been labeled *A* and *B* without numerical designation.

Separate breakers may be operated in each phase, or the relays may control one three-phase breaker which will open all three phases upon operation of any one of the relays.

When a fault occurs at *P* on the lines of Fig. 13.1, increased currents flow from both terminals of the transmission line toward the fault if we assume that sources of power are available beyond both buses 1 and 2. When this assumption is not the case, the protection system becomes somewhat simpler. We will consider the protection of such radial systems later. The increase in current at the line terminals is accompanied by a reduction in voltages. It should be realized that the currents and voltages of the transmission line are at kiloampere and kilovolt levels. These high-level signals are unsuitable for use by the protection system. The power line signals are converted to a lower level (tens of amperes and volts) by the transducers T. Transducers will be discussed at greater length a little later.

The increase in current and reduction in voltage caused by the fault can be used to detect that a fault has occurred on the transmission line. Relays are the logic elements of the protection system. The lower-level signals produced by the transducers are reasonably faithful reproductions of the actual voltages and currents of the transmission line. Relays R12 and R21 process these input signals and make the decision that a fault has in fact occurred on the transmission line 1-2. This decision is reached within a very short time after the occurrence of the fault, typically 8 to 40 milliseconds depending upon the design of the relays.

The decision by relays R12 and R21 that a fault has occurred on the line leads to the tripping of their associated circuit breakers B12 and B21. Circuit breakers are the final link in the fault removal process. They were mentioned briefly in Chap. 10. When the trip circuit of a circuit breaker is energized by its relay the contacts of the circuit breaker—which are in series with the transmission line—begin to move apart very rapidly. As the current through the breaker contacts (the fault current) passes through zero, the space between the contacts becomes a dielectric, and is able to prevent the fault current from flowing again through the circuit breaker. This leads to the disconnection of the transmission line from the rest of the system and the elimination of the fault. The entire process from the time of initiation of the fault to its final clearance takes between 30 and 100 milliseconds depending upon the type of protective system employed.

Certain attributes of a relay are important measures of the quality of its performance. The sequence of events described above indicates that to do its job properly a relay must be fast and reliable (that is, dependable and selective.) The first attribute—speed—is self-explanatory. A relay should make its decision as quickly as possible, consistent with other requirements placed upon it. The attribute of dependability means that the relay should operate consistently for all the faults for which it is designed to operate, and should refrain from operating for any other system condition. Selectivity of a relay refers to the requirement that the smallest possible portion of a system should be isolated following a fault. Selectivity can be illustrated with the help of Fig. 13.1. Consider the relay R23 connected at terminal 2 of line 2-3. The current and voltage inputs to this relay will also change due to the fault at P. This effect of the fault at P upon the relay R23 is often described by saying that the relay R23 also "sees" the fault at P. However, the relay R23 must be selective so that it does not operate for the fault at P if P is outside the area of responsibility known as *reach* of this relay. In general the reliability requirements of a relay are in conflict with its speed requirement, and a compromise must be made in designing the protection system so as to obtain a reasonable measure of these attributes.

13.2 ZONES OF PROTECTION

The idea of an area of responsibility of a protection system mentioned above has been formalized by assigning *zones of protection* to various protection systems. The concept of zones helps define the reliability requirements for different protection systems. We will explain the concept of zones of protection with the help of Fig. 13.2. In this figure, a portion of a power system consisting of a generator, two transformers, two transmission lines, and three buses is represented by a one-line diagram. The closed dashed lines indicate the five zones of protection in which this power system is divided. Each zone contains one or more power system components in addition to two circuit breakers. Each breaker is included in two neighboring zones of protection. Zone 1, for example, contains the generator, its associated transformer, and the connecting leads between the generator and the transformer. Zone 3 contains a transmission line only. Note that zones 1 and 5 contain two power-system components each.

The boundary of each zone defines a portion of the power system such that for a fault anywhere within that zone the protection system responsible for that zone takes action to isolate everything within that zone from the rest of the

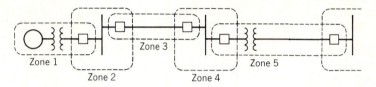

Figure 13.2 Zones of protection indicated by dashed lines enclosing power-system components in each zone.

system. Since the isolation (or deenergization) under faulted conditions is done by circuit breakers, it should be clear that at each point where connection is made between the equipment inside the zone and the rest of the power system, a circuit breaker should be inserted. In other words, the circuit breakers help define the boundaries of the zones of protection.

Another important aspect of zones of protection is that the neighboring zones always overlap. This overlap is necessary, since without it a small part of the system which falls between the neighboring zones, however small it may be, would be left without protection. By overlapping neighboring zones no part of the power system is left without protection, although clearly if a fault should occur within the overlapped region, a much larger portion of the power system (that corresponding to both the zones involved in the overlap) will be isolated and lost from service. To reduce such a possibility to a minimum the region of overlap is made as small as possible.

Example 13.1 (*a*) Consider the power system shown in Fig. 13.3*a* with generating sources beyond buses 1, 3, and 4. What are the zones of protection in which this system should be divided? Which circuit breakers will open for faults at P_1 and P_2?

(*b*) If three circuit breakers are added at the tap point 2, how would the zones of protection be modified? Which circuit breakers will operate for faults at P_1 and P_2 under these conditions?

SOLUTION (*a*) Using the principles of defining the zones of protection, the system of Fig. 13.3*a* can be divided into zones as shown by dashed lines in that figure. For the fault at P_1 breakers *A*, *B*, and *C* will operate. For the fault at P_2 breakers *A*, *B*, *C*, *D*, and *E* will operate.

(*b*) If three circuit breakers *F*, *G*, and *H* are added at bus 2 as shown in Fig. 13.3*b* the zones of protection will be as shown by the dashed lines in that figure. In this case breakers *A* and *F* will operate for the fault at P_1, whereas breakers *G*, *C*, *D*, and *E* will operate for the fault at P_2. Note that in this case a much smaller portion of the power system is deenergized following the two faults. This improved performance is achieved at the expense of three additional circuit breakers and the associated protection equipment.

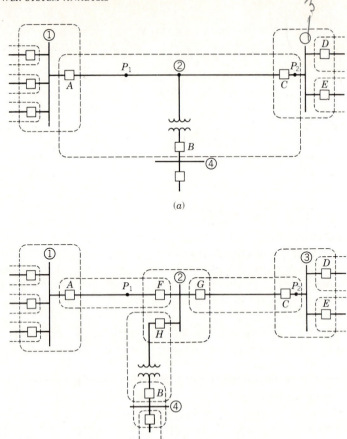

Figure 13.3 One-line diagram for Example 13.1 showing (*a*) original zones of protection and (*b*) modified zones when additional breakers are added at bus 2.

13.3 TRANSDUCERS

Currents and voltages of the protected power equipment are converted by current and voltage transformers to low levels for relay operation. These reduced levels are necessary for two reasons: (1) the lower level input to the relays ensures that the physical hardware used to construct the relays will be quite small and thus less expensive; (2) the personnel who work with the relays will be working in a safe environment. In principle these transducers are no different from the power transformers discussed in Chap. 6. However, the use made of these transformers is rather specialized. For example, it is necessary that a current transformer reproduce in its secondary winding a current which duplicates the primary current waveform as faithfully as possible. It performs this function quite well. Similar considerations hold for a voltage transformer. The amount of power delivered by these transformers is rather modest, since the load connected to them consists only of relays and meters that may be in use at a given time. The load on current transformers (CTs) and voltage transformers (VTs) is commonly known as their *burden*. The term burden usually describes the impedance connected to the transformer secondary winding but may specify the voltamperes delivered to the load. For example, a transformer delivering 5 amperes to a resistive burden of 0.1 ohm may also be said to have a burden of 2.5 voltamperes at 5 amperes.

We will now consider current transformers and voltage transformers separately.

(1) Current Transformers

There are two types of current transformers found in practice. Certain power equipment is of the dead-tank type, having a grounded metal tank in which the power equipment is contained in an insulating medium (usually oil). Examples are power transformers, reactors, and oil circuit breakers. This type of equipment has a bushing through which a terminal of the power equipment is brought out. Current transformers are built within this bushing and are known as bushing CTs. Where such a dead-tank system is not available, for example at an EHV switching station where live-tank circuit breakers are in use, free-standing current transformers are used.

The schematic representation for current transformers is shown in Fig. 13.4. The primary winding of a current transformer usually consists of a single turn, and is represented in Fig. 13.4 by a straight line marked a and b. This single turn is obtained by threading the primary conductor through one or more toroidal steel cores. The secondary windings, the terminals of which are marked as a' and b' in Fig. 13.4, are multiple-turn windings wound on the toroidal cores. The dots placed at the terminals a and a' of the current transformer windings have the same connotation as for a conventional transformer. When the primary current enters terminal a (the terminal with the dot marking) the current leaving the dotted terminal a' of the secondary winding is in phase with the primary current if magnetizing current is neglected.

Current transformers have ratio errors which for some types can be calculated and for other types must be determined by test. The error can be quite high if the impedance burden is too large, but with proper selection of the CT with respect to the burden the error can be maintained at an acceptable value. Since we are mainly concerned with protection methods, we shall not discuss CT errors any further, but we must remain aware of them in our consideration of relays.

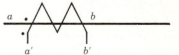

Figure 13.4 Schematic representation to show connection of a current transformer to the line of a power system.

The normal current rating of CT secondaries has been standardized at 5 A, with a second standard of 1 A being used in Europe and, to a lesser extent, in the United States. For short periods of time this rating of the CT secondary windings can be exceeded without damaging the windings. Currents of more than 10 or 20 times normal are often encountered in CT windings during short circuits on the power system.

Standard CT current ratios have been established, and some of these are given in Table 13.1.

(2) Voltage Transformers

Two types of voltage transformers are commonly found in relaying applications. For certain low-voltage applications (system voltages about 12 kV or lower) transformers with a primary winding at the system voltage and secondary windings at 67 V (representing the system line-to-neutral voltage) and $67 \times \sqrt{3} =$ 116 V (representing the system line-to-line voltage) are an industry standard. This type of voltage transformer is quite similar to a multiwinding power transformer, and becomes expensive at higher system voltages. For voltages at HV and EHV levels, a capacitance potential-divider circuit is used as shown in Fig. 13.5. Capacitors C_1 and C_2 are so adjusted that a voltage of a few kilovolts is obtained across C_2 when terminal A is at system potential. In such a coupling-capacitor voltage transformer (CVT) the tapped voltage is further reduced to relaying voltage level by a transformer as shown in Fig. 13.5.

The voltage at point A in the circuit of Fig. 13.5 is essentially that of an infinite bus so far as the connected capacitors are concerned. The Thévenin impedance looking toward the system across the terminals of C_2 is $1/\omega(C_1 + C_2)$. Adjusting L so that ωL equals the Thévenin impedance results in series resonance, and the output of the CVT is in phase with the line potential with no phase-angle error introduced in the CVT output. The CVT is a free-standing device housed in its own supporting insulator structure, and finds application in HV and EHV systems. Whenever a power-system component has a bushing through which a conductor at system voltage passes, such as in a power transformer or in certain types of circuit breakers, a bushing type of CVT built inside the bushing can be made available at little additional cost. In such a CVT, the capacitors C_1 and C_2 are built within the structure of the bushing. In general, bushing type CVTs are capable of supplying smaller burdens than the free-standing CVTs.

Table 13.1 Standard CT ratios

Current ratio	Current ratio	Current ratio
50 : 5	300 : 5	800 : 5
100 : 5	400 : 5	900 : 5
150 : 5	450 : 5	1000 : 5
200 : 5	500 : 5	1200 : 5
250 : 5	600 : 5	

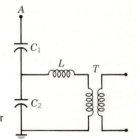

Figure 13.5 Circuit diagram of a capacitor-coupled voltage transformer CVT with its tuning inductance L.

Voltage transformers are generally far more accurate than the current transformers, and their ratio and phase-angle errors are generally neglected. On the other hand it is often necessary to pay attention to the transient response of the CVTs under fault conditions, as errors under these conditions are possible. Transient response of CVTs is beyond the scope of our discussion.

13.4 LOGICAL DESIGN OF RELAYS

The job of a relay is to discriminate between a fault within its zone of protection and all other system conditions. It must act (energize the trip coil of its associated circuit breakers) dependably for faults within its zones of protection, and provide security against false tripping for faults outside those zones. A relay is made secure and dependable by designing into it a logical decision-making capability such that, based upon the condition of its input signals, it is able to produce the correct output for every possible state of its input signals. We will now consider several classes of relays and their logical functional descriptions. In spite of the great variety of relays found on power systems, a majority of them fall into five categories. Their logical performance can be defined in terms of the inputs and outputs of the relay independently of the hardware used in building the relay. For each type of relay, we will specify conditions on their input signals (usually voltages and currents) and the corresponding state of the relay output. The relay output of interest in the present context is the input to the breaker trip coil. Consequently the output state of the relay will be, with its contacts closed, called *trip*, or with its contacts open, called *block* or *block to trip*. The five relay classes to be considered here are:

1. Magnitude relays
2. Directional relays
3. Ratio relays
4. Differential relays
5. Pilot relays

1. Magnitude relays In their most common form these are current magnitude relays—or overcurrent relays. They respond to the magnitude of their input current, and operate to trip whenever the current magnitude exceeds a certain value which is adjustable. If a value $|I_p|$ expressed in terms of the secondary winding of the CT can be found from system short circuit studies such that for all faults within the zone of protection of a relay, the fault current magnitude $|I_f|$ also expressed in terms of the secondary winding will be greater than $|I_p|$, then the following functional description will produce a dependable and secure relay:

$$|I_f| > |I_p| \quad \text{trip}$$
$$|I_f| < |I_p| \quad \text{block}$$

(13.1)

The inequalities expressed by (13.1) are the logical description of an overcurrent relay and can be represented graphically by the phasor diagram shown in Fig. 13.6. The current magnitude $|I_p|$ is known as the pickup value of the relay.

The fault current phasor I_f is drawn in the complex plane with an arbitrary phasor assumed to be the reference. The phase angle of the fault current can lie anywhere between 0 and 360 degrees, since the reference phasor is arbitrary. A circle drawn with the origin as its center and the pickup current magnitude $|I_p|$ as its radius divides the complex phasor plane in two regions labeled trip and block in Fig. 13.6. Any fault current whose phasor representation lies outside the circle in the shaded region will cause the relay to trip. Phasors representing fault currents with magnitude smaller than $|I_p|$ will lie within the circle, and will result in a block decision by the relay. Diagrams such as these are very useful in understanding relay characteristics and are used extensively in relaying literature.

As will be pointed out later, this simplest form of an overcurrent relay is not found to be sufficiently versatile in many cases. It is necessary to introduce another parameter—the time it takes the relay to operate after $|I_f|$ exceeds $|I_p|$. One could supplement the conditions (13.1) with the equation

$$T = \phi(|I_f| - |I_p|) \quad \text{if } |I_f| > |I_p|$$

(13.2)

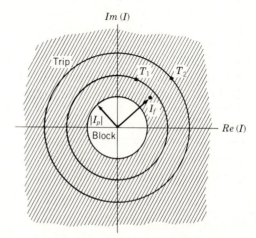

Figure 13.6 Graphical representation of the operating and blocking regions of a time overcurrent relay in the complex plane. Plotting of the phasor current of the relay operating coil on this diagram will show operation or blocking and operating time. Time T_2 is earlier than T_1.

where T is the relay operating time, and ϕ is a function which describes its dependence upon the fault-current level. This functional dependence can be illustrated by adding the time circles such as T_1 and T_2 to the phasor diagram of Fig. 13.6 as shown. The length of phasor $|I_f|$ then falls on a time line (or between two time lines) which represents the operating time of the relay for that fault current. The traditional method of representing the characteristics of a time overcurrent relay is as shown in Fig. 13.7. The pickup setting $|I_p|$ of a relay is adjustable through the taps on its input winding. For example, the relay IFC-53 (General Electric Company) whose characteristic curves are shown in Fig. 13.7 is available with tap settings of 1.0, 1.2, 1.5, 2.0, 3.0, 4.0, 5.0, 6.0, 7.0, 8.0, 10.0, 12.0 A. The function ϕ is usually asymptotic to the pickup value and decreases as some inverse power of the current magnitude for $|I_f| > |I_p|$. The characteristic curves are generally presented with multiples of pickup amperes as the abscissa and operating time as the ordinate. Multiples of pickup amperes means the ratio of relay current to pickup current. The inverse-time characteristic can be shifted up or down by an adjustment known as the time-dial setting. In Fig. 13.7, a time-dial setting of 1/2 produces the fastest operation of the relay, whereas a setting of 10 produces the slowest operation for a given current. Although time-dial adjustments are specified as discrete settings, intermediate values can be obtained by interpolating between the discrete curves.

2. Directional relays In some applications, the zone of a relay includes all of the power system that is situated in only one direction from the relay location. For example, consider the relay R21 shown in Fig. 13.8a. This relay is required to operate for faults to the left of its location, and block for all other conditions. Since the transmission-line impedance is mostly reactive, the faults to the left of R21 have currents flowing from bus 2 toward bus 1 which lag the voltage at bus 2 by an angle of about 90 degrees. On the other hand, for faults to the right of bus 2, the current from bus 2 to bus 1 will lead the voltage at bus 2 by an angle of about 90 degrees. The operation of the relay is described by dividing the complex plane of the phasor diagram of Fig. 13.8b such that for all faults producing current phasors lying in the shaded region (when the voltage at bus 2 is used as a reference) the relay would trip, and for all other faults it would block. Such a relay is called directional, since it depends for its operation upon the direction of the current with respect to the voltage. The quantity that provides the reference phasor is called a polarizing quantity. Thus the directional relay described above uses a polarizing voltage. Sometimes certain current signals may also be used as polarizing signals. The relay can be made more selective by defining a narrower region around the fault-current phasor. In general the operating principle of a directional relay can be described by

$$\theta_{\min} > \theta_{\text{op}} > \theta_{\max} \qquad \text{trip}$$
$$\theta_{\min} < \theta_{\text{op}} < \theta_{\max} \qquad \text{block}$$

$$(13.3)$$

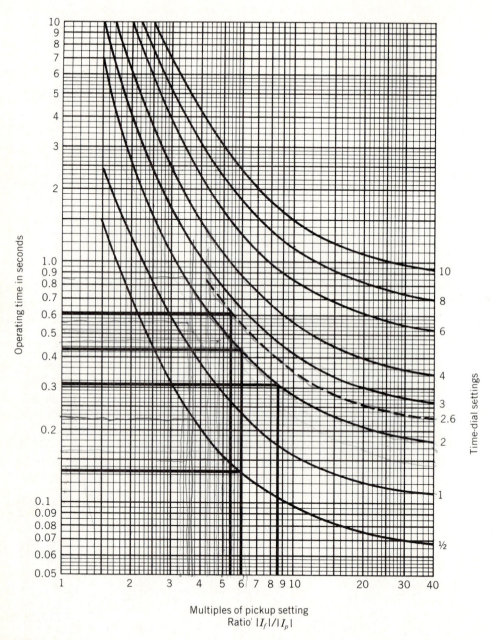

Figure 13.7 Characteristic curves of type IFC-53 time overcurrent relays *(Courtesy General Electric Company).*

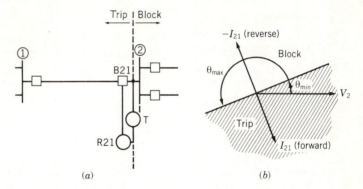

Figure 13.8 Operating principle of a directional relay: (*a*) one-line diagram to show location and (*b*) relay performance characteristic in the complex plane.

where θ_{op} is the phase angle of the operating quantity measured with the polarizing phasor as the reference, and θ_{min} and θ_{max} are the two angles defining the boundary of the operating characteristic.

3. Ratio relays Consider the relay R12 shown in Fig. 13.9*a*. In some applications, it is necessary that the relay operate for faults within a certain distance of its location on any of the lines originating at bus 1. The vicinity is described by the distance along the lines, or equivalently by the impedance between bus 1 and the fault location. The zone of protection is thus a region such that the length of a line originating at bus 1 and having an impedance less than the required setting $|Z_r|$ is included in the zone. This condition can be conveniently expressed as a requirement on the ratio of the voltage and current at the location of R12. Let this ratio (which has the dimensions of an impedance) be

$$Z = \frac{V_1}{I_{12}} \tag{13.4}$$

The relay performance can then be specified by:

$$\begin{aligned} |Z| &< |Z_r| \quad \text{trip} \\ |Z| &> |Z_r| \quad \text{block} \end{aligned} \tag{13.5}$$

and the relay is called an impedance or distance relay. In the complex impedance plane, the locus of constant $|Z_r|$ is a circle as shown in Fig. 13.9*b*. Note that the impedance Z is defined in Eq. (13.4) as a ratio of the voltage and current at relay location 1. During normal system conditions this ratio will be a complex number with some arbitrary phase angle determined by the load power factor. Since the load current is usually much smaller than the fault current, the ratio Z will have a large magnitude (and an arbitrary phase angle) during normal system conditions. Therefore Z plotted in the complex plane under normal system conditions will lie outside the circle of radius $|Z_r|$ and consequently the circuit breaker will not trip during normal system conditions. Under faulted conditions Z appears to the relay to be a load whose impedance is that of the line between the relay location and the fault. The angle associated with Z is θ or $\pi + \theta$ depending upon whether the fault is to the right or left of bus 1 in the circuit of Fig. 13.9a.

A simple modification of an impedance relay is often found to be quite useful. The circle in Fig. 13.9b, which is centered at the origin, can be offset by an amount Z' producing the characteristic of the offset impedance relay shown in Fig. 13.9c. The performance of this type of relay is described by

$$
\begin{aligned}
|Z - Z'| < |Z_r| \qquad &\text{trip} \\
|Z - Z'| > |Z_r| \qquad &\text{block}
\end{aligned}
\tag{13.6}
$$

By selecting $|Z'|$ to be equal to $|Z_r|$, the relay characteristic can be made to pass through the origin. This is the case illustrated in Fig. 13.9c, and the characteristic illustrated here is known as a "mho" characteristic. The impedance relay having a characteristic of Fig. 13.9b is not directional: a fault either to the right or to the left of the relay location (and having $|Z|$ less than $|Z_r|$) will lead to a trip decision by the relay. A mho relay having the characteristic of Fig. 13.9c on the other hand is inherently directional. A fault to the left of the relay at bus 1, no matter how close to bus 1 it may be, will result in a no-trip decision by the relay because Z as defined by Eq. (13.4) will lie in the third quadrant. We will see later that this is a very desirable characteristic in many applications.

4. Differential relays When the entire zone of protection of a relay occupies a relatively small physical space near the relay, it is possible to employ the principle of current continuity to devise a very simple and effective relaying scheme. Consider the zone of protection of one phase of a generator winding shown in Fig. 13.10. Two current transformers having the same turns ratios are placed at the boundaries of the zone of protection (two for each phase of a three-phase

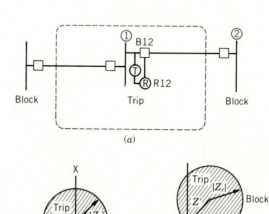

Figure 13.9 Impedance relay characteristics showing (a) zone of protection for R12, (b) complex plane in which measured impedance is plotted for a nondirectional relay, and (c) for a mho relay. In both (b) and (c) the impedance is indicated for a fault to the left of R12.

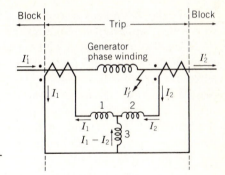

Figure 13.10 Wiring diagram for differential protection of a generator winding.

unit). Then for normal conditions, as well as for faults outside the zone of protection

$$I_1 - I_2 = 0$$

Whereas for a fault inside the protected zone

$$I_1 - I_2 = I_f$$

where I_f is the fault current as seen from the secondary side of the CTs. It should be realized that due to the errors of the current transformers, these equations will not exactly hold in practice. To account for these inaccuracies, a low value of current $|I_p|$ may be chosen such that

$$|I_1 - I_2| < |I_p|$$

for normal system conditions or for faults external to the zone of protection; and

$$|I_1 - I_2| > |I_p|$$

for internal faults. The operating principle of the relay can therefore be defined by

$$
\begin{aligned}
|I_1 - I_2| &> |I_p| \qquad \text{trip} \\
|I_1 - I_2| &< |I_p| \qquad \text{block}
\end{aligned}
\qquad (13.7)
$$

If we connect an overcurrent relay of the type described previously so that its operating coil is coil 3 in the position shown in Fig. 13.10, we see that the current through the coil is $I_1 - I_2$ and the relay will protect the generator winding by tripping a breaker or breakers according to the principle of differential relaying expressed in (13.7). Often the current transformer errors discussed in Sec. 13.3 increase with increasing values of I_1 and I_2. For such cases, it is possible to make the value I_p dependent upon the average of I_1 and I_2. A relay can be designed in this fashion such that the operating principle for the relay becomes

$$
\begin{aligned}
|I_1 - I_2| &> k\,|(I_1 + I_2)|/2 \qquad \text{trip} \\
|I_1 - I_2| &< k\,|(I_1 + I_2)|/2 \qquad \text{block}
\end{aligned}
\qquad (13.8)
$$

Such a relay is known as a percent differential relay. The current $(I_1 + I_2)/2$ is called the restraining current, and the current $(I_1 - I_2)$ is the tripping current of the relay. The relay coils 1 and 2 in Fig. 13.10 have currents I_1 and I_2 flowing through them. If a differential relay is constructed in such a fashion that the currents through coils 1 and 2 oppose the effect of current through 3, then such a relay will exhibit the percent differential characteristic expressed by (13.8). The relative effectiveness of coils 1 and 2 compared to coil 3 is determined by the constant k of the percent differential relay. In an electromechanical percent differential relay, coils 1, 2, and 3 are wound on a common magnetic core in such a direction that currents through 1 and 2 produce a magnetomotive force that is in opposition to that produced by current in 3. In an electronic relay the desired characteristic is obtained by amplification factors in the appropriate signal paths.

A relay of this type can also be used to protect a bus and, of course, a motor.

5. Pilot relays The differential relays discussed above require that the boundary points of the protected zone be physically close to each other so that the signals from these boundary points can be connected to the relay. This is possible only when the zone of protection contains some power equipment of limited size such as a transformer, a generator, or a bus. When transmission lines are to be protected by a relay, their terminals may be hundreds of miles apart, and it becomes impractical to connect the signals from the ends of the transmission line to one relay. Pilot relaying provides a technique of communicating information from a remote zone boundary to the relay at each terminal. Although a direct substitution of pilot communication channels for each differential current signal wire is not economically or technically feasible, equivalent information transmission schemes have been devised. The physical medium used for pilot channels could be conductors of a telephone circuit, high-frequency signals coupled on to the power transmission line itself (known as the power-line carrier), or microwave channels. Application of pilot relaying will be considered in Sec. 13.6.

13.5 PRIMARY AND BACKUP PROTECTION

It was pointed out at the beginning of this chapter that the protection system contains many subsystems, and the successful removal of a fault requires that each subsystem and component function correctly. The protection systems considered so far were primarily responsible for the removal of the fault as soon as possible while deenergizing as little of the system as required. These protection systems are known as the *primary* protection systems. However, it is conceivable that some components or subsystems of the primary protection systems may fail to operate correctly, and it is the usual practice to allow for a *backup* protection system which would take over the job of protection in case the appropriate primary protection system fails to clear the fault. Consider the power system shown in Fig. 13.11. For a fault at P, the primary (main) protection system must

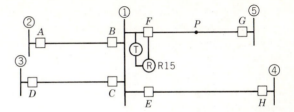

Figure 13.11 One-line diagram of a system having backup protection.

open circuit breakers *F* and *G*. One method of backing up the primary protection system is to duplicate it entirely (or as much of it as is economically feasible) so that the failure of one primary system does not prevent the removal of the fault. Such a backup protection system is known as *duplicate primary*, for obvious reasons, and is used on important circuits where the added cost can be justified. However, there are certain components that are inevitably common to the first primary and the duplicate primary. (Examples are circuit breakers, batteries which operate breaker trip coils, CTs and CVTs.) It is therefore possible that both primaries may be affected by the failure of one of these common components, and provision must be made to furnish backup protection from a remote location where the possibility of a common mode of failure with the primary protection system is slight. This remote backup function is easily incorporated in the primary protection system at the remote location. For example, suppose that the primary protection system at bus 1 in Fig. 13.11 has failed to clear the fault at *P*. (The bus-5 end is assumed to operate correctly.) Recognizing this failure, we can arrange the primary protection systems at buses 2, 3, and 4 to trip circuit breakers *A*, *D*, and *H*, respectively. The protection systems at 2, 3, and 4, in addition to providing their primary protection for lines 2-1, 3-1, and 4-1, will also provide a remote backup protection for the primary protection system at bus 1 for line 1-5. The operation of a remote backup system removes a far greater portion of the power system from service than does the operation of the primary protection system. In the above example, lines 2-1, 3-1, and 4-1 are removed in addition to the originally faulted line 1-5 when the fault is removed by the remote backup protection system. The service to any loads that may be connected at buses 2, 3, and 4 may be affected, and there will be no service to bus 1. Secondly, the backup system must allow a sufficient time for the primary protection system to function normally. A too-hasty operation of the backup protection may lead to unnecessary removal of the larger portion of the power system. The backup system is thus made slower-acting by introducing a delay between the maximum time for fault clearing by the primary system, and the fastest possible response of the backup system. This delay is called the *coordination time delay*, and it is required to help coordinate the operation of the primary and the backup protection system.

Returning once again to Fig. 13.11 we note that, upon the failure of the primary protection at bus 1 associated with line 1-5, a backup system equivalent to the remote backup system described so far may be designed to trip circuit breakers *B*, *C*, and *E* all located at bus 1. Such a protection system is known as the local backup protection, and is generally provided to back up the failure of the circuit breaker responsible for fault clearing (breaker *F* in this example). For this reason, the local backup protection system is also known as breaker-failure protection. Since the primary protection system and the local breaker-failure protection system may share certain subsystems—such as station battery—there are certain modes of failure which are common to the two systems, and some form of remote backup protection is considered essential to a well-designed protection system.

13.6 TRANSMISSION LINE PROTECTION

Transmission line protection has a central role in power-system protection because transmission lines are vital elements of a network which connects the generating plants to the load centers. Also because of the long distances traversed by transmission lines over open countryside, transmission lines are subject to a majority of the faults occurring on the power system. The simplest protection system used at the lowest system voltages consists of fuses which act as relays and circuit breakers combined. We will not consider protection with fuses and reclosers (which are also used in distribution circuits) in this book. Instead, we will concentrate on the protection of medium- and high-voltage transmission lines. The protection system used for medium-voltage transmission lines is somewhat simpler than that used for HV and EHV transmission lines which provide major bulk transmission facilities. Since the consequence of outage of a high-voltage line is far more serious than that of a distribution or subtransmission line, the protection of the bulk power transmission line is generally more elaborate, with greater redundancy, and is also more expensive.

(a) Protection of Subtransmission Lines

The simplest form of protection system can be devised when the generation-load system is radial in nature. Consider the power system shown in Fig. 13.12*a*. The generator at bus 1 (which may be an equivalent representation of one or more transformers feeding bus 1 from a higher-voltage supply point) supplies loads at buses 1, 2, 3, and 4 through three transmission lines. Such a system is known as a radial system because transmission lines radiate from a generating source to supply the loads. Since the source of power is only to the left of each of the transmission lines, it is sufficient to provide only one circuit breaker for each line at the source end. Clearly for any fault on line 1-2, breaker B12 must be opened. In this case all the loads at buses 2, 3, and 4 downstream from breaker 1 will be interrupted.

Overcurrent relays described earlier can be used to protect the transmission lines of this system. Fault current produced by a fault on any of the lines will depend upon the fault location, and since the fault path impedance will increase

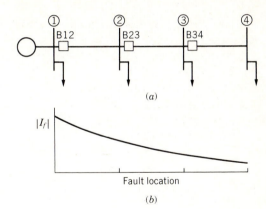

Figure 13.12 Protection of a radial system: (a) one-line diagram of the system and (b) qualitative curve showing fault current $|I_f|$ for faults located along the line.

with the distance to the fault from the generator, the fault current will be inversely proportional to this distance. The fault current I_f as a function of distance from bus 1 is shown qualitatively in Fig. 13.12b. Furthermore, the fault current magnitudes will change depending upon the type of fault and the amount of generation connected at bus 1. For example, if the generator at bus 1 is an equivalent representation of two parallel transformers, then fault currents would be lower when one of the transformers is out of service for any reason. In general, there will be a fault current magnitude curve as shown in Fig. 13.12b for maximum fault-current levels (obtained when maximum generation is in service and a three-phase fault is considered), and one for minimum fault levels (obtained when minimum generation is in service and a line-to-line or a line-to-ground fault, whether or not through impedance to ground, is considered.) The time-overcurrent relays discussed earlier can be set to provide primary protection for a line, as well as the remote backup for a neighboring line for this system. Relays at each of the three buses 1, 2, and 3, are provided to protect their respective lines as primary protection relays, and to provide remote backup protection to one line downstream from the relay location. Thus the relay at 1, in addition to providing primary protection for line 1–2, also provides the remote backup protection for line 2–3. The relay at 3 need provide only primary protection for line 3–4, since there is no other line to the right of line 3–4. When the relay at 1 provides backup protection for line 2–3, it must be so adjusted that it operates with a sufficient time delay (its coordination time delay) such that the relay at bus 2 will always be expected to operate first for faults on line 2–3. It is not considered necessary, nor is it practical, to provide backup protection for any line beyond bus 3 with the relay at bus 1. We will illustrate these concepts with the following numerical examples.

Example 13.2 A portion of a 13.8-kV radial system is shown in Fig. 13.13. The system may be operated with only one rather than two source transformers under certain operating conditions. Assume the high-voltage bus of

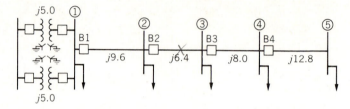

Figure 13.13 One-line diagram of the radial system for Examples 13.2 and 13.3. Line and transformer reactance values are marked in ohms.

the transformer is an infinite bus. The protection system for line-to-line and three-phase faults is to be designed. Transmission line reactances in ohms are shown in Fig. 13.13, and the transformer reactances are in ohms referred to the 13.8-kV side. Neglect resistance and calculate the minimum and maximum fault currents for a fault at bus 5.

SOLUTION Maximum fault current will occur for a three-phase fault with both transformers in service. At bus 5 for this case

$$I_f = \frac{13,800/\sqrt{3}}{j(2.5 + 9.6 + 6.4 + 8.0 + 12.8)} = -j202.75 \text{ A}$$

Minimum fault current will occur with only one transformer in service for a line-to-line fault. For a three-phase fault with just one transformer

$$I_f = \frac{13,800/\sqrt{3}}{j(5.0 + 9.6 + 6.4 + 8.0 + 12.8)} = -j190.6 \text{ A}$$

However, a line-to-line fault would produce a fault current equal to $\sqrt{3}/2$ times the three-phase fault current. This relationship can be verified by solving Prob. 12.2. So the minimum fault current for a fault at bus 5 is

$$I_f = \frac{\sqrt{3}}{2}(-j190.6) = -j165.1 \text{ A}$$

Similar calculations lead to the maximum and minimum fault currents shown in Table 13.2.

Table 13.2 Maximum and minimum fault currents for Example 13.3

Fault at bus	1	2	3	4	5
Maximum fault current, A	3187.2	658.5	430.7	300.7	202.7
Minimum fault current, A	1380.0	472.6	328.6	237.9	165.1

As will be explained in Example 13.3, the principle of backup protection with overcurrent relays for any relay X, backing up the next downstream relay Y, is that X must pick up

(a) for one third of the minimum current seen by Y and
(b) for the maximum current seen by Y but no sooner than 0.3 s after Y should have picked up for that current.

Example 13.3 Select CT ratios, relay tap (pickup) settings, and relay time-dial settings for the system of Example 13.2. Use at all locations the IFC-53 relay whose characteristic curves are given in Fig. 13.7 and whose tap settings are listed in Sec. 13.4. Since each line has a breaker at only one end, simplify the notation by designating the relays at buses 1, 2, 3, and 4 by R1, R2, R3, and R4, respectively. The breaker at each bus will open all three phases when tripped by any of the three associated relays. All three relays at bus 1, for example, will be designated R1.

SOLUTION *Settings for relay R4:* This relay must operate for all currents above 165.1 A, but for reliability a relay would be selected which will operate when current in the line is one third of minimum, or

$$I'_p = \frac{165.1}{3} = 55 \text{ A}$$

For this current a CT ratio of 50/5 (Table 13.1) will result in a relay current of

$$I_p = 55 \times \frac{5}{50} = 5.5 \text{ A}$$

So a relay tap setting of 5.0 A is the proper value.

Since this relay is at the end of a radial system, no coordination with any other relay is necessary. Consequently, the fastest possible operation is desirable. The time-dial setting is therefore chosen to be 1/2.

Settings for relay R3: This relay must back up relay R4, and therefore it must pick up reliably for the smallest fault current seen by R4 which is 165.1 A. So, just as for R4, we use a CT ratio of 50/5 and a relay tap of 5 A for R3 also.

To determine the time-dial setting it is customary to require that the backup relay (in this case R3) operate at least 0.3 s after the time the relay being backed up (R4) should have operated. This interval is the coordination delay time. As we shall soon see we must provide for R3 a delay of 0.3 s for the highest fault current seen by R4 (rather than the lowest fault current). Then R3 will operate no less than 0.3 s after R4 for every possible fault seen by R4.

The highest fault current seen by R4 is the current for a fault just beyond R4 toward bus 5, or 300.7 A according to Table 13.2. The relay current of both R3 and R4 is then

$$300.7 \times \frac{5}{50} = 30.1 \text{ A}$$

and for a relay tap setting of 5 the ratio of relay current to tap setting for both is $30.1/5 = 6.0$. Figure 13.7 tells us the operating time for R4 is 0.135 s since the time-dial setting is 1/2. So in the event of failure of R4, relay R3 must operate in

$$0.135 + 0.3 = 0.435 \text{ s}$$

and Fig. 13.7 tells us the required time-dial setting for R3 is 2.0.

If we had provided for R3 a delay of 0.3 s for the lowest rather than the highest fault current seen by R4, we would have determined a time-dial setting of less than 2.0, which would not have provided a delay of 0.3 s for the highest current seen by R4.

Setting for relay R2: The smallest fault current for which R2 must pick up to provide backup for R3 is 237.9 A as given in Table 13.2. We might choose a CT ratio of 100/5. Then with the required reliability which causes us to design for pickup at one third of the minimum fault current we compute a pickup setting of

$$\frac{1}{3} \times 237.9 \times \frac{5}{100} = 3.9 \text{ A}$$

and we specify the 4.0 tap.

To find the time-dial setting for R2 we see that the maximum fault current at bus 3 is 430.7 A. Relay R3 for this current will have a ratio of relay current to pickup setting of

$$430.7 \times \frac{5}{50} \times \frac{1}{5} = 8.6$$

Since the time-dial setting of R3 is 2.0 that relay will operate in 0.31 s as read from Fig. 13.7. So for proper coordination with R3 relay R2 must operate in

$$0.31 + 0.3 = 0.61 \text{ s}$$

In backing up R3 relay R2 also sees the fault current of 430.7 A for which this ratio of relay current to pickup setting is

$$430.7 \times \frac{5}{100} \times \frac{1}{4} = 5.4$$

We read from Fig. 13.7 a time-dial setting of 2.6.

The relay R1 is set similarly. The final CT ratios, pickup values, and time-dial settings for all the relays are given in Table 13.3.

Table 13.3 Relay settings for Example 13.3

	R1	R2	R3	R4
CT ratio	100 : 5	100 : 5	50 : 5	50 : 5
Pickup setting, A	5	4	5	5
Time-dial setting	2.9	2.6	2.0	$\frac{1}{2}$

It is worth pointing out that there is a slight danger of relay R3 operating before R4 if the current in line 4-5 happens to be close to the relay pickup value during a heavy load or light fault condition. Both R3 and R4 see the same current (since they have the same CT ratios and pickup settings) and there may be just enough of an error in the CTs or relays that R3 may see this as a trip condition (current slightly greater than its pickup value) while R4 sees a current slightly below its pickup value. To avoid possible problems of this nature, the pickup tap setting of R3 should be set at a value somewhat greater than that for R4.

The system shown in Fig. 13.13 can be protected with time-over-current relays (which are simple and relatively inexpensive) because it is a radial system. Consider the system shown in Fig. 13.14a which has multiple sources, and the system of Fig. 13.14b, both of which are similar with respect to protection methods since both are loop systems. In such systems the fault current will flow from both ends of a transmission line for a fault on the line. Therefore to remove a faulted line from the system circuit breakers must be provided at both ends of each line. If, however, every relay responds only to the flow of currents in the forward direction (toward its zone of protection) as shown by the arrows in Fig. 13.14 and does nothing for currents in the reverse direction, then the loop system can be protected much like a radial system. Relays associated with circuit breakers A, C, E must be coordinated among themselves, and relays F, D, B are coordinated together. The overcurrent relays are made directional by using an additional directional relay at each location, and arranging the outputs of the

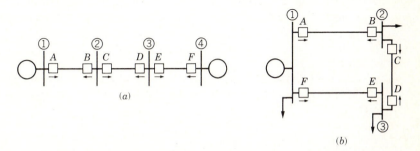

Figure 13.14 One-line diagrams for loop systems. Heavy arrows beside each circuit breaker show the direction to the fault for which the relay will respond. Relays having all arrows pointing in the same direction around the loop coordinate with each other.

directional and overcurrent units in such a manner that a logical "and" operation between their outputs is performed. Their associated breakers will not operate unless both relays provide a trip signal.

(b) Protection of HV and EHV transmission lines

In a bulk power network there are no radial or single-loop systems. Many generating stations and subtransmission feed points are interconnected to form a network, so that no simple loops can be identified. In such a system it becomes impossible to coordinate directional overcurrent relays to provide protection for the transmission lines, since for a given fault location the current seen by the relay varies over a very wide range depending upon the system operating conditions.

The impedance relay described earlier provides a method of protecting transmission lines connected in a network. The relay is made to respond to the impedance between the relay location and the fault point. This impedance is proportional to the distance to the fault, hence the name *distance relay*, and does not depend upon the fault current levels. Consider the system shown in Fig. 13.15a to be a portion of a large system. For a fault at P_1 the relay R12, whose forward direction is in the direction from bus 1 to bus 2, is designed to respond to the positive sequence impedance (or distance) between bus 1 and P_1. Similarly, a relay designated R21 is located at bus 2 with a forward direction from bus 2 to bus 1.

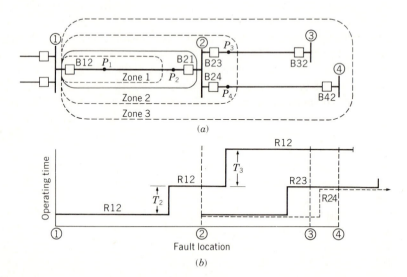

(a)

(b)

Figure 13.15 Coordination of distance (impedance) relays. The zone of protection shown by the solid line in (a) is replaced by zones 1 and 2 identified by dashed lines. Zone 3 provides backup protection for neighboring protection systems. Time delay and operating time is shown in (b) for R12, R23, and R24.

The impedance relays which respond to line-to-line voltages (such as $V_a - V_b$) and the difference between line currents (such as $I_a - I_b$, called *delta currents*) are known as phase relays. They detect the *positive-sequence impedance* between the fault point and the relay location. Three such relays respond correctly to all possible line-to-line faults, double line-to-ground faults, and three-phase faults. These relays, however, do not respond correctly to line-to-ground faults. Three additional relays which utilize line-to-neutral voltages V_a, V_b, V_c, line currents I_a, I_b, I_c and the zero-sequence current I_0 are provided, and which detect the *positive-sequence impedance* between the fault and the relay location for all line faults involving ground.

The distance relays are made directional by incorporating a directional unit similar to that in a directional overcurrent relay. There are distance relays which do not need an added directional unit because they have directionality inherent in their design. The principal example of such a relay is the mho relay described earlier. The directionality is necessary so that a relay will respond to the distance in the forward direction (that is, looking into its zone of protection) and block its operation for all faults in the reverse direction.

Consider the application of the directional distance relays to the protection of line 1-2 in Fig. 13.15a. The zone of protection for the relays R12 and R21 which protect this line is shown by a solid line. For the relay R12 the faults at P_2, P_3, and P_4 all appear to be at the same distance from bus 1; yet faults at P_3 and P_4 are clearly outside the zone of protection of R12. Consequently if the distance relay is set to respond to the fault at P_2, it will also respond to the faults at P_3 and P_4. This is an improper operation of relay R12. To avoid this basic problem, the zone of the distance relay is modified as shown for R12 by the dashed lines in Fig. 13.15a. The single zone of protection shown by the solid line is replaced by two zones: zone 1 and zone 2. Zone 1 extends a shorter distance than the zone shown by the solid line and is usually about 80% of the line length. For faults within this zone, the distance relay at bus 1 operates normally (that is, as quickly as possible). This shortened zone 1 is commonly known as an *underreaching* zone. Zone 2, on the other hand, extends beyond the line terminal well into the lines connected to the remote bus and is said to be *overreaching*. The relay R12 responds to a zone-2 fault with a time delay so that it may coordinate with R23 and R24.

Similar zone 1 and zone 2 settings are available for the relay R21 at bus 2. For a fault at P_1, both relays R12 and R21 operate at highest possible speed since this fault is in zone 1 of both relays. A fault at P_2 is in zone 1 of R21 and consequently breaker B21 will be tripped at high speed. Relay R12 however will not clear the fault at high speed, since this fault lies in zone 2 of R12. After its zone 2 time delay has expired, the relay R12 will operate and trip circuit breaker B12. There is thus a delayed fault clearing from bus 1 while the bus 2 end is cleared at high speed for faults such as P_2.

Now consider a fault at P_3. This fault, lying in zone 1 of relay R23 will be cleared by R23 and circuit breaker B23 at a high speed. Relay R12 will trip circuit breaker B12 at bus 1 to isolate the fault in the zone-2 time of relay R12 if

B23 has failed to operate. Clearly the zone-2 clearing time must be slower than the slowest possible zone-1 clearing time of relay R23, so that R12 does not trip for the fault at P_3 prematurely. A similar zone-2 setting also exists at bus 4 for relay R42 and circuit breaker B42. The fault at P_3 lies in zone 2 of relay R42.

The response time of relays R12, R23, and R24 to their respective zone-1 and zone-2 faults is shown schematically in Fig. 13.15*b*. The abscissa in the response-time diagram is the fault location along the appropriate lines, and the ordinate is the relay operating time. The zone-1 operating time is of the order of 1 cycle, whereas the zone-2 operating time varies between 15 and 30 cycles.

In most cases, the distance relays are provided with another zone of protection known as their zone 3 to provide remote backup for the neighboring lines. The third zone of a relay must reach beyond the longest line emanating from the bus at the remote end of its protected line. The remote backup function must coordinate with the primary protection which it backs up. Thus the third zone of relay R12 must be coordinated with zone 2 of the relays at bus 2 (R23 and R24). Notice also that the Fig. 13.15*b* points to an important principle of relay coordination. The coordination is by time and also by distance. A faster zone of protection must reach in distance beyond the reach of its slower backup. Thus the zone-2 reach of relay R12 is shorter than the zone-1 reach of relay R23 or R24. Similarly zone-3 reach of relay R12 is shorter than the zone-2 reach of R23 and R24. Unless this coordination by distance was employed, for certain faults, instead of high-speed clearing an unnecessarily slow backup time clearing would result. The zone-3 coordinating time is generally of the order of one second. The three zones of protection of relay R12 and the operating times for the three zones are shown schematically in Fig. 13.15*b*.

The characteristic of a directional distance relay in the complex *R-X* plane is shown in Fig. 13.16*a*. A straight line called the line impedance locus is shown in the figure. Along this line the positive-sequence impedances of the protected line

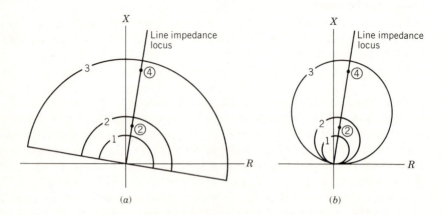

Figure 13.16 Characteristics of (*a*) directional impedance relay and (*b*) mho relay for Example 13.4.

as seen by the relay between its location and different points along the protected line can be plotted. The directional unit of the relay causes separation of the trip and block regions of the relay characteristic in Fig. 13.16a by a line drawn perpendicular to the line impedance locus. The impedance measured from bus 1 to buses 2 and 4 along the line impedance locus are indicated by numerals enclosed in circles. Zone circles whose centers are at the origin of the R-X plane have radii equal to the magnitudes of the impedances of the protected line seen by R12 from bus 1 to the end of the zone identified by the numbers of the circles. Thus, the intersection of a zone circle with the line impedance locus is the line impedance between the relay and the end of the zone. After a fault occurs the impedance seen by the relay is very small compared to the load impedance seen during normal operation. The relays operate when the impedance seen by the relay lies within a zone circle. For impedances inside the zone-1 circle, operation occurs in minimum time. Time delays are set for later operation for zone-2 and then zone-3 faults.

The characteristic of a mho relay is shown in Fig. 13-16b with the centers of the zone circles lying on the line impedance locus. Note that the radii of the zone circles are half of the radii of the corresponding zone circles of the directional distance relay for the same impedances since the intersection of a zone circle with the line impedance locus must still be the impedance of the line seen by the relay between its location and the end of the zone.

The three-step distance relaying scheme described here provides a versatile protection system for high-voltage transmission lines. With minor modifications to accommodate the nature of a given system this protection is used almost universally to protect transmission lines of a modern power network. In many cases ground faults are provided for by the directional time overcurrent relays of the type described earlier, while three-phase and line-to-line faults are covered by distance relays. Occasionally it becomes very difficult to coordinate the third zone times of a relay with the second zones of the neighboring lines; especially at EHV levels. In such cases, the remote backup (zone-3) function of the distance relays may be omitted.

During emergency load-flow conditions when the load is quite high the impedance seen by the relay is low and must be checked to make sure that it does not fall within one of the zone circles of the relay characteristic.

Example 13.4 Consider again the portion of a 138-kV transmission system shown in Fig. 13.15a. Lines 1-2, 2-3, and 2-4 are respectively 64, 64, and 96 km (40, 40, and 60 mi) long. The positive-sequence impedance of the transmission lines is $0.05 + j0.5$ ohm per kilometer. The maximum load carried by line 1-2 under emergency conditions is 50 MVA. Design a three-zone step distance relaying system to the extent of determining for R12 the zone settings which are the impedance values in terms of CT and CVT secondary quantities. The zone settings give points on the R-X plane through which the zone circles of the relay characteristic must pass.

SOLUTION The positive sequence impedances of the three lines are

Line 1-2 $3.2 + j32.0 \ \Omega$
Line 2-3 $3.2 + j32.0 \ \Omega$
Line 2-4 $4.8 + j48.0 \ \Omega$

Since distance relays depend on the ratio of voltage to current, both a CT and CVT will be needed for each phase. The maximum load current is

$$\frac{50 \times 10^6}{\sqrt{3} \times 138 \times 10^3} = 209.2 \text{ A}$$

and we select a CT ratio of 200/5 which will produce about 5 A in the secondary winding under maximum loading conditions.

System voltage to neutral is

$$\frac{138 \times 10^3}{\sqrt{3}} = 79.67 \times 10^3 \text{ V}$$

It was pointed out in Sec. 13.3 that the industry standard for CVT secondary voltage is 67 V for line-to-neutral voltages. Consequently we select a CVT ratio of

$$\frac{79.67 \times 10^3}{67} = 1189.1/1$$

in which case a normal system voltage will produce 67 volts for each phase on the CVT secondary. Denoting primary and secondary voltages of the CVT at bus 1 as V_p and the primary current of the CT as I_p, we have for the impedance measured by the relay

$$\frac{V_p/1189.1}{I_p/40} = Z_{\text{line}} \times 0.0336$$

Thus the impedances of the three lines as seen by the relay R12 are approximately

Line 1-2 $0.11 + j1.1 \ \Omega$, secondary
Line 2-3 $0.11 + j1.1 \ \Omega$, secondary
Line 2-4 $0.16 + j1.6 \ \Omega$, secondary

The maximum load current of 209.2 A appears to the relay, assuming a power factor of 0.8 lagging, to be

$$Z_{\text{load}} = \frac{67}{209.2(5/200)} (0.8 + j0.6)$$

$$= 10.2 + j7.7 \ \Omega, \text{ secondary}$$

The zone-1 setting of the relay R12 must underreach the line 1-2, so that the setting should be

$$0.8 \times (0.11 + j1.1) = 0.088 + j0.88 \ \Omega, \text{ secondary}$$

The zone-2 setting should reach past terminal 2 of the line 1-2. To allow for the various possible inaccuracies of the transducer-relay system, zone 2 is usually set at about 1.2 times the length of the line being protected. Zone 2 for R12 is therefore set at

$$1.2 \times (0.11 + j1.1) = 0.13 + j1.32 \; \Omega, \text{ secondary}$$

The zone-3 setting should reach beyond the longest line connected to bus 2. Thus the zone-3 setting must be

$$(0.11 + j1.1) + 1.2 \times (0.16 + j1.6) = 0.302 + j3.02 \; \Omega, \text{ secondary}$$

Note that once again a factor of 1.2 is used as a multiplier on the impedance of the longest line connected to bus 2 to make sure that zone 3 of relay R12 will reach past bus 4 even in the presence of inaccuracies introduced in the relaying system. A directional impedance relay with the characteristic shown in Fig. 13.16a can be used.

Both Figs. 13.16a and 13.16b apply to this example, and points 2 and 4 on the line impedance locus correspond to the calculated impedances from bus 1 to buses 2 and 4. As seen by the relay the load impedance Z_{load} is more than 3 times the line impedance from bus 1 to bus 4 and lies well beyond the zone-3 circle of both the directional impedance relay and the mho relay. Consequently there is no danger of tripping the line during any load swings that may occur on the transmission line. If the maximum load had been too close to the zone-3 setting of the directional impedance relay, it might have been necessary to replace it by the mho relay whose zone-3 circle encloses a smaller area of the *R-X* plane, as is well apparent in Fig. 13.16.

(c) Line protection with pilot relays

It was pointed out in the previous section that a line (such as 1-2 in Fig. 13.17) is protected by zone 1 and zone 2 of a distance relay and that the normal reach for zone 1 is about 80 percent of the line length. This zone is often called the direct-trip zone or a high-speed zone, and relay operating times of the order of one cycle are quite common for zone-1 faults. A fault such as that at P_2 in Fig. 13.17 falls in zone 2 of relay R12 and will be cleared by relay R12 in its zone-2 time. Of course an identical set of relays exists at bus 2, and these relays will see the fault P_2 as being in their zone 1. Consequently the breaker at bus 2 will operate in high-speed relaying time for the fault at P_2. Thus for faults in the middle 60 percent of the line, both ends clear the fault with high speed, whereas for faults in the 20 percent line length at either end, the nearest end will clear the fault at high speed while the distant end will clear the fault with a zone-2 time delay.

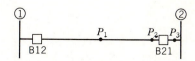

Figure 13.17 Line to be protected by pilot relaying.

In modern HV and EHV systems, this delayed clearing from the remote end is often found to be unacceptable because of the complex nature of the modern interconnected network and the tighter stability margins. It then becomes necessary to provide high-speed protection for the entire line (instead of the middle 60 percent). High-speed protection for the entire line is provided by pilot relaying of the type described in Sec. 13.4. For a fault anywhere on the protected line, the directional relays R12 and R21 see a unique condition: both relays see the fault current flowing in the forward direction. This information, when communicated to the remote ends over a pilot channel, confirms that a fault is indeed on the protected line. Consider faults at P_2 and P_3 on line 1-2 shown in Fig. 13.15. Although the relay R12 sees no difference between the two faults, the relay R21 sees P_2 as an internal fault and P_3 as an external fault (being in the reverse direction from its zone of protection). Upon receiving this directional information at R12, that relay will be able to block (prevent) tripping for the fault at P_3. (Note that the fault at P_3 is in the zone of protection of bus 2 but not of the line 1-2). The fault at P_2 is tripped simultaneously at high speed from both ends. Such a system is called a "directional comparison" pilot scheme. Equivalent performance can be obtained by comparing the phase angles of the fault currents seen at the two ends of a line, and exchanging this phase angle information over the pilot channel. Such a scheme is known as a "phase comparison" scheme. A discussion of the merits and demerits of the two schemes is beyond the scope of this book, although it may be mentioned that both types of protection schemes are used in practice.

13.7 PROTECTION OF POWER TRANSFORMERS

As in the case of transmission lines, the type of protection used for power transformers depends upon their size, voltage rating, and the nature of their application. For small transformers (smaller than about 2 MVA) protection with fuses may be adequate; whereas for transformers of greater than 10-MVA capacity differential relays with harmonic restraint may be used.

Consider first the differential protection of a single-phase two-winding power transformer shown in Fig. 13.18. If the power transformer is carrying load

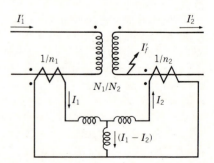

Figure 13.18 Connection diagram for differential protection of a transformer.

currents I'_1 and I'_2 in the primary and secondary windings, we know that with magnetizing current neglected

$$\frac{I'_1}{I'_2} = \frac{N_2}{N_1} \tag{13.9}$$

where N_1 and N_2 are the turns of the primary and secondary windings of the power transformer, respectively. The CT secondary currents are I_1 and I_2, and the turns ratio of the CTs on the primary and secondary side of the power transformer are n_1 and n_2, respectively (one turn on the primary of a CT means n turns on the secondary). So

$$I_1 = \frac{I'_1}{n_1} \qquad I_2 = \frac{I'_2}{n_2} \tag{13.10}$$

To prevent tripping under normal conditions when Eq. (13.9) must be true the current $I_1 - I_2$ through the trip coil must be zero; that is I_1 must equal I_2. So from Eqs. (13.9) and (13.10) we require that

$$\frac{n_1}{n_2} = \frac{N_2}{N_1} \tag{13.11}$$

With an internal fault on the secondary side of the power transformer and with current I'_f in the fault

$$I_1 - I_2 = \frac{I'_f}{n_2} \tag{13.12}$$

For a fault on the primary side of the transformer, the right-hand side of Eq. (13.12) would be I'_f/n_1. If the relay is set with a sufficiently small value of pickup current $|I_p|$ the differential relay would trip for an internal fault, while blocking for external faults or normal load.

As we discussed in Sec. 8.10, a power transformer is usually equipped with variable tap settings, which allow its secondary voltage to be adjusted over a certain range. The adjustments usually vary in small steps over a range ± 10 percent from the nominal turns ratio of N_1/N_2. If tap settings result in an off-normal turns ratio, the relay will see a differential current during normal load conditions. To avoid improper operation in this case a percent differential relay must be used.

Three-phase transformers with Y-Δ windings require further discussion. As pointed out in Sec. 11.4 the primary and secondary currents of such transformers differ in magnitude and phase angles during normal operating conditions. The current transformers must therefore be connected in such a manner that the CT secondary line currents as seen by the relay are in phase, and the CT ratios must also be adjusted so that the current magnitudes as seen by the relay are equal under normal (unfaulted) conditions. The correct phase-angle relationship is obtained by connecting the CTs on the wye side of the power transformer in delta, and those on the delta side of the power transformer in wye. In this manner, the CT connections compensate for the phase shift created by the Y-Δ-connected power transformer. These considerations are illustrated by the following example.

Example 13.5 A three-phase 345/34.5 kV transformer is rated at 50 MVA and its short-term emergency rating is 60 MVA. Using standard CT ratios available, determine the CT ratios, CT connections, and the currents in the power transformer and the CTs. The 345-kV side is Y-connected, while the 34.5-kV side is Δ-connected.

SOLUTION The currents on the 345-kV side and 34.5-kV side of the transformer when it is carrying its maximum expected load are

$$\frac{60 \times 10^6}{\sqrt{3} \times 345 \times 10^3} = 100.4 \text{ A} \qquad \text{and} \qquad \frac{60 \times 10^6}{\sqrt{3} \times 34.5 \times 10^3} = 1004.1 \text{ A}$$

We will use CT ratio of 1000/5 on the 34.5-kV side. Since the CTs on this side are connected in wye, the current flowing to the differential relay from this side will be

$$1004 \times \frac{5}{1000} \cong 5.0 \text{ A}$$

To balance this current, the line currents produced from the Δ-connected CTs on the 345-kV side must also be 5.0 A. This will require that each of the secondary windings of the Δ-connected CTs have a current of

$$\frac{5.0}{\sqrt{3}} \cong 2.9 \text{ A}$$

This current in the CT secondary windings would require CT ratios of

$$\frac{100.4}{2.9} = 34.64$$

for the CTs on the 345-kV side. The nearest available standard CT ratio is 200/5. If we use this ratio, the CT secondary currents would be

$$100.4 \times \frac{5}{200} = 2.51 \text{ A}$$

and the line currents from the Δ-connected CTs to the differential relays would be

$$2.51 \times \sqrt{3} = 4.35 \text{ A}$$

Clearly this current cannot balance the 5.0 A produced by the 34.5-kV side.

This situation of mismatched currents when standard CT ratios are used is quite commonly encountered in designing the protection system for Y-Δ-connected transformers. A convenient solution is provided by auxiliary current transformers which provide a wide range of turn ratios. These auxiliary

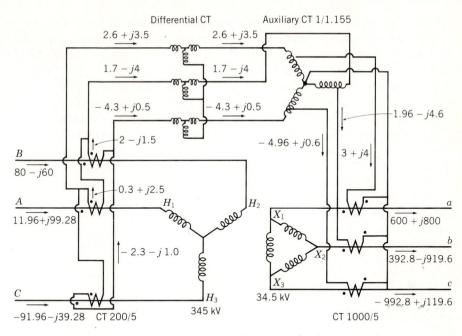

Figure 13.19 Wiring diagram showing currents in amperes for the power and relaying circuits of Example 13.5.

CTs are small and inexpensive devices since their primary and secondary windings are low-voltage low-current circuits. Using a set of three auxiliary CTs with turns ratios of

$$\frac{5.0}{4.35} = 1.155$$

would produce a balanced set of currents in the differential relay when the power transformer is carrying normal load. For an assumed power factor 0.8 lagging, the currents in the power transformer and the various CTs are as shown in Fig. 13.19.

Although the auxiliary transformers can be used as discussed here, in general it is a good practice to use these only as a last resort. The auxiliary CTs add their own burden to the main CTs, and also add to the total errors of transformation. A much more suitable approach is to use tap settings on relay coils themselves, which serve the same purpose as a variable turns ratio auxiliary CT. In most cases, relay coil tap settings provide adequate margin to correct for most practical ratio mismatches.

Notice that in assuming the primary and secondary currents of power transformers are related by their turns ratio, we have assumed that their magnetizing currents are negligible. This is a good approximation when the transformer is in normal operation and the magnetizing current is very small. However, when a transformer is energized, it can draw very heavy magnetizing currents (known as the magnetizing inrush currents) which decay with time to the very small steady-state value. If this high inrush current does flow during energization of the transformer, it appears as differential current since it flows in the primary winding only. It is essential to detect such a condition and stop the differential relay from tripping the transformer. One of the more common methods used to accomplish this function is based upon the fact that the magnetizing inrush current is rich in harmonics whereas a fault current is a purer fundamental frequency sinusoid. To take advantage of this fact, in addition to the fundamental frequency restraining current of $(I_1 + I_2)/2$, another restraining signal proportional to the harmonic content of the differential current is generated in the differential relay. By selecting a proper weighting coefficient for the harmonic component, it is possible to prevent the differential relay from tripping the transformer during its energization, even though a substantial tripping current may be produced due to the magnetizing inrush current.

13.8 RELAY HARDWARE

In Sec. 13.4 we described the logical design of certain types of relays. Most of the relays discussed so far have been built with electromechanical devices. Some of the more common types include plunger type relays, balance-beam type relays, and rotating cup (or disc) relays which are similar to the watt-hour meters found on most residential circuits. The electromechanical relays have served well since the earliest days, and are an important part of the present protection system design practices. These relays are robust, inexpensive, and relatively immune to the harsh environment in an electric power substation. Their response time is somewhat slow in terms of modern power system needs, and also their design is somewhat inflexible in terms of available characteristics, burden capability, and tap settings.

In the late 1950s relays using solid-state circuitry were introduced. These relays use analog circuits in conjunction with logic gates to produce the desired relay characteristic. Solid-state relays are capable of providing characteristics similar to those of electromechanical relays, and in fact a number of newer characteristic shapes are available with solid-state relays. Although earlier models of these relays were prone to frequent component failures under the harsh operating environment of a substation, newer models do perform very well and are providing excellent service on HV and EHV power systems. Quite recently, relays based on microcomputers have been proposed, and are under active investigation at this time.

13.9 SUMMARY

In this chapter we have concentrated on the discussion of protective relaying systems for modern high-voltage power networks. We did not study the application of fuses and reclosers, which are more commonly used on distribution

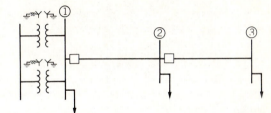

Figure 13.20 One-line diagram for Prob. 13.4.

systems. In many ways, the coordination of these devices is similar to that of time overcurrent relays discussed in Secs. 13.4 and 13.6. In terms of equipment protection we discussed transformer protection in Sec. 13.7. Protection of generators has been mentioned only in connection with a differential relay which can also be used to protect bus bars. Protective systems for equipment and lines offer unique challenges worthy of extensive treatment.†

PROBLEMS

13.1 In the system whose one-line diagram and zones of protection are shown in Fig. 13.3b where is the fault located if the breakers tripped are (a) G and C, (b) F, G, and H, (c) F, G, H, and B, and (d) D, C, and E?

13.2 Determine the time-dial setting for R3 of Example 13.3 if the setting had been determined for the lowest fault current seen by R4? Why is the setting made for the highest rather than the lowest fault current seen by R4?

13.3 In Example 13.3, assume that a line-to-line fault occurs at the midpoint of line 2-3. Which relay will operate for this fault? What will be its operating time? Assume that this relay fails to clear the fault, which relay will operate to clear the fault now? How long will it take?

13.4 An 11-kV radial system is shown in Fig. 13.20. The positive- and zero-sequence impedances of line 1-2 are 0.8 Ω and 2.5 Ω, respectively. Impedances of line 2-3 are three times as large. The positive- and zero-sequence impedances of each of the two transformers are $j2.0$ ohms and $j3.5$ ohms, respectively. Under emergency conditions, the system may be operated with one transformer out of service. Determine the CT ratios, pickup values, and time-dial settings for IFC-53 relays designed to provide single line-to-ground fault protection for this system. Assume the high-tension bus is an infinite bus whose voltage will provide 11 kV at the low-tension bus at no load.

13.5 A portion of a 765-kV network is shown in Fig. 13.21. The positive and zero sequence impedances of the transmission lines are $0.01 + j0.6$ ohm and $0.1 + j1.8$ ohms per mile respectively. Assume that the generator impedance is $j10.0$ ohms for positive and negative sequence, and $j20.0$ ohms for the zero sequence.

(a) Relays R12, R23, and R34 use inverse-time directional IFC-53 overcurrent relays for single line-to-ground fault protection of this system. Determine the fault current needed for setting the pickup value of R12 ground relay. You may neglect line resistances for this calculation.

† See, for instance, C. R. Mason, *The Art and Science of Protective Relaying*, John Wiley and Sons, Inc., New York, 1956, and Westinghouse Electric Corporation, *Applied Protective Relaying*, Relay Instruments Division, Newark, N.J., 1976.

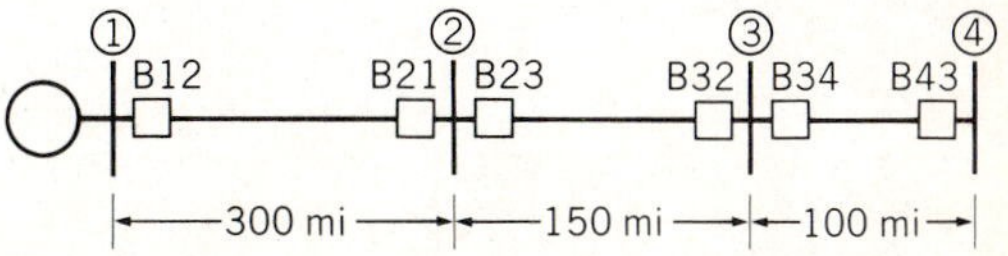

Figure 13.21 One-line diagram for Prob. 13.6.

(b) Select CT and CVT ratios for phase distance relays at bus 1. You may assume that the relay current coils can carry 10 A continuously, and that the emergency line loading limit is 3000 MVA. Use standard CT ratios.

(c) Determine and show on the (secondary) R-X diagram the three zones of a directional impedance relay at bus 1 for phase fault protection.

(d) Also show the location of the equivalent impedance of the emergency load on the R-X diagram. Do you see any problems with line operating at emergency load? What solution would you propose?

13.6 A three-phase fault occurs at the terminals of the delta winding in Fig. 13.19 within the zone of protection of the differential relay. Assume that the positive-sequence impedance of the transformer as seen from the 345-kV side is $j250$ ohms, and that the power system feeding the 345-kV side is of infinite short-circuit capacity. What are the currents flowing in all the leads shown in Fig. 13.19 in this case? Neglect prefault current and assume no fault current originates in the low-tension part of the system.

FOURTEEN

POWER SYSTEM STABILITY

When ac generators were driven by reciprocating steam engines, one of the major problems in the operation of machinery was hunting. The periodic variations in the torque applied to the generators caused periodic variations in speed. The resulting periodic variations in voltage and frequency were transmitted to the motors connected to the system. Oscillations of the motors caused by the variations in voltage and frequency sometimes caused the motors to lose synchronism entirely if their natural frequency of oscillation coincided with the frequency of oscillation caused by the engines driving the generators. Damper windings were first used to minimize hunting by the damping action of the losses resulting from the currents induced in the damper windings by any relative motion between the rotor and the rotating field set up by the armature current. The use of turbines has reduced the problem of hunting, although it is still present where the prime mover is a diesel engine. Maintaining synchronism between the various parts of a power system becomes increasingly difficult, however, as the systems and interconnections between systems continue to grow.

14.1 THE STABILITY PROBLEM

Power system stability may be defined as that property of the system which enables the synchronous machines of the system to respond to a disturbance from a normal operating condition so as to return to a condition where their operation is again normal. Stability studies are usually classified into three types depending upon the nature and order of magnitude of the disturbance. These are *transient*, *dynamic*, and *steady-state* stability studies.

Today, transient stability studies constitute the major analytical approach to the study of power-system electromechanical dynamic behavior. Transient stability studies are aimed at determining if the system will remain in synchronism following *major* disturbances such as transmission system faults, sudden load changes, loss of generating units, or line switching. Such studies began more than 50 years ago but were then confined to consideration of the dynamic problems of not more than two machines. Present-day power systems are vast, heavily interconnected systems with many hundreds of machines which can dynamically interact through the medium of their extra-high voltage and ultra-high voltage networks. These machines have associated excitation systems and turbine-governing control systems which, in some but not all cases, must be modeled in order to properly reflect the correct dynamic response of the power system to certain system disturbances.

Dynamic and steady-state stability studies are less extensive in scope and involve one or just a few machines undergoing *slow* or *gradual* changes in operating conditions. Therefore, dynamic and steady-state stability studies concern the stability of the locus of essentially steady-state operating points of the system. The distinction made between steady-state and dynamic stability studies is really artificial since the stability problems are the same in nature; they differ only in the degree of detail used to model the machines. In dynamic stability studies, the excitation system and turbine-governing system are represented along with synchronous machine models which provide for flux-linkage variation in the machine air-gap. Steady-state stability problems use a very simple generator model which treats the generator as a constant voltage source. The solution technique of steady-state and dynamic stability problems is to examine the stability of the system under incremental variations about an equilibrium point. The nonlinear differential and algebraic equations for the system can be replaced by a set of linear equations which are then solved by methods of linear analysis to determine whether the machine or machines will remain in synchronism following small changes from the operating point.

Transient stability studies are much more commonly undertaken thereby reflecting their greater importance in practice. Such problems involve large disturbances which do not allow the linearization process to be used and the nonlinear differential and algebraic equations must be solved by direct methods or by iterative step-by-step procedures. Transient stability problems can be subdivided into first-swing and multiswing stability problems. First-swing stability is based on a reasonably simple generator model without representation of control systems. Usually the time period under study is the first second following a system fault. If the machines of the system are found to remain in synchronism within the first second, the system is said to be stable. Multiswing stability problems extend over a longer study period and therefore must consider effects of generator control systems which affect machine performance during the extended time period. Machine models of greater sophistication must be represented to reflect proper behavior.

In all stability studies, the objective is to determine whether or not the rotors of the machines being perturbed return to constant speed operation. Obviously, this means that the rotor speeds must depart at least temporarily from synchronous speed. To facilitate computation, three fundamental assumptions are made in *all* stability studies:

1. Only synchronous frequency currents and voltages are considered in the stator windings and the power system. Consequently, dc offset currents and harmonic components are neglected.
2. Symmetrical components are used in the representation of unbalanced faults.
3. Generated voltage is considered unaffected by machine speed variations.

These assumptions permit the use of phasor algebra for the transmission network and solution by load-flow techniques using 60-Hz parameters. Also, negative- and zero-sequence networks can be incorporated into the positive-sequence network at the fault point. As we shall see, three-phase balanced faults are generally considered. However, in some special studies, circuit-breaker clearing operations may be such that consideration of unbalanced conditions is unavoidable.†

14.2 ROTOR DYNAMICS AND THE SWING EQUATION

The equation governing the motion of the rotor of a synchronous machine is based on the elementary principle in dynamics which states that accelerating torque is the product of the moment of inertia of the rotor times its angular acceleration. In the MKS system of units, this equation can be written for the synchronous generator in the form

$$J\frac{d^2\theta_m}{dt^2} = T_a = T_m - T_e \qquad \text{N-m} \qquad (14.1)$$

where the symbols have the following meanings:

  J the total moment of inertia of the rotor masses, in kg-m^2
  θ_m the angular displacement of the rotor with respect to a stationary axis, in mechanical radians
  t time, in seconds
  T_m the mechanical or shaft torque supplied by the prime mover less retarding torque due to rotational losses, in N-m
  T_e the net electrical or electromagnetic torque, in N-m
  T_a the net accelerating torque, in N-m

The mechanical torque T_m and the electrical torque T_e are considered positive for the synchronous generator. This means that T_m is the resultant shaft torque

† For information beyond the scope of this book, see P. M. Anderson and A. A. Fouad, *Power System Control and Stability*, The Iowa State University Press, Ames, Iowa, 1977.

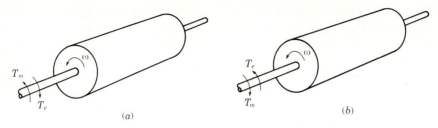

Figure 14.1 Representation of a machine rotor comparing direction of rotation and mechanical and electrical torques for (*a*) a generator and (*b*) a motor.

which tends to accelerate the rotor in the positive θ_m direction of rotation as shown in Fig. 14.1*a*. Under steady-state operation of the generator T_m and T_e are equal and the accelerating torque T_a is zero. In this case there is no acceleration or deceleration of the rotor masses and the resultant constant speed is the *synchronous speed*. The rotating masses which include the rotor of the generator and the prime mover are said to be in synchronism with the other machines operating at synchronous speed in the power system. The prime mover may be a hydro turbine or a steam turbine for which models of different levels of complexity exist to represent their effect on T_m. In this text T_m is considered constant at any given operating condition. This assumption is a fair one for generators even though input from the prime mover is controlled by governors. Governors do not act until after a change in speed is sensed and so are not considered effective during the time-period in which rotor dynamics are of interest in our stability studies here. The electrical torque T_e corresponds to the net air-gap power in the machine and thus accounts for the total output power of the generator plus $|I|^2R$ losses in the armature winding. In the synchronous motor the direction of power flow is opposite to that in the generator. Accordingly, for a motor both T_m and T_e in Eq. (14.1) are reversed in sign as shown in Fig. 14.1*b*. T_e then corresponds to the air-gap power supplied by the electrical system to drive the rotor while T_m represents the counter-torque of the load and rotational losses tending to retard the rotor.

Since θ_m is measured with respect to a stationary reference axis on the stator, it is an absolute measure of rotor angle. Consequently, it continuously increases with time even at constant synchronous speed. Since the rotor speed relative to synchronous speed is of interest it is more convenient to measure the rotor angular position with respect to a reference axis which rotates at synchronous speed. Therefore, we define

$$\theta_m = \omega_{sm}t + \delta_m \tag{14.2}$$

where ω_{sm} is the synchronous speed of the machine in mechanical radians per second and δ_m is the angular displacement of the rotor, in mechanical radians, from the synchronously rotating reference axis. The derivatives of Eq. (14.2) with respect to time are

$$\frac{d\theta_m}{dt} = \omega_{sm} + \frac{d\delta_m}{dt} \qquad (14.3)$$

and

$$\frac{d^2\theta_m}{dt^2} = \frac{d^2\delta_m}{dt^2} \qquad (14.4)$$

Equation (14.3) shows that the rotor angular velocity $d\theta_m/dt$ is constant and equals the synchronous speed only when $d\delta_m/dt$ is zero. Therefore $d\delta_m/dt$ represents the deviation of the rotor speed from synchronism and the units of measure are mechanical radians per second. Equation (14.4) represents the rotor acceleration measured in mechanical radians per second-squared.

Substituting Eq. (14.4) into Eq. (14.1), we obtain

$$J\frac{d^2\delta_m}{dt^2} = T_a = T_m - T_e \qquad \text{N-m} \qquad (14.5)$$

It is convenient for notational purposes to introduce

$$\omega_m = \frac{d\theta_m}{dt} \qquad (14.6)$$

for the angular velocity of the rotor. We recall from elementary dynamics that power equals torque times angular velocity and so, multiplying Eq. (14.5) by ω_m, we obtain

$$J\omega_m\frac{d^2\delta_m}{dt^2} = P_a = P_m - P_e \qquad \text{W} \qquad (14.7)$$

where P_m is the shaft power input to the machine less rotational losses, P_e is the electrical power crossing its air-gap and P_a is the accelerating power which accounts for any unbalance between those two quantities. Usually we will neglect rotational losses and armature $|I|^2R$ losses and think of P_m as power supplied by the prime mover and P_e as the electrical power output.

The coefficient $J\omega_m$ is the angular momentum of the rotor; at synchronous speed ω_{sm}, it is denoted by M and called the *inertia constant* of the machine. Obviously the units in which M is expressed must correspond to those of J and ω_m. A careful check of the units in each term of Eq. (14.7) shows that M is expressed in joule-seconds per mechanical radian and we write

$$M\frac{d^2\delta_m}{dt^2} = P_a = P_m - P_e \qquad \text{W} \qquad (14.8)$$

While we have used M in this equation, the coefficient is not a constant in the strictest sense because ω_m does not equal synchronous speed under all conditions of operation. However, in practice ω_m does not differ significantly from synchronous speed when the machine is stable and since power is more convenient in calculations than torque, Eq. (14.8) is preferred. In machine data supplied for stability studies, another constant related to inertia is very often encountered. This is the so-called H *constant* which is defined by

$$H = \frac{\text{stored kinetic energy in megajoules at synchronous speed}}{\text{machine rating in MVA}}$$

and

$$H = \frac{\frac{1}{2}J\omega_{sm}^2}{S_{mach}} = \frac{\frac{1}{2}M\omega_{sm}}{S_{mach}} \quad \text{MJ/MVA} \tag{14.9}$$

where S_{mach} is the three-phase rating of the machine in MVA. Solving for M in Eq. (14.9), we obtain

$$M = \frac{2H}{\omega_{sm}} S_{mach} \quad \text{MJ/mech rad} \tag{14.10}$$

and substituting this equation in Eq. (14.8), we find

$$\frac{2H}{\omega_{sm}} \frac{d^2\delta_m}{dt^2} = \frac{P_a}{S_{mach}} = \frac{P_m - P_e}{S_{mach}} \tag{14.11}$$

This equation leads to a very simple result.

Note that δ_m is expressed in mechanical radians in the numerator of Eq. (14.11) while ω_{sm} is expressed in mechanical radians per second in the denominator. Therefore, we can write the equation in the form

$$\frac{2H}{\omega_s} \frac{d^2\delta}{dt^2} = P_a = P_m - P_e \quad \text{per unit} \tag{14.12}$$

provided both δ and ω_s have consistent units which may be mechanical or electrical degrees or radians. H and t have consistent units since megajoules per megavoltampere is in units of time in seconds and P_a, P_m, and P_e must be in per unit on the same base as H. When the subscript m is associated with ω, ω_s, and δ it means mechanical units are used; otherwise electrical units are implied. Accordingly, ω_s is the synchronous speed in electrical units. For a system with an electrical frequency of f hertz, Eq. (14.12) becomes

$$\frac{H}{\pi f} \frac{d^2\delta}{dt^2} = P_a = P_m - P_e \quad \text{per unit} \tag{14.13}$$

when δ is in electrical radians while

$$\frac{H}{180f} \frac{d^2\delta}{dt^2} = P_a = P_m - P_e \quad \text{per unit} \tag{14.14}$$

applies when δ is in electrical degrees.

Equation (14.12), called the *swing equation* of the machine, is the fundamental equation which governs the rotational dynamics of the synchronous machine in stability studies. We note that it is a second-order differential equation which can be written as the two first-order differential equations

$$\frac{2H}{\omega_s} \frac{d\omega}{dt} = P_m - P_e \qquad \text{per unit} \tag{14.15}$$

$$\frac{d\delta}{dt} = \omega - \omega_s \tag{14.16}$$

in which ω, ω_s, and δ involve electrical radians or electrical degrees. We will use the various equivalent forms of the swing equation throughout this chapter to determine the stability of a machine within a power system. When the swing equation is solved we obtain the expression for δ as a function of time. A graph of the solution is called the *swing curve* of the machine and inspection of the swing curves of all the machines of the system will show whether the machines remain in synchronism after a disturbance.

14.3 FURTHER CONSIDERATIONS OF THE SWING EQUATION

The MVA base used in Eq. (14.11) is the machine rating which is introduced by the definition of H. In a stability study of a power system with many synchronous machines, only one MVA base common to all parts of the system can be chosen. Since the right-hand side of the swing equation for each machine must be expressed in per unit on this common system base, it is clear that H on the left-hand side of each swing equation must be also consistent with the system base. This is accomplished by converting the H constant for each machine based on its own individual rating, to a value determined by the system base, S_{system}. Equation (14.11), multiplied on each side by the ratio $(S_{\text{mach}}/S_{\text{system}})$, leads to the conversion formula

$$H_{\text{system}} = H_{\text{mach}} \frac{S_{\text{mach}}}{S_{\text{system}}} \tag{14.17}$$

in which the subscript for each term indicates the corresponding base being used. In industry studies the system base usually chosen is 100 MVA.

The inertia constant M is rarely used in practice and the forms of the swing equation involving H are more often encountered. This is because the value of M varies widely with the size and type of the machine whereas H assumes a much narrower range of values as shown in Table 14.1. Machine manufacturers also use the symbol WR^2 to specify for the rotating parts of a generating unit (including the prime mover) the weight in pounds multiplied by the square of the radius of gyration in feet. Hence $WR^2/32.2$ is the moment of inertia of the machine in slug-feet squared.

Table 14.1 Typical inertia constants of synchronous machines†

Type of machine	Inertia constant H‡ MJ/MVA
Turbine generator:	
Condensing, 1800 r/min	9–6
3600 r/min	7–4
Noncondensing, 3600 r/min	4–3
Waterwheel generator:	
Slow-speed, <200 r/min	2–3
High-speed, >200 r/min	2–4
Synchronous condenser:§	
Large	1.25
Small	1.00
Synchronous motor with load varies from 1.0 to 5.0 and higher for heavy flywheels	2.00

† Reprinted by permission of the Westinghouse Electric Corporation from "Electrical Transmission and Distribution Reference Book."
‡ Where range is given, the first figure applies to machines of smaller megavoltampere rating.
§ Hydrogen-cooled, 25% less.

Example 14.1 Develop a formula to calculate the H constant from WR^2 and evaluate H for a nuclear generating unit rated at 1333 MVA, 1800 r/min with $WR^2 = 5,820,000$ lb-ft^2.

SOLUTION The kinetic energy of rotation in foot-pounds at synchronous speed is

$$\text{KE} = \frac{1}{2}\frac{WR^2}{32.2}\left[\frac{2\pi(\text{r/min})}{60}\right]^2 \qquad \text{ft-lb}$$

Since 550 ft-lb/s equals 746 W, it follows that 1 ft-lb equals 746/550 J. Hence, converting foot-pounds to megajoules and dividing by the machine rating in megavoltamperes, we obtain

$$H = \frac{\left(\dfrac{746}{550}\times 10^{-6}\right)\dfrac{1}{2}\dfrac{WR^2}{32.2}\left[\dfrac{2\pi(\text{r/min})}{60}\right]^2}{S_{\text{mach}}}$$

which yields upon simplification

$$H = \frac{2.31\times 10^{-10}WR^2(\text{r/min})^2}{S_{\text{mach}}}$$

Inserting the given machine data in this equation, we obtain

$$H = \frac{2.31 \times 10^{-10}(5.82 \times 10^6)(1800)^2}{1333}$$

$$= 3.27 \text{ MJ/MVA}$$

Converting H to a 100-MVA system base, we obtain

$$H = 3.27 \times \frac{1333}{100} = 43.56 \text{ MJ/MVA}$$

In a stability study for a large system with many machines geographically dispersed over a wide area, it is desirable to minimize the number of swing equations to be solved. This can be done if the transmission line fault, or other disturbance on the system, affects the machines within a power plant so that their rotors swing together. In such cases, the machines within the plant can be combined into a single equivalent machine just as if their rotors were mechanically coupled and only one swing equation need be written for them. Consider a power plant with two generators connected to the same bus which is electrically remote from the network disturbances. The swing equations on the common system base are

$$\frac{2H_1}{\omega_s} \frac{d^2\delta_1}{dt^2} = P_{m1} - P_{e1} \qquad \text{per unit} \qquad (14.18)$$

$$\frac{2H_2}{\omega_s} \frac{d^2\delta_2}{dt^2} = P_{m2} - P_{e2} \qquad \text{per unit} \qquad (14.19)$$

Adding the equations together, and denoting δ_1 and δ_2 by δ since the rotors swing together, we obtain

$$\frac{2H}{\omega_s} \frac{d^2\delta}{dt^2} = P_m - P_e \qquad \text{per unit} \qquad (14.20)$$

where $H = (H_1 + H_2)$, $P_m = (P_{m1} + P_{m2})$ and $P_e = (P_{e1} + P_{e2})$. This single equation, which is in the form of Eq. (14.12), can be solved to represent the plant dynamics.

Example 14.2 Two 60-Hz generating units operate in parallel within the same power plant and have the following ratings:

Unit 1: 500 MVA, 0.85 power factor, 20 kV, 3600 r/min

$$H_1 = 4.8 \text{ MJ/MVA}$$

Unit 2: 1333 MVA, 0.9 power factor, 22 kV, 1800 r/min

$$H_2 = 3.27 \text{ MJ/MVA}$$

Calculate the equivalent H constant for the two units on a 100-MVA base.

SOLUTION The total kinetic energy of rotation of the two machines is

$$KE = (4.8 \times 500) + (3.27 \times 1333) = 6759 \text{ MJ}$$

Therefore the H constant for the equivalent machine on 100-MVA base is

$$H = 67.59 \text{ MJ/MVA}$$

and this value can be used in a single swing equation provided the machines swing together so that their rotor angles are in step at each instant of time.

Machines which swing together are called *coherent* machines. It is noted that, when both ω_s and δ are expressed in electrical degrees or radians, the swing equations for coherent machines can be combined together even though, as in the example, the rated speeds are different. This fact is often used in stability studies involving many machines in order to reduce the number of swing equations which need to be solved.

For *any pair* of non-coherent machines in a system, swing equations similar to Eqs. (14.18) and (14.19) can be written. Dividing each equation by its left-hand-side coefficient and subtracting the resultant equations, we obtain

$$\frac{d^2\delta_1}{dt^2} - \frac{d^2\delta_2}{dt^2} = \frac{\omega_s}{2}\left(\frac{P_{m1} - P_{e1}}{H_1} - \frac{P_{m2} - P_{e2}}{H_2}\right) \qquad (14.21)$$

Multiplying each side by $H_1 H_2/(H_1 + H_2)$ and rearranging, we find that

$$\frac{2}{\omega_s}\left(\frac{H_1 H_2}{H_1 + H_2}\right)\frac{d^2(\delta_1 - \delta_2)}{dt^2} = \frac{P_{m1} H_2 - P_{m2} H_1}{H_1 + H_2} - \frac{P_{e1} H_2 - P_{e2} H_1}{H_1 + H_2} \qquad (14.22)$$

which also may be written more simply in the form of the basic swing equation, Eq. (14.12), as follows

$$\frac{2}{\omega_s} H_{12} \frac{d^2\delta_{12}}{dt^2} = P_{m12} - P_{e12} \qquad (14.23)$$

Here the relative angle δ_{12} equals $\delta_1 - \delta_2$ and an equivalent inertia and weighted input and output powers are defined by

$$H_{12} = \frac{H_1 H_2}{H_1 + H_2} \qquad (14.24)$$

$$P_{m12} = \frac{P_{m1} H_2 - P_{m2} H_1}{H_1 + H_2} \qquad (14.25)$$

$$P_{e12} = \frac{P_{e1} H_2 - P_{e2} H_1}{H_1 + H_2} \qquad (14.26)$$

A noteworthy application of these equations concerns a two-machine system having only one generator (machine one) and a synchronous motor (machine two) connected by a network of pure reactances. Whatever change occurs in the generator output is thus absorbed by the motor and we can write

$$P_{m1} = -P_{m2} = P_m$$
$$P_{e1} = -P_{e2} = P_e$$

(14.27)

Under these conditions, $P_{m12} = P_m$, $P_{e12} = P_e$ and Eq. (14.22) reduces to

$$\frac{2H_{12}}{\omega_s}\frac{d^2\delta_{12}}{dt^2} = P_m - P_e$$

which is also the format of Eq. (14.12) which applies for a single machine.

Equation (14.22) demonstrates that stability of a machine within a system is a relative property associated with its dynamic behavior with respect to the other machines of the system. The rotor angle of one machine, say δ_1, can be chosen for comparison with the rotor angle of each other machine, symbolized by δ_2. In order to be stable the angular differences between all machines must decrease after the final switching operation such as the opening of a circuit-breaker to clear a fault. While we may choose to plot the angle between a machine's rotor and a synchronously rotating reference axis, it is the relative angles between machines which are important. Our discussion above emphasizes the relative nature of the system stability property and shows that the essential features of a stability study are revealed by consideration of two-machine problems. Such problems are of two types; those having one machine of finite inertia swinging with respect to an infinite bus and those having two finite-inertia machines swinging with respect to each other. An infinite bus may be considered for stability purposes as a bus at which is located a machine of constant internal voltage, having zero impedance and infinite inertia. The point of connection of a generator to a large power system may be regarded as such a bus. In all cases the swing equation assumes the form of Eq. (14.12), each term of which must be explicitly described before it can be solved. The equation for P_e is essential to this description and we now proceed to its characterization for a general two-machine system.

14.4 THE POWER-ANGLE EQUATION

In the swing equation for the generator, the input mechanical power from the prime mover P_m will be considered constant. As we have mentioned previously, this is a reasonable assumption because conditions in the electrical network can be expected to change before the control governor can cause the turbine to react. Since P_m in Eq. (14.12) is constant, the electrical power output P_e will determine whether the rotor accelerates, decelerates, or remains at synchronous speed. When P_e equals P_m the machine operates at steady-state synchronous speed; when P_e changes from this value the rotor deviates from synchronous speed.

Changes in P_e are determined by conditions on the transmission and distribution networks and the loads on the system to which the generator supplies power. Electrical network disturbances resulting from severe load changes, network faults, or circuit breaker operations, may cause the generator output P_e to change rapidly in which case electromechanical transients exist. Our fundamental assumption is that the effect of machine speed variations upon the generated voltage is negligible so that the manner in which P_e changes is determined by the load flow equations applicable to the state of the electrical network and by the model chosen to represent the electrical behavior of the machine. Each synchronous machine is represented for transient stability studies by its transient internal voltage E' in series with the transient reactance X'_d as shown in Fig. 14.2a in which V_t is the terminal voltage. This corresponds to the steady-state representation in which synchronous reactance X_d is in series with the synchronous internal or no-load voltage E. Armature resistance is negligible in most cases so that the phasor diagram of Fig. 14.2b applies. Since each machine must be considered relative to the system of which it is part, the phasor angles of the machine quantities are measured with respect to the common system reference.

Figure 14.3 schematically represents a generator supplying power through a transmission system to a receiving-end system at bus 2. The rectangle represents the transmission system of linear passive components such as transformers, transmission lines, and capacitors, and includes the transient reactance of the generator. Therefore, the voltage E'_1 represents the transient internal voltage of the generator at bus 1. The voltage E'_2 at the receiving end is regarded here as that of an infinite bus or as the transient internal voltage of a synchronous motor whose transient reactance is included in the network. Later we shall consider the case of two generators supplying constant-impedance loads within

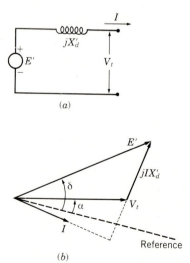

(a)

(b)

Figure 14.2 Phasor diagram of a synchronous machine for transient stability studies.

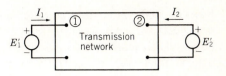

Figure 14.3 Schematic diagram for stability studies. Transient reactances associated with E_1' and E_2' are included in the transmission network.

the network. The elements of the bus admittance matrix for the network reduced to two nodes in addition to the reference node is

$$\mathbf{Y}_{bus} = \begin{vmatrix} Y_{11} & Y_{12} \\ Y_{21} & Y_{22} \end{vmatrix} \tag{14.28}$$

Repeating Eq. (8.7) we have

$$P_k - jQ_k = V_k^* \sum_{n=1}^{N} Y_{kn} V_n \tag{14.29}$$

which upon letting k and N equal 1 and 2, respectively, and substituting E_2' for V, can be written

$$P_1 + jQ_1 = E_1'(Y_{11}E_1')^* + E_1'(Y_{12}E_2')^* \tag{14.30}$$

or where

$$E_1' = |E_1'| \underline{/\delta_1} \qquad\qquad E_2' = |E_2'| \underline{/\delta_2}$$

$$Y_{11} = G_{11} + jB_{11} \qquad Y_{12} = |Y_{12}| \underline{/\theta_{12}}$$

we obtain

$$P_1 = |E_1'|^2 G_{11} + |E_1'||E_2'||Y_{11}| \cos(\delta_1 - \delta_2 - \theta_{12}) \tag{14.31}$$

$$Q_1 = -|E_1'|^2 B_{11} + |E_1'||E_2'||Y_{12}| \sin(\delta_1 - \delta_2 - \theta_{12}) \tag{14.32}$$

Similar equations apply at bus 2 by interchanging the subscripts in the two equations above.

If we let

$$\delta = \delta_1 - \delta_2$$

and define a new angle γ such that

$$\gamma = \theta_{12} - \frac{\pi}{2}$$

we obtain from Eqs. (14.31) and (14.32)

$$P_1 = |E_1'|^2 G_{11} + |E_1'||E_2'||Y_{12}| \sin(\delta - \gamma) \tag{14.33}$$

$$Q_1 = -|E_1'|^2 B_{11} - |E_1'||E_2'||Y_{12}| \cos(\delta - \gamma) \tag{14.34}$$

Equation (14.33) may be written more simply as

$$P_e = P_c + P_{max} \sin(\delta - \gamma) \tag{14.35}$$

where

$$P_c = |E_1'|^2 G_{11} \qquad P_{max} = |E_1'||E_2'||Y_{12}| \tag{14.36}$$

Since P_1 represents the electric power output of the generator (armature loss neglected) we have replaced it by P_e in Eq. (14.35) which is often called the *power-angle equation*; its graph as a function of δ is called the *power-angle curve*. The parameters P_c, P_{max}, and γ are constants for a given network configuration and constant voltage magnitudes $|E'_1|$ and $|E'_2|$. When the network is considered without resistance all the elements of $\mathbf{Y}_{bus}$ are susceptances and so both G_{11} and γ are zero. The power-angle equation which then applies for the pure reactance network is simply the familiar equation

$$P_e = P_{max} \sin \delta \tag{14.37}$$

where $P_{max} = |E'_1||E'_2|/X$ and X is the transfer reactance between E'_1 and E'_2.

Example 14.3 The single-line diagram of Fig. 14.4 shows a generator connected through parallel transmission lines to a large metropolitan system considered as an infinite bus. The machine is delivering 1.0 per unit power and both the terminal voltage and the infinite-bus voltage are 1.0 per unit. Numbers on the diagram indicate the values of the reactances on a common system base. The transient reactance of the generator is 0.20 per unit as indicated. Determine the power-angle equation for the system applicable to the operating conditions.

SOLUTION The reactance diagram for the system is shown in Fig. 14.5a. The series reactance between the terminal voltage and the infinite bus is

$$X = 0.10 + \frac{0.4}{2} = 0.3 \text{ per unit}$$

and therefore the 1.0 per unit power output of the generator is determined by the power-angle equation

$$\frac{|V_t||V|}{X} \sin \alpha = \frac{(1.0)(1.0)}{0.3} \sin \alpha = 1.0$$

where V is the voltage of the infinite bus and α is the angle of the terminal voltage relative to the infinite bus. Solving for α, we obtain

$$\alpha = \sin^{-1} 0.3 = 17.458°$$

so that the terminal voltage is

$$V_t = 1.0\underline{/17.458°} = 0.954 + j0.300 \text{ per unit}$$

Figure 14.4 One-line diagram for Examples 14.3 and 14.4. Point P is at the center of the line.

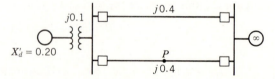

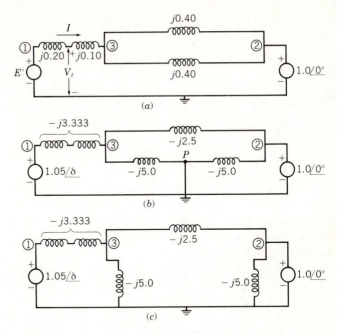

Figure 14.5 Reactance diagram (*a*) for prefault network for Example 14.3 with impedances in per unit and (*b*) and (*c*) for the faulted network for Example 14.4 with the same impedances converted to admittances and marked in per unit.

The output current from the generator is now calculated as

$$I = \frac{1.0\underline{/17.458°} - 1.0\underline{/0°}}{j0.3}$$

$$= 1.0 + j0.1535 = 1.012\underline{/8.729°} \text{ per unit}$$

and the transient internal voltage is then found to be

$$E' = (0.954 + j0.30) + j(0.2)(1.0 + j0.1535)$$

$$= 0.923 + j0.5 = 1.050\underline{/28.44°} \text{ per unit}$$

The power-angle equation relating the transient internal voltage E' and the infinite bus voltage V is determined by the *total* series reactance

$$X = 0.2 + 0.1 + \frac{0.4}{2} = 0.5 \text{ per unit}$$

Hence, the desired equation is

$$P_e = \frac{(1.050)(1.0)}{0.5} \sin \delta = 2.10 \sin \delta \qquad \text{per unit}$$

where δ is the machine rotor angle with respect to the infinite bus.

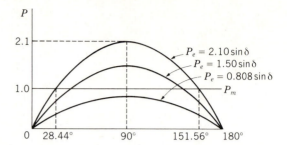

Figure 14.6 Plot of power-angle curves found in Examples 14.3 to 14.5.

This power-angle equation is plotted in Fig. 14.6. Note that the mechanical input power P_m is constant and intersects the sinusoidal power-angle curve at the operating angle $\delta_0 = 28.44°$. This is the initial angular position of the generator rotor corresponding to the given operating conditions. The swing equation for the machine may be written

$$\frac{H}{180f}\frac{d^2\delta}{dt^2} = 1.0 - 2.10\sin\delta \qquad \text{per unit} \qquad (14.38)$$

where H is in megajoules per megavoltampere, f is the electrical frequency of the system, and δ is in electrical degrees. We can easily check the example results since, under the operating conditions, $P_e = 2.10\sin 28.44° = 1.0$ per unit which corresponds exactly to the mechanical power input P_m and the acceleration is zero.

In the next example, we determine the power-angle equation for the same system with a three-phase fault at P, the midpoint of one of the transmission lines. Positive acceleration is shown to exist while the fault is on.

Example 14.4 The system of Example 14.3 is operating under the indicated conditions when a three-phase fault occurs at point P in Fig. 14.4. Determine the power-angle equation for the system with the fault on and the corresponding swing equation. Take $H = 5$ MJ/MVA.

SOLUTION The reactance diagram is shown in Fig. 14.5b with the fault on the system at point P. Values shown are admittances in per unit. The effect of the short circuit caused by the fault is clearly shown by redrawing the reactance diagram as in Fig. 14.5c. As calculated in Example 14.3, transient internal voltage of the generator remains at $E' = 1.05\underline{/28.44°}$ based on the assumption of constant flux linkages in the machine. The net transfer admittance connecting the voltage sources remains to be determined. The buses are numbered as shown and the $\mathbf{Y}_{bus}$ is formed by inspection of Fig. 14.5c as follows

$$\mathbf{Y}_{bus} = j\begin{bmatrix} -3.333 & 0 & 3.333 \\ 0 & -7.50 & 2.50 \\ 3.333 & 2.50 & -10.833 \end{bmatrix}$$

Bus 3 has no external source connection and it may be removed by the node elimination procedure of Sec. 7.4 to yield the reduced bus admittance matrix

$$\begin{bmatrix} Y_{11} & Y_{12} \\ Y_{21} & Y_{22} \end{bmatrix} = j \begin{bmatrix} -2.308 & 0.769 \\ 0.769 & -6.923 \end{bmatrix}$$

The magnitude of the transfer admittance is 0.769 and therefore

$$P_{max} = |E_1'||E_2'||Y_{12}| = (1.05)(1.0)(0.769) = 0.808 \text{ per unit}$$

The power-angle equation with the fault on the system is therefore

$$P_e = 0.808 \sin \delta \qquad \text{per unit}$$

and the corresponding swing equation is

$$\frac{5}{180f} \frac{d^2\delta}{dt^2} = 1.0 - 0.808 \sin \delta \qquad \text{per unit} \qquad (14.39)$$

For later reference note that, because of its inertia, the rotor cannot change position instantly upon occurrence of the fault. Therefore, the rotor angle δ is initially 28.44° as in Example 14.3 and the electrical power output is $P_e = 0.808 \sin 28.44° = 0.385$. The initial accelerating power is

$$P_a = 1.0 - 0.385 = 0.615 \text{ per unit}$$

and the initial acceleration is positive with the value given by

$$\frac{d^2\delta}{dt^2} = \frac{180f}{5}(0.615) = 22.14f \qquad \text{elec deg/s}^2$$

where f is the system frequency.

The line-relaying schemes will sense the fault on the line and will act to clear the fault by simultaneous opening of the line-end breakers. When this occurs another power-angle equation applies since a network change has occurred.

Example 14.5 The fault on the system of Example 14.4 is cleared by simultaneous opening of the circuit breakers at each end of the affected line. Determine the power-angle equation and the swing equation for the post-fault period.

SOLUTION Inspection of Fig. 14.5a shows that, upon removal of the faulted line, the net transfer admittance across the system is

$$y_{12} = \frac{1}{j(0.2 + 0.1 + 0.4)} = -j1.429 \text{ per unit}$$

or in the bus admittance matrix

$$Y_{12} = j1.429$$

Therefore the postfault power-angle equation is

$$P_e = (1.05)(1.0)(1.429) \sin \delta = 1.500 \sin \delta$$

and the swing equation is

$$\frac{5}{180f} \frac{d^2\delta}{dt^2} = 1.0 - 1.500 \sin \delta$$

The acceleration at the instant of clearing the fault depends upon the angular position of the rotor at that time. The power-angle curves for Examples 14.3 to 14.5 are compared in Fig. 14.6.

14.5 SYNCHRONIZING POWER COEFFICIENTS

In Example 14.3 the operating point on the sinusoidal P_e curve of Fig. 14.6 was found to be at δ_0 equal to 28.44° where the mechanical power input P_m equals the electrical power output P_e. In the same figure it is also seen that P_e equals P_m at $\delta = 151.56°$ and this might appear to be an equally acceptable operating point. However, this is now shown not to be the case.

A common-sense requirement for an acceptable operating point is that the generator shall not lose synchronism when small temporary changes occur in the electrical power output from the machine. To examine this requirement, for fixed mechanical input power P_m, consider small incremental changes in the operating point parameters, that is, consider

$$\delta = \delta_0 + \delta_\Delta \qquad P_e = P_{e0} + P_{e\Delta} \qquad (14.40)$$

where the subscript zero denotes the steady-state operating point values and the subscript delta identifies the incremental variations from those values. Substituting Eqs. (14.40) into Eq. (14.37), we obtain the power-angle equation for the general two-machine system in the form

$$P_{e0} + P_{e\Delta} = P_{max} \sin (\delta_0 + \delta_\Delta)$$
$$= P_{max} (\sin \delta_0 \cos \delta_\Delta + \cos \delta_0 \sin \delta_\Delta)$$

Since δ_Δ is a small incremental displacement from δ_0,

$$\sin \delta_\Delta \cong \delta_\Delta \qquad \text{and} \qquad \cos \delta_\Delta \cong 1 \qquad (14.41)$$

and the previous equation becomes

$$P_{e0} + P_{e\Delta} = P_{max} \sin \delta_0 + (P_{max} \cos \delta_0)\delta_\Delta \qquad (14.42)$$

where strict equality is now used. At the initial operating point δ_0,

$$P_m = P_{e0} = P_{max} \sin \delta_0 \qquad (14.43)$$

and it therefore follows from the last two equations that

$$P_m - (P_{e0} + P_{e\Delta}) = -(P_{max} \cos \delta_0)\delta_\Delta \qquad (14.44)$$

Substituting the incremental variables of Eq. (14.40) into the basic swing equation, Eq. (14.12), we obtain

$$\frac{2H}{\omega_s} \frac{d^2(\delta_0 + \delta_\Delta)}{dt^2} = P_m - (P_{e0} + P_{e\Delta}) \tag{14.45}$$

Replacing the right-hand side of this equation by Eq. (14.44) and transposing terms, we obtain

$$\frac{2H}{\omega_s} \frac{d^2\delta_\Delta}{dt^2} + (P_{max} \cos \delta_0)\delta_\Delta = 0 \tag{14.46}$$

since δ_0 is a constant value. Noting that $P_{max} \cos \delta_0$ is the slope of the power-angle curve at the angle δ_0, we denote this slope as S_P and define it as

$$S_P = \left.\frac{dP_e}{d\delta}\right|_{\delta = \delta_0} = P_{max} \cos \delta_0 \tag{14.47}$$

where S_P is called the *synchronizing power coefficient*. When S_P is used in Eq. (14.46), the swing equation governing the incremental rotor-angle variations may be rewritten in the form

$$\frac{d^2\delta_\Delta}{dt^2} + \frac{\omega_s S_P}{2H} \delta_\Delta = 0 \tag{14.48}$$

This is a linear, second-order differential equation, the solution to which depends upon the algebraic sign of S_P. When S_P is positive, the solution $\delta_\Delta(t)$ corresponds to that of simple harmonic motion; such motion is represented by the oscillations of an undamped swinging pendulum.† When S_P is negative, the solution $\delta_\Delta(t)$ increases exponentially without limit. Therefore, in Fig. 14.6 the operating point $\delta = 28.44°$ is a point of stable equilibrium, in the sense that the rotor angle swing is bounded following a small perturbation. In the physical situation, damping will restore the rotor angle to $\delta_0 = 28.44°$ following the temporary electrical perturbation. On the other hand, the point $\delta = 151.56°$ is a point of unstable equilibrium since S_P is negative there. So this point is not a valid operating point.

The changing position of the generator rotor swinging with respect to the infinite bus may be visualized by an analogy. Consider a pendulum swinging from a pivot on a stationary frame, as shown in Fig. 14.7a. Points *a* and *c* are the maximum points of the oscillation of the pendulum about the equilibrium point *b*. Damping will eventually bring the pendulum to rest at *b*. Now imagine a disk rotating in a clockwise direction about the pivot of the pendulum, as shown in Fig. 14.7b, and superimpose the motion of the pendulum on the motion of the disk. When the pendulum is moving from *a* to *c*, the combined angular velocity

---

† The equation of simple harmonic motion is $d^2x/dt^2 + \omega_n^2 x = 0$ which has the general solution $A \cos \omega_n t + B \sin \omega_n t$ with constants A and B determined by the initial conditions. The solution when plotted is an undamped sinusoid of angular frequency ω_n.

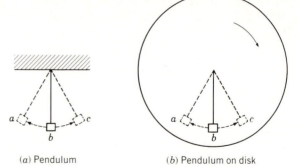

Figure 14.7 Pendulum and rotating disk to illustrate a rotor swinging with respect to an infinite bus.

(a) Pendulum   (b) Pendulum on disk

is slower than that of the disk. When the pendulum is moving from c to a, the combined angular velocity is faster than that of the disk. At points a and c, the angular velocity of the pendulum alone is zero and the combined angular velocity equals that of the disk. If the angular velocity of the disk corresponds to the synchronous speed of the rotor, and if the motion of the pendulum alone represents the swinging of the rotor with respect to an infinite bus, the superimposed motion of the pendulum on that of the disk represents the actual angular motion of the rotor.

From the above discussion, we conclude that the solution to Eq. (14.48) represents sinusoidal oscillations provided the synchronizing power coefficient S_P is positive. The angular frequency of the undamped oscillations is given by

$$\omega_n = \sqrt{\frac{\omega_s S_P}{2H}} \quad \text{elec rad/s} \tag{14.49}$$

which corresponds to a frequency of oscillation given by

$$f_n = \frac{1}{2\pi} \sqrt{\frac{\omega_s S_P}{2H}} \quad \text{Hz} \tag{14.50}$$

Example 14.6 The machine of Example 14.3 is operating at $\delta = 28.44°$ when it is subjected to a slight temporary electrical-system disturbance. Determine the frequency and period of oscillation of the machine rotor if the disturbance is removed before the prime mover responds. $H = 5$ MJ/MVA.

SOLUTION The applicable swing equation is Eq. (14.48) and the synchronizing power coefficient at the operating point is

$$S_P = 2.10 \cos 28.44° = 1.8466$$

The angular frequency of oscillation is therefore

$$\omega_n = \sqrt{\frac{\omega_s S_P}{2H}} = \sqrt{\frac{377 \times 1.8466}{2 \times 5}} = 8.343 \text{ elec rad/s}$$

The corresponding frequency of oscillation is

$$f_n = \frac{8.343}{2\pi} = 1.33 \text{ Hz}$$

and the period of oscillation is

$$T = \frac{1}{f_n} = 0.753 \text{ s}$$

This example is an important one from the practical viewpoint since it indicates the order of magnitude of the frequencies which can be superimposed upon the nominal 60-Hz frequency in a large power system having many interconnected machines. As load on the system changes randomly throughout the day, inter-machine oscillations involving frequencies of the order of 1 Hz tend to arise but these are quickly damped out by the various damping influences caused by the prime mover, the system loads, and the machine itself. It is worthwhile to note that even if the transmission system in our example contains resistance, nonetheless, the swinging of the rotor is harmonic and undamped. Problem 14.8 examines the effect of resistance on the synchronizing power coefficient and the frequency of oscillations. In a later section we again discuss the concept of synchronizing coefficients. In the next section we examine a method of determining stability under transient conditions caused by large disturbances.

14.6 EQUAL-AREA CRITERION OF STABILITY

In Sec. 14.4 we developed swing equations which are nonlinear in nature. Formal solution of such equations cannot be explicitly found. Even in the case of a single machine swinging with respect to an infinite bus it is very difficult to obtain literal-form solutions and therefore digital computer methods are normally used. To examine the stability of a two-machine system without solving the swing equation, a direct approach is possible and will now be discussed.

The system shown in Fig. 14.8 is the same as that considered previously except for the addition of a short transmission line. Initially circuit breaker A is closed but circuit breaker B at the opposite end of the short line is open. Therefore, the initial operating conditions of Example 14.3 may be considered

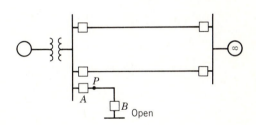

Figure 14.8 One line-diagram of the system of Fig. 14.4 with the addition of a short transmission line.

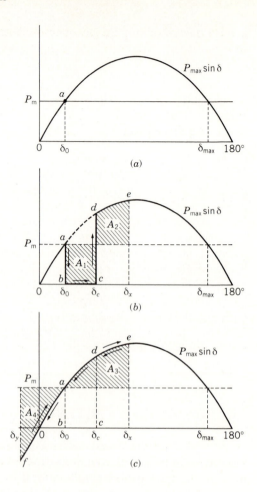

Figure 14.9 Power-angle curves for the generator shown in Fig. 14.8. Areas A_1 and A_2 are equal as are areas A_3 and A_4.

unaltered. At point P close to the bus, a three-phase fault occurs and is cleared by circuit breaker A after a short period of time. Therefore, the effective transmission system is unaltered except while the fault is on. The short circuit caused by the fault is effectively at the bus and so the electrical power output from the generator is zero until the fault is cleared. The physical conditions before, during, and after the fault can be understood by analyzing the power-angle curves in Fig. 14.9.

Originally the generator is operating at synchronous speed with a rotor angle of δ_0 and the input mechanical power P_m equals the output electrical power P_e as shown at point a in Fig. 14.9a. When the fault occurs at $t = 0$, the electrical power output is suddenly zero while the input mechanical power is unaltered as shown in Fig. 14.9b. The difference in power must be accounted for by a rate of change of stored kinetic energy in the rotor masses. This can be accomplished only by an increase in speed which results from the constant accelerating power P_m. If we denote the time to clear the fault by t_c, then for time t less than t_c the acceleration is constant and is given by

$$\frac{d^2\delta}{dt^2} = \frac{\omega_s}{2H} P_m \tag{14.51}$$

While the fault is on, the velocity increase above synchronous speed is found by integrating this equation to obtain

$$\frac{d\delta}{dt} = \int_0^t \frac{\omega_s}{2H} P_m \, dt = \frac{\omega_s}{2H} P_m t \tag{14.52}$$

A further integration with respect to time yields

$$\delta = \frac{\omega_s P_m}{4H} t^2 + \delta_0 \tag{14.53}$$

for the rotor angular position.

Equations (14.52) and (14.53) show that the velocity of the rotor relative to synchronous speed increases linearly with time while the rotor angle advances from δ_0 to the angle at clearing δ_c; that is, in Fig. 14.9 the angles δ goes from b to c. At the instant of fault clearing, the increase in rotor speed and the angle separation between the generator and the infinite bus are given, respectively, by

$$\left.\frac{d\delta}{dt}\right|_{t=t_c} = \frac{\omega_s P_m}{2H} t_c \tag{14.54}$$

and

$$\left.\delta(t)\right|_{t=t_c} = \frac{\omega_s P_m}{4H} t_c^2 + \delta_0 \tag{14.55}$$

When the fault is cleared at the angle δ_c, the electrical power output abruptly increases to a value corresponding to point d on the power-angle curve. At d the electrical power output exceeds the mechanical power input and thus the accelerating power is negative. As a consequence the rotor slows down as P_e goes from d to e in Fig. 14.9c. At e the rotor speed is again synchronous although the rotor angle has advanced to δ_x. The angle δ_x is determined by the fact that areas A_1 and A_2 must be equal, as will be explained later. The accelerating power at e is still negative (retarding), and so the rotor cannot remain at synchronous speed but must continue to slow down. The relative velocity is negative and the rotor angle moves back from δ_x at e along the power-angle curve of Fig. 14.9c to point a at which the rotor speed is less than synchronous. From a to f the mechanical power exceeds the electrical power and rotor increases speed again until it reaches synchronism at f. Point f is located so that areas A_3 and A_4 are equal. In the absence of damping the rotor would continue to oscillate in the sequence f-a-e, e-a-f, etc., with synchronous speed occurring at e and f.

As noted, we shall soon show that the shaded areas A_1 and A_2 in Fig. 14.9b must be equal, and similarly, areas A_3 and A_4 in Fig. 14.9c must be equal. In a system where one machine is swinging with respect to an infinite bus we may use this principle of equality of areas, called the *equal-area criterion,* to determine the stability of the system under transient conditions without solving the swing equation. Although not applicable to multimachine systems, the method helps in understanding how certain factors influence the transient stability of any system.

The derivation of the equal-area criterion is made for one machine and an infinite bus although the considerations in Sec. 14.3 show that the method can be readily adapted to general two-machine systems. The swing equation for the machine connected to the bus is

$$\frac{2H}{\omega_s}\frac{d^2\delta}{dt^2} = P_m - P_e \qquad (14.56)$$

Define the angular velocity of the rotor relative to synchronous speed by

$$\omega_r = \frac{d\delta}{dt} = \omega - \omega_s \qquad (14.57)$$

Differentiating Eq. (14.57) with respect to t and substituting in Eq. (14.56) we obtain

$$\frac{2H}{\omega_s}\frac{d\omega_r}{dt} = P_m - P_e \qquad (14.58)$$

It is clear that when the rotor speed is synchronous, ω equals ω_s and ω_r is zero. Multiplying both sides of Eq. (14.58) by $\omega_r = d\delta/dt$, we have

$$\frac{H}{\omega_s}2\omega_r\frac{d\omega_r}{dt} = (P_m - P_e)\frac{d\delta}{dt} \qquad (14.59)$$

The left-hand side of the equation can be rewritten to give

$$\frac{H}{\omega_s}\frac{d(\omega_r^2)}{dt} = (P_m - P_e)\frac{d\delta}{dt} \qquad (14.60)$$

Multiplying by dt and integrating, we obtain

$$\frac{H}{\omega_s}(\omega_{r2}^2 - \omega_{r1}^2) = \int_{\delta_1}^{\delta_2}(P_m - P_e)\,d\delta \qquad (14.61)$$

The subscripts for the ω_r terms correspond to those for the δ limits, that is, the rotor speed ω_{r1} corresponds to that at the angle δ_1 and ω_{r2} corresponds to δ_2. Since ω_r represents the *departure* of the rotor speed from synchronous speed, we readily see that if the rotor speed is synchronous at δ_1 *and* δ_2, then, correspondingly, $\omega_{r1} = \omega_{r2} = 0$. Under this condition Eq. (14.61) becomes

$$\int_{\delta_1}^{\delta_2}(P_m - P_e)\,d\delta = 0 \qquad (14.62)$$

This equation applies to any two points δ_1 and δ_2 on the power-angle diagram, provided they are points at which the rotor speed is synchronous. In Fig. 14.9*b* two such points are a and e corresponding to δ_0 and δ_x. If we perform the integration in two steps, we can write

$$\int_{\delta_0}^{\delta_c} (P_m - P_e)\, d\delta + \int_{\delta_c}^{\delta_x} (P_m - P_e)\, d\delta = 0 \qquad (14.63)$$

or

$$\int_{\delta_0}^{\delta_c} (P_m - P_e)\, d\delta = \int_{\delta_c}^{\delta_x} (P_e - P_m)\, d\delta \qquad (14.64)$$

The left integral applies to the fault period while the right integral corresponds to the immediate postfault period up to the point of maximum swing δ_x. In Fig. 14.9*b* P_e is zero during the fault. The shaded area A_1 is given by the left-hand side of Eq. (14.64) and the shaded area A_2 is given by the right-hand side. So the two areas A_1 and A_2 are equal.

Since the rotor speed is synchronous at δ_x and also at δ_y in Fig. 14.9*c* the same reasoning as above shows that A_3 equals A_4. The areas A_1 and A_4 are directly proportional to the increase in kinetic energy of the rotor while it is accelerating, whereas areas A_2 and A_3 are proportional to the decrease in kinetic energy of the rotor while it is decelerating. This can be seen by inspection of both sides of Eq. (14.61). Therefore the equal-area criterion merely states that whatever kinetic energy is added to the rotor following a fault must be removed after the fault to restore the rotor to synchronous speed.

The shaded area A_1 is dependent upon the time taken to clear the fault. If there is delay in clearing, the angle δ_c is increased; likewise the area A_1 increases and the equal-area criterion requires that area A_2 also increase to restore the rotor to synchronous speed at a larger angle of maximum swing δ_x. If the delay in clearing is prolonged so that the rotor angle δ swings beyond the angle $\delta_{\max}$ in Fig. 14.9 then the rotor speed at that point on the power-angle curve is above synchronous speed when positive accelerating power is again encountered. Under the influence of this positive accelerating power the angle δ will increase without limit and instability results. Therefore there is a critical angle for clearing the fault in order to satisfy the requirements of the equal-area criterion for stability. This angle, called the *critical clearing angle* δ_{cr}, is shown in Fig. 14.10. The corresponding critical time for removing the fault is called the *critical clearing time* t_{cr}.

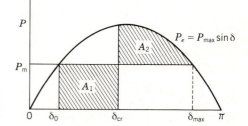

Figure 14.10 Power-angle curve showing the critical-clearing angle δ_{cr}. Areas A_1 and A_2 are equal.

In the particular case of Fig. 14.10, *both* the critical clearing angle *and* the critical clearing time can be calculated as follows. The rectangular area A_1 is

$$A_1 = \int_{\delta_0}^{\delta_{cr}} P_m \, d\delta = P_m(\delta_{cr} - \delta_0) \tag{14.65}$$

while the area A_2 is

$$A_2 = \int_{\delta_{cr}}^{\delta_{max}} (P_{max} \sin \delta - P_m) \, d\delta$$

$$= P_{max}(\cos \delta_{cr} - \cos \delta_{max}) - P_m(\delta_{max} - \delta_{cr}) \tag{14.66}$$

Equating the expressions for A_1 and A_2, and transposing terms, yields

$$\cos \delta_{cr} = (P_m/P_{max})(\delta_{max} - \delta_0) + \cos \delta_{max} \tag{14.67}$$

We see from the sinusoidal power-angle curve that

$$\delta_{max} = \pi - \delta_0 \qquad \text{elec rad} \tag{14.68}$$

and

$$P_m = P_{max} \sin \delta_0 \tag{14.69}$$

Substituting for δ_{max} and P_m in Eq. (14.67), simplifying the result and solving for δ_{cr}, we obtain

$$\delta_{cr} = \cos^{-1}[(\pi - 2\delta_0) \sin \delta_0 - \cos \delta_0] \tag{14.70}$$

for the critical clearing angle. The value for δ_{cr} calculated from this equation, when substituted for the left-hand side of Eq. (14.55), yields

$$\delta_{cr} = \frac{\omega_s P_m}{4H} t_{cr}^2 + \delta_0 \tag{14.71}$$

from which is found

$$t_{cr} = \sqrt{\frac{4H(\delta_{cr} - \delta_0)}{\omega_s P_m}} \tag{14.72}$$

for the critical clearing time.

Example 14.7 Calculate the critical clearing angle and the critical clearing time for the system of Fig. 14.8 when the system is subjected to a three-phase fault at point P on the short transmission line. The initial conditions are the same as those in Example 14.3 and H is 5 MJ/MVA.

SOLUTION From Example 14.3 the power-angle equation is

$$P_e = P_{max} \sin \delta = 2.10 \sin \delta$$

the initial rotor angle is

$$\delta_0 = 28.44° = 0.496 \text{ elec rad}$$

and mechanical input power P_m is 1.0 per unit. Therefore, from Eq. (14.70) we obtain

$$\delta_{cr} = \cos^{-1}[(\pi - 2 \times 0.496) \sin 28.44° - \cos 28.44°]$$
$$= 81.697° = 1.426 \text{ elec rad}$$

for the critical clearing angle. Inserting this value with the other known quantities in Eq. (14.72) we obtain

$$t_{cr} = \sqrt{\frac{4 \times 5(1.426 - 0.496)}{377 \times 1}}$$
$$= 0.222 \text{ s}$$

This value is equivalent to a critical clearing time of 13.3 cycles based on a frequency of 60 Hz.

This example serves to establish the concept of critical clearing time which is essential to the design of proper relaying schemes for fault clearing. In more general cases, the critical clearing time cannot be explicitly found without solving the swing equations by digital computer simulation.

14.7 FURTHER APPLICATIONS OF THE EQUAL-AREA CRITERION

Although the equal-area criterion can be applied only for the case of two machines or one machine and an infinite bus, it is a very useful means for beginning to see what happens when a fault occurs. The digital computer is the only practical way to determine the stability of a large system. However, because the equal-area criterion is so helpful in understanding transient stability, we shall continue to examine it briefly before discussing the determination of swing curves and the digital-computer approach.

When a generator is supplying power to an infinite bus over two parallel transmission lines, opening one of the lines may cause the generator to lose synchronism even though the load could be supplied over the remaining line under steady-state conditions. If a three-phase short circuit occurs on the bus to which two parallel lines are connected, no power can be transmitted over either line. This is essentially the case in Example 14.7. However, if the fault is at the end of one of the lines, opening breakers at both ends of the line will isolate the fault from the system and allow power to flow through the other parallel line. When a three-phase fault occurs at some point on a double-circuit line other than on the paralleling buses or at the extreme ends of the line, there is some impedance between the paralleling buses and the fault. Therefore, some power is transmitted while the fault is still on the system. The power-angle equation in Example 14.4 demonstrates this fact.

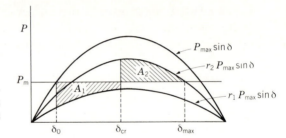

Figure 14.11 Equal-area criterion applied to fault clearing when power is transmitted during the fault. Areas A_1 and A_2 are equal.

When power is transmitted during a fault, the equal-area criterion is applied as shown in Fig. 14.11 which is similar to the power-angle diagram of Fig. 14.6. Before the fault, $P_{max} \sin \delta$ is the power which can be transmitted; during the fault, $r_1 P_{max} \sin \delta$ is the power which can be transmitted; and $r_2 P_{max} \sin \delta$ is the power which can be transmitted after the fault is cleared by switching at the instant when $\delta = \delta_{cr}$. Examination of Fig. 14.11 shows that δ_{cr} is the critical clearing angle in this case. By evaluating the areas A_1 and A_2 using the procedural steps of the previous section, we would find

$$\cos \delta_{cr} = \frac{(P_m/P_{max})(\delta_{max} - \delta_0) + r_2 \cos \delta_{max} - r_1 \cos \delta_0}{r_2 - r_1} \qquad (14.73)$$

A literal form solution for the critical clearing time t_{cr} is not possible in this case. For the system and fault location shown in Fig. 14.8, the values are $r_1 = 0$, $r_2 = 1$ and the equation then reduces to Eq. 14.67.

Regardless of their location, short-circuit faults not involving all three phases allow the transmission of some power, because they are represented by connecting some impedance rather than a short circuit between the fault point and the reference bus in the positive-sequence impedance diagram. The larger the impedance shunted across the positive-sequence network to represent the fault, the larger the power transmitted during the fault. The amount of power transmitted during the fault affects the value of A_1 for any given clearing angle. Thus, smaller values of r_1 result in greater disturbances to the system, as low r_1 means low power transmitted during the fault and larger A_1. In order of increasing severity (decreasing $r_1 P_{max}$) the various faults are:

1. single line-to-ground fault
2. line-to-line fault
3. double line-to-ground fault
4. three-phase fault

The single line-to-ground fault occurs most frequently, and the three-phase fault is least frequent. For complete reliability a system should be designed for transient stability for three-phase faults at the worst locations, and this is virtually the universal practice.

Example 14.8 Determine the critical clearing angle for the three-phase fault described in Examples 14.4 and 14.5 when the initial system configuration and prefault operating conditions are as described in Example 14.3.

SOLUTION The power-angle equations obtained in the previous examples are

$$\text{Before the fault:} \qquad P_{\max} \sin \delta = 2.100 \sin \delta$$

$$\text{During the fault:} \quad r_1 P_{\max} \sin \delta = 0.808 \sin \delta$$

$$\text{After the fault:} \qquad r_2 P_{\max} \sin \delta = 1.500 \sin \delta$$

Hence

$$r_1 = \frac{0.808}{2.100} = 0.385 \qquad r_2 = \frac{1.500}{2.100} = 0.714$$

From Example 14.3, we have

$$\delta_0 = 28.44° = 0.496 \text{ rad}$$

and from Fig. 14.11, we calculate

$$\delta_{\max} = 180° - \sin^{-1} \frac{1.000}{1.500} = 138.190° = 2.412 \text{ rad}$$

Therefore, inserting numerical values in Eq. (14.73), we obtain

$$\cos \delta_{cr} = \frac{(1.0/2.10)(2.412 - 0.496) + 0.714 \cos (138.19°) - 0.385 \cos (28.44°)}{0.714 - 0.385}$$

$$= 0.127$$

Hence

$$\delta_{cr} = 82.726°$$

To determine the critical clearing time we must obtain the swing curve of δ versus t for this example. In Sec. 14.9 we shall discuss one method of computing such swing curves.

14.8 MULTIMACHINE STABILITY STUDIES: CLASSICAL REPRESENTATION

The equal area criterion cannot be used directly in systems where three or more machines are represented. Although the physical phenomena observed in the two-machine problems basically reflect that of the multimachine case, nonetheless, the complexity of the numerical computations increases with the number of machines considered in a transient stability study. When a multimachine system operates under electromechanical transient conditions, intermachine oscillations occur between the machines through the medium of the transmission system which connects them. If any one machine could be considered to act alone as the single oscillating source, it would send into the interconnected system an electromechanical oscillation determined by its inertia and synchronizing power. A typical frequency of such oscillation is of the order of 1 to 2 Hz and this is superimposed upon the nominal 60-Hz frequency of the system. When many machine rotors are simultaneously undergoing transient oscillation, the swing curves will reflect the combined presence of many such oscillations. Therefore, the transmission system frequency is not unduly perturbed from nominal frequency, and the assumption is made that the 60-Hz network parameters are still applicable. To ease the complexity of system modeling and, thereby, the computational burden, the following additional assumptions are commonly made in transient stability studies:

(a) The mechanical power input to each machine remains constant during the entire period of the swing curve computation.
(b) Damping power is negligible.
(c) Each machine may be represented by a constant transient reactance in series with a constant transient internal voltage.
(d) The mechanical rotor angle of each machine coincides with δ, the electrical phase angle of the transient internal voltage.
(e) All loads may be considered as shunt impedances to ground with values determined by conditions prevailing immediately prior to the transient conditions.

The system stability model based on these assumptions is called the *classical stability model* and studies which use this model are called *classical stability studies*. These assumptions, which we shall adopt, are in addition to the fundamental assumptions set forth in Sec. 14.1 for *all* stability studies. Of course, detailed computer programs with more sophisticated machine and load models are available to modify one or more of assumptions (a) to (e). However, throughout this chapter the classical model is used to study system disturbances originating from three-phase faults.

As already seen, in any transient stability study, the system conditions before the fault and the network configuration during and after its occurrence must be known. Consequently, in the multimachine case, two preliminary steps are required:

1. The steady-state prefault conditions for the system are calculated using a production-type load-flow program.
2. The prefault network representation is determined and then modified to account for the fault and for the postfault conditions.

From the first preliminary step we know the values for power, reactive power, and voltage at each generator terminal and load bus with all angles measured with respect to the swing bus. The transient internal voltage of each generator is then calculated using the equation

$$E' = V_t + jX'_d I \tag{14.74}$$

where V_t is the corresponding terminal voltage and I, the output current. Each load is converted into a constant admittance to ground at its bus using the equation

$$Y_L = \frac{P_L - jQ_L}{|V_L|^2} \tag{14.75}$$

where $P_L + jQ_L$ is the load and $|V_L|$ is the magnitude of the corresponding bus voltage. The bus-admittance matrix used for the prefault load-flow calculation is augmented to include the transient reactance of each generator and the shunt load admittances as suggested in Fig. 14.12. Note that the injected current is zero at all buses except the three-generator internal buses. The second preliminary step determines the modified bus-admittance matrices corresponding to the faulted and postfault conditions. Since only the generator internal buses have injections, all other buses can be eliminated to reduce the dimensions of the modified matrices to correspond to the number of generators. During and after the fault, the power flow into the network from each generator is calculated by the corresponding power-angle equations. For example, in Fig. 14.12 the power out of generator 1 is given by

$$P_{e1} = |E_1'|^2 G_{11} + |E_1'||E_2'||Y_{12}| \cos(\delta_{12} - \theta_{12}) \\ + |E_1'||E_3'||Y_{13}| \cos(\delta_{13} - \theta_{13}) \tag{14.76}$$

where δ_{12} equals $\delta_1 - \delta_2$. Similar equations are written for P_{e2} and P_{e3} with the Y_{ij} values chosen from the 3×3 bus-admittance matrices appropriate to the fault or postfault condition. The power-angle equations form part of the swing equations

$$\frac{2H_i}{\omega_s} \frac{d^2\delta_i}{dt^2} = P_{mi} - P_{ei} \qquad i = 1, 2, 3 \tag{14.77}$$

to represent the motion of each rotor for the fault and postfault periods. The solutions depend upon the location and duration of the fault, and the $\mathbf{Y}_{bus}$ which

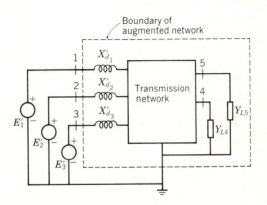

Figure 14.12 Augmented network of a power system.

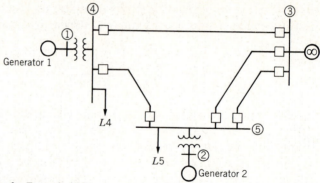

Figure 14.13 One-line diagram for Example 14.9.

results when the faulted line is removed. The basic procedures used in digital computer programs for classical stability studies are revealed in the following examples.

Example 14.9 A 60-Hz, 230-kV transmission line has two generators and an infinite bus as shown in Fig. 14.13. The transformer and line data are given in Table 14.2. A three-phase fault occurs on line 4–5 near bus 4. Using the prefault load-flow solution shown in Table 14.3 determine the swing equation for each machine during the fault period. The generators, with reactances and H values expressed on a 100-MVA base, are described as follows:

Generator 1   400 MVA, 20 kV, $X'_d = 0.067$ per unit, $H = 11.2$ MJ/MVA
Generator 2   250 MVA, 18 kV, $X'_d = 0.10$ per unit, $H = 8.0$ MJ/MVA

Table 14.2 Line and transformer data for Example 14.9, all values in per unit on 230-kV, 100-MVA base.

Bus to bus	Series Z		Shunt Y
	R	X	B
Trans. 1–4	$\cdots\cdots$	0.022	
Trans. 2–5	$\cdots\cdots$	0.040	
Line 3–4	0.007	0.040	0.082
Line 3–5(1)	0.008	0.047	0.098
Line 3–5(2)	0.008	0.047	0.098
Line 4–5	0.018	0.110	0.226

Table 14.3 Bus data and prefault load-flow values in per unit on 230-KV, 100-MVA base.

Bus	Voltage	Generation		Load	
		P	Q	P	Q
1	$1.030\underline{/8.88°}$	3.500	0.712		
2	$1.020\underline{/6.38°}$	1.850	0.298		
3	$1.000\underline{/0°}$	......	......		
4	$1.018\underline{/4.68°}$	......	......	1.00	0.44
5	$1.011\underline{/2.27°}$	......	......	0.50	0.16

SOLUTION To determine the swing equations we need to find transient internal voltages. The current into the network at bus 1 based on the data in Table 14.3 is

$$I_1 = \frac{(P_1 + jQ_1)^*}{V_1^*} = \frac{3.50 - j0.712}{1.030 \underline{/-8.88°}} = 3.468 \underline{/-2.619°}$$

and similarly

$$I_2 = \frac{1.850 - j0.298}{1.020 \underline{/-6.38°}} = 1.837 \underline{/-2.771}$$

so,

$$E_1' = 1.030\underline{/8.88°} + j0.067 \times 3.468 \underline{/-2.619} = 1.100 \underline{/20.82°}$$

$$E_2' = 1.020\underline{/6.38°} + j0.10 \times 1.837 \underline{/-2.771} = 1.065 \underline{/16.19°}$$

At the infinite bus

$$E_3' = E_3 = 1.000 \underline{/0.0°}$$

and so

$$\delta_{13} = \delta_1 \quad \text{and} \quad \delta_{23} = \delta_2$$

The loads at buses 4 and 5 are represented by the admittances calculated by Eq. (14.75) to yield

$$Y_{L4} = \frac{1.00 - j0.44}{(1.018)^2} = 0.9649 - j0.4246$$

$$Y_{L5} = \frac{0.50 - j0.16}{(1.011)^2} = 0.4892 - j0.1565$$

These admittances, together with the transient reactances, are used with the line and transformer parameters to form for the prefault system the augmented bus admittance matrix which includes the transient reactances of the

Table 14.4 Elements of prefault bus admittance matrix for Example 14.9, admittances in per unit.

Bus	1	2	3	4	5
1	$-j11.2360$	0.0	0.0	$j11.2360$	0.0
2	0.0	$-j7.1429$	0.0	0.0	$j7.1429$
3	0.0	0.0	11.2841 $-j65.4731$	-4.2450 $+j24.2571$	-7.0392 $+j41.3550$
4	$j11.2360$	0.0	-4.2450 $+j24.2571$	6.6588 $-j44.6175$	-1.4488 $+j8.8538$
5	0.0	$j7.1429$	-7.0392 $+j41.3550$	-1.4488 $+j8.8538$	8.9772 $-j57.2972$

machines. Therefore, we will now designate as buses 1 and 2 the fictitious internal nodes between the internal voltages and the transient reactances of the machines. So, in the matrix, for example:

$$Y_{11} = \frac{1}{j0.067 + j0.022} = -j11.236$$

$$Y_{34} = -\frac{1}{0.007 + j0.040} = -4.2450 + j24.2571$$

The sum of the admittances connected to nodes 3, 4, and 5 must include the shunt capacitances of the transmission lines. So

$$Y_{44} = -j11.236 + \frac{j0.082}{2} + \frac{j0.226}{2} + 4.2450$$

$$-j24.2571 + \frac{1}{0.018 + j0.110} + 0.9649 - j0.4246$$

$$= 6.6587 - j44.6175$$

The entire augmented matrix is displayed as Table 14.4.

During the fault, bus 4 must be short circuited to ground. Row and column 4 of Table 14.4 disappear because node 4 is merged with the reference node. The new row 4 and column 4 (node 5) are eliminated by Eq. (7.30) to reduce the bus-admittance matrix for the faulted network to that shown in the upper half of Table 14.5. The faulted-system Y_{bus} shows that bus 1 decouples from the other buses during the fault and that bus 2 is connected directly to bus 3. This reflects the physical fact that the short circuit at bus 4 reduces to zero the power injected into the system from generator 1 and causes generator 2 to deliver its power radially to bus 3. Under fault conditions we find by using values from Table 14.5 for the per-unit power-angle equations

Table 14.5 Elements of faulted and postfault bus admittance matrices for Example 14.9, admittances in per unit.

	Faulted network		
Bus	1	2	3
1	$0.0000 - j11.2360$ $(11.2360 \,\underline{/-90°})$	$0.0 + j0.0$	$0.0 + j0.0$
2	$0.0 + j0.0$	$0.1362 - j6.2737$ $(6.2752 \,\underline{/-88.7563°})$	$-0.0681 + j5.1661$ $(5.1665 \,\underline{/90.7552})$
3	$0.0 + j0.0$	$-0.681 + j5.1661$ $(5.1665 \,\underline{/90.7552°})$	$5.7986 - j35.6299$ $(36.0987 \,\underline{/-80.7564})$
	Postfault network		
1	$0.5005 - j7.7897$ $(7.8058 \,\underline{/-86.3237°})$	$0.0 + j0.0$	$-0.2216 + j7.6291$ $(7.6323 \,\underline{/91.6638°})$
2	$0.0 + j0.0$	$0.1591 - j6.1168$ $(6.1189 \,\underline{/-88.5101°})$	$-0.0901 + j6.0975$ $(6.0982 \,\underline{/90.8466°})$
3	$-0.2216 + j7.6291$ $(7.6323 \,\underline{/91.6638°})$	$-0.0901 - j6.0975$ $(6.0982 \,\underline{/90.8466°})$	$1.3927 - j13.8728$ $(13.9426 \,\underline{/-84.2672})$

$$P_{e1} = 0$$

$$P_{e2} = |E_2'|^2 G_{22} + |E_2'||E_3||Y_{23}| \cos(\delta_{23} - \theta_{23})$$

$$= (1.065)^2(0.1362) + (1.065)(1.0)(5.1665) \cos(\delta_2 - 90.755°)$$

$$= 0.1545 + 5.5023 \sin(\delta_2 - 0.755°)$$

Therefore, while the fault is on the system, the desired swing equations (values of P_{m1} and P_{m2} from Table 14.3) are

$$\frac{d^2\delta_1}{dt^2} = \frac{180f}{H_1}(P_{m1} - P_{e1}) = \frac{180f}{H_1} P_{a1}$$

$$= \frac{180f}{11.2}(3.5) \qquad \text{elec deg/s}^2$$

$$\frac{d^2\delta_2}{dt^2} = \frac{180f}{H_2}(P_{m2} - P_{e2}) = \frac{180f}{H_2} P_{a2}$$

$$= \frac{180f}{8.0} \{ \overset{P_m}{\overbrace{1.85}} - [\overset{P_c}{\overbrace{0.1545}} + \overset{P_{max}}{\overbrace{5.5023}} \sin(\delta_2 - \overset{\gamma}{\overbrace{0.755°}})]\}$$

$$= \frac{180f}{8.0} [\underset{P_m - P_c}{\underbrace{1.6955}} - \underset{P_{max}}{\underbrace{5.5023}} \sin(\delta_2 - \underset{\gamma}{\underbrace{0.755°}})] \qquad \text{elec deg/s}^2$$

Example 14.10 The three-phase fault in Example 14.9 is cleared by simultaneously opening the circuit breakers at the ends of the faulted line. Determine the swing equations for the postfault period.

SOLUTION When the fault is cleared by removing line 4-5, the prefault $\mathbf{Y}_{bus}$ of Table 14.4 must again be modified. This is accomplished by substituting zero for Y_{45} and Y_{54} and subtracting the series admittance of line 4-5 and the capacitive susceptance of one-half the line from elements Y_{44} and Y_{55} of Table 14.4. The reduced bus-admittance matrix applicable to the postfault network is shown in the lower half of Table 14.5 and it is noted that a zero element appears in the first and second rows. This reflects the fact that, physically, the generators are not interconnected when line 4-5 is removed. Accordingly, each generator is connected radially to the infinite bus. Therefore, the per-unit power-angle equations for postfault conditions are

$$P_{e1} = |E_1'|^2 G_{11} + |E_1'||E_3||Y_{13}| \cos(\delta_{13} - \theta_{13})$$
$$= (1.100)^2(0.5005) + (1.100)(1.0)(7.6323) \cos(\delta_1 - 91.664°)$$
$$= 0.6056 + 8.3955 \sin(\delta_1 - 1.664°)$$

and

$$P_{e2} = |E_2'|^2 G_{22} + |E_2'||E_3||Y_{23}| \cos(\delta_{23} - \theta_{23})$$
$$= (1.065)^2(0.1591) + (1.065)(1.0)(6.0982) \cos(\delta_2 - 90.847°)$$
$$= 0.1804 + 6.4934 \sin(\delta_2 - 0.847°)$$

For the postfault period the applicable swing equations are

$$\frac{d^2\delta_1}{dt^2} = \frac{180f}{11.2}\{3.5 - [0.6056 + 8.3955 \sin(\delta_1 - 1.664°)]\}$$

$$= \frac{180f}{11.2}[2.8944 - 8.3955 \sin(\delta_1 - 1.664°)] \qquad \text{elec deg/s}^2$$

and

$$\frac{d^2\delta_2}{dt^2} = \frac{180f}{8.0}\{1.85 - [0.1804 + 6.4934 \sin(\delta_2 - 0.847°)]\}$$

$$= \frac{180f}{8.0}[1.6696 - 6.4934 \sin(\delta_2 - 0.847°)] \qquad \text{elec deg/s}^2$$

Each of the power-angle equations obtained in Example 14.9 and in this example is of the form of Eq. (14.35). The resultant swing equation in each case assumes the form

$$\frac{d^2\delta}{dt^2} = \frac{180f}{H}[P_m - P_c - P_{max} \sin(\delta - \gamma)] \qquad (14.78)$$

where the bracketed right-hand term represents the accelerating power on the rotor. Accordingly, we may write the equation in the form

$$\frac{d^2\delta}{dt^2} = \frac{180f}{H} P_a \qquad \text{elec deg/s}^2 \qquad (14.79)$$

where

$$P_a = P_m - P_c - P_{\max} \sin (\delta - \gamma) \qquad (14.80)$$

In the next section we shall discuss how to solve equations of the form of Eq. (14.79) to obtain δ as a function of time for specified clearing times.

14.9 STEP-BY-STEP SOLUTION OF THE SWING CURVE

For large systems we depend on the digital computer which determines δ versus t for all the machines in which we are interested; and δ may be plotted versus t for a machine to obtain the swing curve of that machine. The angle δ is calculated as a function of time over a period long enough to determine whether δ will increase without limit or reach a maximum and start to decrease. Although the latter result usually indicates stability, on an actual system where a number of variables are taken into account it may be necessary to plot δ versus t over a long enough interval to be sure δ will not increase again without returning to a low value.

By determining swing curves for various clearing times the length of time permitted before clearing a fault can be determined. Standard interrupting times for circuit breakers and their associated relays are commonly 8, 5, 3, or 2 cycles after a fault occurs, and thus breaker speeds may be specified. Calculations should be made for a fault in the position which will allow the least transfer of power from the machine and for the most severe type of fault for which protection against loss of stability is justified.

A number of different methods are available for the numerical evaluation of second-order differential equations in step-by-step computations for small increments of the independent variable. The more elaborate methods are practical only when the computations are performed on a digital computer. The step-by-step method used for hand calculation is necessarily simpler than some of the methods recommended for digital computers. In the method for hand calculation the change in the angular position of the rotor during a short interval of time is computed by making the following assumptions:

1. The accelerating power P_a computed at the beginning of an interval is constant from the middle of the preceding interval to the middle of the interval considered.
2. The angular velocity is constant throughout any interval at the value computed for the middle of the interval. Of course, neither of the assumptions is true, since δ is changing continuously and both P_a and ω are functions

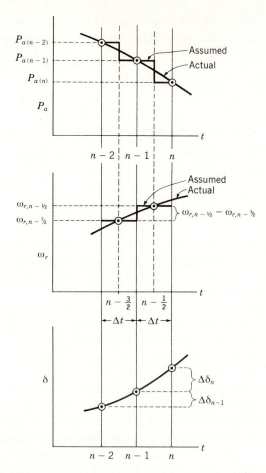

Figure 14.14 Actual and assumed values of P_a, ω_r, and δ as functions of time.

of δ. As the time interval is decreased, the computed swing curve approaches the true curve.

Figure 14.14 will help in visualizing the assumptions. The accelerating power is computed for the points enclosed in circles at the ends of the $n-2$, $n-1$, and n intervals, which are the beginnings of the $n-1$, n, and $n+1$ intervals. The step curve of P_a in Fig. 14.14 results from the assumption that P_a is constant between midpoints of the intervals. Similarly, ω_r, the *excess* of the angular velocity ω over the synchronous angular velocity ω_s, is shown as a step curve that is constant throughout the interval at the value computed for the midpoint. Between the ordinates $n-\frac{3}{2}$ and $n-\frac{1}{2}$ there is a change of speed caused by the constant accelerating power. The change in speed is the product of the acceleration and the time interval, and so

$$\omega_{r,\,n-1/2} - \omega_{r,\,n-3/2} = \frac{d^2\delta}{dt^2}\,\Delta t = \frac{180f}{H}\,P_{a,\,n-1}\,\Delta t \qquad (14.81)$$

The change in δ over any interval is the product of ω_r for the interval and the time of the interval. Thus, the change in δ during the $n - 1$ interval is

$$\Delta\delta_{n-1} = \delta_{n-1} - \delta_{n-2} = \Delta t\omega_{r,\,n-3/2} \qquad (14.82)$$

and during the nth interval

$$\Delta\delta_n = \delta_n - \delta_{n-1} = \Delta t\omega_{r,\,n-1/2} \qquad (14.83)$$

Subtracting Eq. (14.82) from Eq. (14.83) and substituting Eq. (14.81) in the resulting equation to eliminate all values of ω_r yields

$$\Delta\delta_n = \Delta\delta_{n-1} + kP_{a,\,n-1} \qquad (14.84)$$

where

$$k = \frac{180f}{H}(\Delta t)^2 \qquad (14.85)$$

Equation (14.84) is the important one for the step-by-step solution of the swing equation with the necessary assumptions enumerated, for it shows how to calculate the change in δ during an interval if the change in δ for the previous interval and the accelerating power for the interval in question are known. Equation (14.84) shows that, subject to the stated assumptions, the change in torque angle during a given interval is equal to the change in torque angle during the preceding interval plus the accelerating power at the beginning of the interval times k. The accelerating power is calculated at the beginning of each new interval. The solution progresses through enough intervals to obtain points for plotting the swing curve. Greater accuracy is obtained when the duration of the intervals is small. An interval of 0.05 s is usually satisfactory.

The occurrence of a fault causes a discontinuity in the accelerating power P_a which is zero before the fault and a definite amount immediately following the fault. The discontinuity occurs at the beginning of the interval, when $t = 0$. Reference to Fig. 14.14 shows that our method of calculation assumes that the accelerating power computed at the beginning of an interval is constant from the middle of the preceding interval to the middle of the interval considered. When the fault occurs, we have two values of P_a at the beginning of an interval, and we must take the average of these two values as our constant accelerating power. The procedure is illustrated in the following example.

Example 14.11 Prepare a table showing the steps taken to plot the swing curve for machine 2 for the fault on the 60-Hz system of Examples 14.9 and 14.10. The fault is cleared by simultaneous opening of the circuit breakers at the ends of the faulted line at 0.225 s.

SOLUTION Without loss of generality, we will consider the detailed computations for machine 2. Computations to plot the swing curve for machine 1 are left to the student. Accordingly we drop the subscript 2 as the indication of the machine number from all symbols in what follows. All our calculations are made in per unit on 100 MVA base. For the time interval $\Delta t = 0.05$ s the parameter k applicable to machine 2 is

$$k = \frac{180f}{H}(\Delta t)^2 = \frac{180 \times 60}{8.0} \times 25 \times 10^{-4} = 3.375 \text{ elec deg}$$

When the fault occurs at $t = 0$ the rotor angle of machine 2 cannot change instantly. Hence, from Example 14.9,

$$\delta_0 = 16.19°$$

and, during the fault,

$$P_e = 0.1545 + 5.5023 \sin(\delta - 0.755°)$$

Therefore, as already seen in Example 14.9

$$P_a = P_m - P_e = 1.6955 - 5.5023 \sin(\delta - 0.755°)$$

At the beginning of the first interval there is a discontinuity in the accelerating power of each machine. Just before the fault occurs, $P_a = 0$ and just after the fault occurs

$$P_a = 1.6955 - 5.5023 \sin(16.19° - 0.755°) = 0.231 \text{ per unit}$$

The average value of P_a at $t = 0$ is $\frac{1}{2} \times 0.2310 = 0.1155$ per unit. We then find

$$kP_a = 3.375 \times 0.1155 = 0.3898°$$

and consequently where we now identify the interval by numerical subscripts

$$\Delta\delta_1 = 0 + 0.3898 = 0.3898°$$

is the change in rotor angle of machine 2 as time advances over the *first interval* from 0 to Δt. Therefore at the end of the first time interval

$$\delta_1 = \delta_0 + \Delta\delta_1 = 16.19 + 0.3898 = 16.5798°$$

and

$$\delta_1 - \gamma = 16.5798 - 0.755 = 15.8248°$$

We then find at $t = \Delta t = 0.05$ s

$$kP_{a,1} = 3.375[(P_m - P_c) - P_{max} \sin(\delta_1 - \gamma)]$$
$$= 3.375[1.6955 - 5.5023 \sin(15.8248)] = 0.6583°$$

and it follows that the increase in rotor angle over the second time interval is

$$\Delta\delta_2 = \Delta\delta_1 + kP_{a,1} = 0.3898 + 0.6583 = 1.0481°$$

Hence, at the end of the second time interval

$$\delta_2 = \delta_1 + \Delta\delta_2 = 16.5798 + 1.0481 = 17.6279°$$

The subsequent steps in the computations are shown in Table 14.6. Note that the postfault equation found in Example 14.10 is needed.

In Table 14.6 the terms $P_{max} \sin(\delta - \gamma)$, P_a, and δ_n are values computed at the time t shown in the first column but $\Delta\delta_n$ is the *change* in rotor angle *during* the interval that begins at the time indicated. For example, in the row for $t = 0.10$ s the angle 17.6279° is the first value calculated and is found by adding the change in angle during the preceding time-interval

Table 14.6 Computation of swing curve for machine 2 of Example 14.11 for clearing at 0.225 s.

$k = (180f/H)(\Delta t)^2 = 3.375$ elec deg. Before clearing $P_m - P_c = 1.6955$ p.u., $P_{max} = 5.5023$ p.u., and $\gamma = 0.755°$. After clearing these values become 1.6696, 6.4934, and 0.847, respectively.

t, s	$\delta_n - \gamma$ elec deg	$P_{max} \sin(\delta_n - \gamma)$ per unit	P_a per unit	kP_a elec deg	$\Delta\delta_n$ elec deg	δ_n, elec deg
0−	....	....	0.00	....		16.19
0+	15.435	1.4644	0.2310	....		16.19
0 av	....	....	0.1155	0.3898	....	16.19
					0.3898	
0.05	15.8248	1.5005	0.1950	0.6583	....	16.5798
					1.0481	
0.10	16.8729	1.5970	0.0985	0.3323	....	17.6279
					1.3804	
0.15	18.2533	1.7234	−0.0279	−0.0942	....	19.0083
					1.2862	
0.20	19.5395	1.8403	−0.1448	−0.4886	....	20.2945
					0.7976	
0.25	20.2451	2.2470	−0.5774	−1.9487	....	21.0921
					−1.1511	
0.30	19.0940	2.1241	−0.4545	−1.534	....	19.9410
					−2.6852	
0.35	16.4088	1.8343	−0.1647	−0.5559	....	17.2558
					−3.2410	
0.40	13.1678	1.4792	0.1904	0.6425	....	14.0148
					−2.5985	
0.45	10.5693	1.1911	0.4785	1.6151	....	11.4163
					−0.9833	
0.50	9.5860	1.0813	0.5883	1.9854	....	10.4330
					1.0020	
0.55	10.5880	1.1931	0.4765	1.6081	....	11.4350
					2.6101	
0.60	13.1981	1.4826	0.1870	0.6312	....	14.0451
					3.2414	
0.65	16.4395	1.8376	−0.1680	−0.5672	....	17.2865
					2.6742	
0.70	19.1137	2.1262	−0.4566	−1.5411	....	19.9607
					1.1331	
0.75	20.2468	2.2471	−0.5775	−1.9492	....	21.0938
					−0.8161	
0.80	19.4307	2.1601	−0.4905	−1.6556	....	20.2777
					−2.4716	
0.85	....	....	....	....	....	17.8061

(0.05 to 0.10 s) to the angle at $t = 0.05$ s. Next $P_{max} \sin (\delta - \gamma)$ is calculated for $\delta = 17.6279°$. Then, $P_a = (P_m - P_c) - P_{max} \sin (\delta - \gamma)$ and kP_a are calculated. The value of kP_a is $0.3323°$, which is added to the angular change of $1.0481°$ during the preceding interval to find the change of $1.3804°$ during the interval beginning at $t = 0.10$ s. This value added to $17.6279°$ gives the value $\delta = 19.0083$ at $t = 0.15$ s. Note that at 0.25 s the value of $P_m - P_c$ has changed because the fault was cleared at 0.225 s. The angle γ has also changed from 0.755 to $0.847°$.

Whenever a fault is cleared, a discontinuity occurs in the accelerating power P_a. When clearing is at 0.225 s, as is the case for the calculations in Table 14.6, no special approach is required since our procedure assumes a discontinuity at the middle of an interval. At the beginning of the interval following clearing the assumed constant value of P_a is that determined for δ at the beginning of the interval following clearing.

When clearing is at the beginning of an interval such as at three cycles (0.05 s), two values of accelerating power result from the two expressions (one during the fault and one after clearing) for the power output of the generator at the beginning of the interval. For the system of Example 14.11, if the discontinuity occurs at 0.05 s, the average of the two values is assumed as the constant value of P_a from 0.025 to 0.075 s. The procedure is the same as that followed upon occurrence of the fault at $t = 0$ as demonstrated in Table 14.6.

In the same manner as followed in preparing Table 14.6, we could determine δ versus t for machine 1 for clearing at 0.225 s and for both machines for clearing at 0.05 s. In the next section, we shall see computer printouts of δ versus t for both machines calculated for clearing at 0.05 and 0.225 s. The swing curves are plotted for the two machines in Fig. 14.15 for clearing at 0.225 s. Evidently, machine 1 is unstable in this case.

Swing curves for clearing at 0.20 s would show the system to be stable. Since the output of the unstable machine is zero during the fault the equal-area criterion can be applied by solving Prob. 14.16 to find the actual critical clearing time which will be between 0.20 and 0.225 s.

It is noted from the swing curves of Fig. 14.15 that, even though clearing does not occur until 13.5 cycles after the onset of the fault, the change in the rotor angle of machine 2 is quite small. Consequently, it is interesting to calculate the approximate frequency of oscillation of the rotor based on the linearization procedure presented in Sec. 14.5. The synchronizing power coefficient calculated from the postfault power-angle equation for machine 2 is given by

$$S_P = \frac{dP_e}{d\delta} = \frac{d}{d\delta} [0.1804 + 6.4934 \sin (\delta - 0.847°)]$$

$$= 6.4934 \cos (\delta - 0.847°)$$

We note that for the points calculated for Table 14.6, the angle of δ for machine 2 varies between 10.43 and $21.09°$. Using either angle makes little difference in the

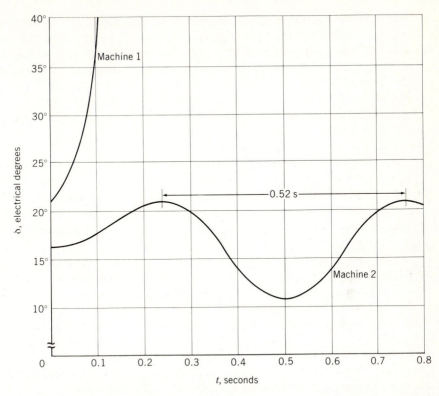

Figure 14.15 Swing curves for machines 1 and 2 of Examples 14.9 to 14.11 for clearing at 0.225 s.

value found for S_P. If we used the average value of 15.76°, we find

$$S_P = 6.274 \text{ per-unit power/elec rad}$$

and by Eq. (14.50) the frequency of oscillation is

$$f_n = \frac{1}{2\pi}\sqrt{\frac{377 \times 6.274}{2 \times 8}} = 1.935 \text{ Hz}$$

and the period of oscillation is

$$T = \frac{1}{1.935} = 0.517 \text{ s}$$

Examination of Fig. 14.15 or Table 14.6 confirms the applicability of these calculations for machine 2 when the fault is on for 0.225 s. For faults of shorter duration, similar results can be expected since the rotor-angle swing is correspondingly smaller.

It is possible to calculate the swing curves separately for each machine in the above examples because of the fault location considered. When other fault locations are chosen, the intermachine oscillations between the two generators occur because there is no decoupling of the machines. The swing-curve computations are more unwieldy. For such cases, manual calculations are time consuming and should be avoided. Digital computer programs of great versatility are generally available and should be used.

14.10 DIGITAL-COMPUTER PROGRAMS FOR TRANSIENT STABILITY STUDIES

Present-day digital computer programs for transient stability studies have evolved from two basic needs (*a*) the requirement to study very large inter-connected systems with very many machines and (*b*) the need to represent machines and their associated control systems by more detailed models. The classical machine representation is suitable for many studies. However, more elaborate models may be required to represent modern turboalternators with dynamic characteristics determined by the many technological advances in design of machine and control systems.

The simplest possible synchronous machine model is that used in classical stability studies. Much more complicated two-axis machine models are available which provide for direct- and quadrature-axis flux conditions during the subtransient and transient periods following a system disturbance. For example, unless the machine model explicitly provides for varying flux linkages of the field winding in the direct axis, it is not possible to represent the action of the continuously acting automatic voltage regulator and excitation system with which all modern machines are equipped. Turbine control systems, which automatically govern the mechanical power input to the generating unit, also have dynamic response characteristics which can influence rotor dynamics. If these control schemes are to be represented, the generating unit model must be further extended. The more complex generator models give rise to a larger number of differential and algebraic equations for each machine. In large system studies, many generators are interconnected with widely dispersed load centers by an extensive transmission system whose performance also must be represented by a very large number of algebraic equations. Therefore, two sets of equations must be solved simultaneously for each interval following the occurrence of a system disturbance. One set consists of the *algebraic* equations for the *steady-state* behavior of the network and its loads and the algebraic equations relating V_t and E' of the synchronous machines. The other set consists of the *differential* equations which describe the *dynamic* performance of the machines and associated control systems.

The Newton-Raphson load-flow procedure described in Chap. 8 is probably the most commonly used solution technique for the network equations. Any one of several well known step-by-step procedures may be chosen for numerical integration of the differential equations. The fourth-order Runge-Kutta method

Table 14.7 Computer printout of swing curves for machines 1 and 2 of Examples 14.9 to 14.11 for clearing at 0.225 and 0.05 s.

	CLEARING AT 0.225 S			CLEARING AT 0.05 S	
	MACH 1	MACH 2		MACH 1	MACH 2
TIME	ANGLE	ANGLE	TIME	ANGLE	ANGLE
0.00	20.8	16.2	0.00	20.8	16.2
0.05	25.1	16.6	0.05	25.1	16.6
0.10	37.7	17.6	0.10	32.9	17.2
0.15	58.7	19.0	0.15	37.3	17.2
0.20	88.1	20.3	0.20	36.8	16.7
0.25	123.1	20.9	0.25	31.7	15.9
0.30	151.1	19.9	0.30	23.4	15.0
0.35	175.5	17.4	0.35	14.6	14.4
0.40	205.1	14.3	0.40	8.6	14.3
0.45	249.9	11.8	0.45	6.5	14.7
0.50	319.3	10.7	0.50	10.1	15.6
0.55	407.0	11.4	0.55	17.7	16.4
0.60	489.9	13.7	0.60	26.6	17.1
0.65	566.0	16.8	0.65	34.0	17.2
0.70	656.4	19.4	0.70	37.6	16.8
0.75	767.7	20.8	0.75	36.2	16.0

is very often used in production-type transient-stability programs. Other methods known as the Euler method, the modified-Euler method, the trapezoidal method, and predictor-corrector methods like the step-by-step method developed in Sec. 14.9 are alternatives. Each of these methods has advantages and disadvantages associated with numerical stability, time-step size, computational effort per integration step, and accuracy of solutions obtained.†

Table 14.7 shows the computer printout for plotting the swing curves of machines 1 and 2 of Example 14.11 for clearing at 0.225 and at 0.05 s obtained by use of a production-type stability program which couples a Newton-Raphson load-flow program with a fourth-order Runge-Kutta procedure. It is interesting to compare the closeness of our hand-calculated values in Table 14.6 with those for machine 2 in Table 14.7 when the fault is cleared at 0.225 s.

† For further information, see G. W. Stagg and A. H. El-Abiad, *Computer Methods in Power System Analysis*, chaps. 9 and 10, McGraw-Hill Book Company, New York, 1968.

Our assumption of constant admittances of the loads allowed us to absorb these admittances into $\mathbf{Y}_{bus}$ and avoid load-flow calculations required when the more accurate solutions using Runge-Kutta calculations are desired. The latter, being fourth order, require four iterative load-flow computations per time step.

14.11 FACTORS AFFECTING TRANSIENT STABILITY

There are two factors which can act as guideline criteria for the relative stability of a generating unit within a power system. These are the angular swing of the machine during and following fault conditions and the critical clearing time. It is apparent from the equations which we have developed in this chapter that the H constant and the transient reactance X'_d of the generating unit have a direct effect on both of these criteria.

Inspection of Eqs. (14.84) and (14.85) indicates that the smaller the H constant, the larger the angular swing during any time interval. On the other hand, Eq. (14.36) shows that P_{max} decreases as the transient reactance of the machine increases. This is so because the transient reactance forms part of the overall series reactance of the system which is the reciprocal of the transfer admittance. Examination of Fig. 14.11 shows that all three power curves are lowered when P_{max} is decreased. Accordingly, for a given shaft power P_m, the initial rotor angle δ_0 is increased, δ_{max} is decreased, and a smaller difference between δ_0 and δ_{cr} exists for a smaller P_{max}. The net result is that a decreased P_{max} constrains a machine to swing through a smaller angle from its original position before it reaches the critical clearing angle. Thus, any developments which lower the H constant and increase the transient reactance of a machine cause the critical clearing time to decrease and lessens the probability of maintaining stability under transient conditions. As power systems continually increase in size, there is a corresponding need for higher-rated generating units. These larger units have advanced cooling systems which allow higher-rated capacities without comparable increase in rotor size. As a result, H constants continue to decrease with potential adverse impact on generating unit stability. At the same time, this uprating process tends to result in higher transient and synchronous reactances which makes the job of designing a reliable and stable system even more challenging.

Fortunately, stability control techniques and transmission system designs have also been evolving to increase overall system stability. The control schemes include

Excitation systems
Turbine valve control
Single-pole operation of circuit breakers
Faster fault clearing times

The system design strategies, aimed at lowering system reactance, include

Minimum transformer reactance
Series capacitor compensation of lines
Additional transmission lines

When a fault occurs on a system the voltages at all buses are reduced. At generator terminals the reduced voltages are sensed by the automatic voltage regulators which act within the excitation system to restore generator terminal voltages. The general effect of the excitation system is to reduce the initial rotor angle swing following the fault. This is accomplished by boosting the voltage applied to the field winding of the generator through action of the amplifiers in the forward path of the voltage regulators. The increased air-gap flux exerts a restraining torque on the rotor which tends to slow down its motion. Modern excitation systems employing thyristor controls rapidly respond to bus voltage reduction and can effect from one-half to one-and-one-half cycles gain in critical clearing times for three-phase faults on the high-side bus of the generator step-up transformer.

Modern electrohydraulic turbine governing systems have the ability to close turbine valves to reduce unit acceleration during severe system faults near the unit. Immediately upon detecting differences between mechanical input and electrical output, control action initiates the valve closing which reduces the power input. A gain of one to two cycles in critical clearing time can be achieved.

Reducing the reactance of the system during fault conditions increases $r_1 P_{max}$ decreasing the acceleration area of Fig. 14.11, and thereby enhances the possibility of maintaining stability. Since single-phase faults occur more often than three-phase faults, relaying schemes, allowing independent or selective circuit-breaker pole operation, can be used to clear the faulted phase while keeping the unfaulted phases intact. Separate relay systems, trip coils, and operating mechanisms can be provided for each pole so that stuck breaker contingencies following three-phase faults can be mitigated in effect. Independent-pole operation of critical circuit breakers can extend the critical clearing time by 2 to 5 cycles depending upon whether 1 or 2 poles fail to open under fault conditions. Such gain in critical clearing time can be important especially if backup clearing times are a problem for system stability.

Reducing the reactance of a transmission line is another way of raising P_{max}. Compensation for line reactance by series capacitors is often economical for increasing stability. Increasing the number of parallel lines between two points is a common means of reducing reactance. When parallel transmission lines are used instead of a single line, some power is transferred over the remaining line even during a three-phase fault on one of the lines unless the fault occurs at a paralleling bus. For other types of faults on one line, more power is transferred during the fault if there are two lines in parallel than is transferred over a single faulted line. For more than two lines in parallel the power transferred during the fault is even greater. Power transferred is subtracted from power input to obtain accelerating power. Thus increased power transferred during a fault means lower accelerating power for the machine and increased chance of stability.

PROBLEMS

14.1 A 60-Hz four-pole turbogenerator rated 500 MVA, 22 kV has an inertia constant of $H = 7.5$ MJ/MVA. Find (a) the kinetic energy stored in the rotor at synchronous speed and (b) the angular acceleration if the electrical power developed is 400 MW when the input less the rotational losses is 740,000 hp.

14.2 If the acceleration computed for the generator described in Prob. 14.1 is constant for a period of 15 cycles, find the change in δ in electrical degrees in that period and the speed in revolutions per minute at the end of 15 cycles. Assume that the generator is synchronized with a large system and has no accelerating torque before the 15-cycle period begins.

14.3 The generator of Prob. 14.1 is delivering rated megavolt-amperes at 0.8 power factor lag when a fault reduces the electric power output by 40%. Determine the accelerating torque in newton-meters at the time the fault occurs. Neglect losses and assume constant power input to the shaft.

14.4 Determine the WR^2 of the generator of Prob. 14.1.

14.5 A generator having $H = 6$ MJ/MVA is connected to a synchronous motor having $H = 4$ MJ/MVA through a network of reactances. The generator is delivering power of 1.0 per unit to the motor when a fault occurs which reduces the delivered power. At the time when the reduced power delivered is 0.6 per unit determine the angular acceleration of the generator with respect to the motor.

14.6 A power system is identical to that of Example 14.3 except that the impedance of each of the parallel transmission lines is $j0.5$ and the delivered power is 0.8 per unit when both the terminal voltage of the machine and the voltage of the infinite bus are 1.0 per unit. Determine the power-angle equation for the system during the specified operating conditions.

14.7 If a three-phase fault occurs on the power system of Prob. 14.6 at a point on one of the transmission lines at a distance of 30% of the line length away from the sending-end terminal of the line, determine (a) the power-angle equation during the fault and (b) the swing equation. Assume the system is operating under the conditions specified in Prob. 14.6 when the fault occurs. Let $H = 5.0$ MJ/MVA as in Example 14.4.

14.8 Series resistance in the transmission network results in positive values for P_c and γ in Eq. (14.80). For a given electrical power output, show the effects of resistance on the synchronizing coefficient S_P, the frequency of rotor oscillations, and the damping of these oscillations.

14.9 A generator having $H = 6.0$ MJ/MVA is delivering power of 1.0 per unit to an infinite bus through a purely reactive network when the occurrence of a fault reduces the generator output power to zero. The maximum power that could be delivered is 2.5 per unit. When the fault is cleared the original network conditions again exist. Determine the critical clearing angle and critical clearing time.

14.10 A 60-Hz generator is supplying 60% of P_{max} to an infinite bus through a reactive network. A fault occurs which increases the reactance of the network between the generator internal voltage and the infinite bus by 400%. When the fault is cleared the maximum power that can be delivered is 80% of the original maximum value. Determine the critical clearing angle for the condition described.

14.11 If the generator of Prob. 14.10 has an inertia constant of $H = 6$ MJ/MVA and P_m (equal to $0.6 P_{max}$) is 1.0 per-unit power, find the critical clearing time for the condition of Prob. 14.10. Use $\Delta t = 0.05$ s to plot the necessary swing curve.

14.12 For the system and fault conditions described in Probs. 14.6 and 14.7 determine the power-angle equation if the fault is cleared by the simultaneous opening of breakers at both ends of the faulted line at 4.5 cycles after the fault occurs. Then plot the swing curve of the generator through $t = 0.25$ s.

14.13 Extend Table 14.6 to find δ at $t = 1.00$ s.

14.14 Calculate the swing curve for machine 2 of Examples 14.9 to 14.11 for fault clearing at 0.05 s by the method described in Sec. 14.9. Compare the results with the values obtained by the production-type program and listed in Table 14.7.

14.15 If the three-phase fault on the system of Example 14.9 occurs on line 4-5 at bus 5 and is cleared by simultaneous opening of breakers at both ends of the line at 4.5 cycles after the fault occurs prepare a table like that of Table 14.6 to plot the swing curve of machine 2 through $t = 0.30$ s.

14.16 By applying the equal-area criterion to the swing curves obtained in Examples 14.9 and 14.10 for machine 1, (*a*) derive an equation for the critical clearing angle, (*b*) solve the equation by trial and error to evaluate δ_{cr}, and (*c*) use Eq. (14.72) to find the critical clearing time.

APPENDIX

X am =

^{1}in = $x$am $\left(\dfrac{1\,am}{5\,secm}\right)$

$$^{10}\sqrt{\left(\cdot\,6965\right)^{4}\left(\tfrac{4}{3}\right)^{4}\left(\theta\sqrt{2}\right)}$$

$.85|7,28$

Table A.1 Electrical characteristics of bare aluminum conductors steel-reinforced (ACSR)†

Code word	Aluminum area, cmil	Stranding Al/St	Layers of aluminum	Outside diameter, in	Resistance Dc, 20°C, Ω/1,000 ft	Ac, 60 Hz 20°C, Ω/mi	Ac, 60 Hz 50°C, Ω/mi	GMR D_s, ft	Inductive X_a, Ω/mi	Capacitive X_a', MΩ·mi
Waxwing	266,800	18/1	2	0.609	0.0646	0.3488	0.3831	0.0198	0.476	0.1090
Partridge	266,800	26/7	2	0.642	0.0640	0.3452	0.3792	0.0217	0.465	0.1074
Ostrich	300,000	26/7	2	0.680	0.0569	0.3070	0.3372	0.0229	0.458	0.1057
Merlin	336,400	18/1	2	0.684	0.0512	0.2767	0.3037	0.0222	0.462	0.1055
Linnet	336,400	26/7	2	0.721	0.0507	0.2737	0.3006	0.0243	0.451	0.1040
Oriole	336,400	30/7	2	0.741	0.0504	0.2719	0.2987	0.0255	0.445	0.1032
Chickadee	397,500	18/1	2	0.743	0.0433	0.2342	0.2572	0.0241	0.452	0.1031
Ibis	397,500	26/7	2	0.783	0.0430	0.2323	0.2551	0.0264	0.441	0.1015
Pelican	477,000	18/1	2	0.814	0.0361	0.1957	0.2148	0.0264	0.441	0.1004
Flicker	477,000	24/7	2	0.846	0.0359	0.1943	0.2134	0.0284	0.432	0.0992
Hawk	477,000	26/7	2	0.858	0.0357	0.1931	0.2120	0.0289	0.430	0.0988
Hen	477,000	30/7	2	0.883	0.0355	0.1919	0.2107	0.0304	0.424	0.0980
Osprey	556,500	18/1	2	0.879	0.0309	0.1679	0.1843	0.0284	0.432	0.0981
Parakeet	556,500	24/7	2	0.914	0.0308	0.1669	0.1832	0.0306	0.423	0.0969
Dove	556,500	26/7	2	0.927	0.0307	0.1663	0.1826	0.0314	0.420	0.0965
Rook	636,000	24/7	2	0.977	0.0269	0.1461	0.1603	0.0327	0.415	0.0950
Grosbeak	636,000	26/7	2	0.990	0.0268	0.1454	0.1596	0.0335	0.412	0.0946
Drake	795,000	26/7	2	1.108	0.0215	0.1172	0.1284	0.0373	0.399	0.0912
Tern	795,000	45/7	3	1.063	0.0217	0.1188	0.1302	0.0352	0.406	0.0925
Rail	954,000	45/7	3	1.165	0.0181	0.0997	0.1092	0.0386	0.395	0.0897
Cardinal	954,000	54/7	3	1.196	0.0180	0.0988	0.1082	0.0402	0.390	0.0890
Ortolan	1,033,500	45/7	3	1.213	0.0167	0.0924	0.1011	0.0402	0.390	0.0885
Bluejay	1,113,000	45/7	3	1.259	0.0155	0.0861	0.0941	0.0415	0.386	0.0874
Finch	1,113,000	54/19	3	1.293	0.0155	0.0856	0.0937	0.0436	0.380	0.0866
Bittern	1,272,000	45/7	3	1.345	0.0136	0.0762	0.0832	0.0444	0.378	0.0855
Pheasant	1,272,000	54/19	3	1.382	0.0135	0.0751	0.0821	0.0466	0.372	0.0847
Bobolink	1,431,000	45/7	3	1.427	0.0121	0.0684	0.0746	0.0470	0.371	0.0837
Plover	1,431,000	54/19	3	1.465	0.0120	0.0673	0.0735	0.0494	0.365	0.0829
Lapwing	1,590,000	45/7	3	1.502	0.0109	0.0623	0.0678	0.0498	0.364	0.0822
Falcon	1,590,000	54/19	3	1.545	0.0108	0.0612	0.0667	0.0523	0.358	0.0814
Bluebird	2,156,000	84/19	4	1.762	0.0080	0.0476	0.0515	0.0586	0.344	0.0776

† Most used multilayer sizes.

‡ Data, by permission, from Aluminum Association, "Aluminum Electrical Conductor Handbook," New York, September 1971.

Table A.2 Inductive reactance spacing factor X_d at 60 Hz† (ohms per mile per conductor)

Feet	Inches 0	1	2	3	4	5	6	7	8	9	10	11
0	......	−0.3015	−0.2174	−0.1682	−0.1333	−0.1062	−0.0841	−0.0654	−0.0492	−0.0349	−0.0221	−0.0106
1	0.0841	0.0097	0.0187	0.0271	0.0349	0.0423	0.0492	0.0558	0.0620	0.0679	0.0735	0.0789
2	0.1333	0.0891	0.0938	0.0984	0.1028	0.1071	0.1112	0.1152	0.1190	0.1227	0.1264	0.1299
3	0.1682	0.1366	0.1399	0.1430	0.1461	0.1491	0.1520	0.1549	0.1577	0.1604	0.1631	0.1657
4	0.1953	0.1707	0.1732	0.1756	0.1779	0.1802	0.1825	0.1847	0.1869	0.1891	0.1912	0.1933
5	0.2174	0.1973	0.1993	0.2012	0.2031	0.2050	0.2069	0.2087	0.2105	0.2123	0.2140	0.2157
6	0.2361	0.2191	0.2207	0.2224	0.2240	0.2256	0.2271	0.2287	0.2302	0.2317	0.2332	0.2347
7	0.2523	0.2376	0.2390	0.2404	0.2418	0.2431	0.2445	0.2458	0.2472	0.2485	0.2498	0.2511
8	0.2666											
9	0.2794											
10	0.2910											
11	0.3015											
12	0.3112											
13	0.3202											
14	0.3286											
15	0.3364											
16	0.3438											
17	0.3507											
18	0.3573											
19	0.3635											
20	0.3694											
21	0.3751											
22	0.3805											
23	0.3856											
24	0.3906											
25	0.3953											
26	0.3999											
27	0.4043											
28	0.4086											
29	0.4127											
30	0.4167											
31	0.4205											
32	0.4243											
33	0.4279											
34	0.4314											
35	0.4348											
36	0.4382											
37	0.4414											
38	0.4445											
39	0.4476											
40	0.4506											
41	0.4535											
42	0.4564											
43	0.4592											
44	0.4619											
45	0.4646											
46	0.4672											
47	0.4697											
48	0.4722											

At 60 Hz, in Ω/mi per conductor
$X_d = 0.2794 \log d$
d = separation, ft
For three-phase lines
$d = D_{eq}$

† From "Electrical Transmission and Distribution Reference Book," by permission of the Westinghouse Electric Corporation.

Table A.3 Shunt capacitive-reactance spacing factor X_d at 10 Hz† (megohm-miles per conductor)

| | | Separation — Inches | | | | | | | | | | |
Feet	0	1	2	3	4	5	6	7	8	9	10	11
0	......	-0.0737	-0.0532	-0.0411	-0.0326	-0.0260	-0.0206	-0.0160	-0.0120	-0.0085	-0.0054	-0.0026
1	0.0206	0.0024	0.0046	0.0066	0.0085	0.0103	0.0120	0.0136	0.0152	0.0166	0.0180	0.0193
2	0.0326	0.0218	0.0229	0.0241	0.0251	0.0262	0.0272	0.0282	0.0291	0.0300	0.0309	0.0318
3	0.0411	0.0334	0.0342	0.0350	0.0357	0.0365	0.0372	0.0379	0.0385	0.0392	0.0399	0.0405
4	0.0478	0.0417	0.0423	0.0429	0.0435	0.0441	0.0446	0.0452	0.0457	0.0462	0.0467	0.0473
5	0.0532	0.0482	0.0487	0.0492	0.0497	0.0501	0.0506	0.0510	0.0515	0.0519	0.0523	0.0527
6	0.0577	0.0536	0.0540	0.0544	0.0548	0.0552	0.0555	0.0559	0.0563	0.0567	0.0570	0.0574
7	0.0617	0.0581	0.0584	0.0588	0.0591	0.0594	0.0598	0.0601	0.0604	0.0608	0.0611	0.0614
8	0.0652											
9	0.0683											
10	0.0711											
11	0.0737											
12	0.0761											
13	0.0783											
14	0.0803											
15	0.0823											
16	0.0841											
17	0.0858											
18	0.0874											
19	0.0889											
20	0.0903											
21	0.0917											
22	0.0930											
23	0.0943											
24	0.0955											
25	0.0967											
26	0.0978											
27	0.0989											
28	0.0999											
29	0.1009											
30	0.1019											
31	0.1028											
32	0.1037											
33	0.1046											
34	0.1055											
35	0.1063											
36	0.1071											
37	0.1079											
38	0.1087											
39	0.1094											
40	0.1102											
41	0.1109											
42	0.1116											
43	0.1123											
44	0.1129											
45	0.1136											
46	0.1142											
47	0.1149											
48	0.1155											
49												

At 60 Hz, in MΩ·mi per conductor
$X_d' = 0.06831 \log d$
d = separation, ft
For three-phase lines
$d = D_{eq}$

† From "Electrical Transmission and Distribution Reference Book," by permission of the Westinghouse Electric Corporation.

Table A.4 Typical reactances of three-phase synchronous machines.†

Values are per unit. For each reactance a range of values is listed below the typical value.‡

	Turbine-generators				Salient-pole generators	
	2-pole		4-pole			
	Conventional cooled	Conductor cooled	Conventional cooled	Conductor cooled	With dampers	Without dampers
X_d	1.76	1.95	1.38	1.87	1	1
	1.7–1.82	1.72–2.17	1.21–1.55	1.6–2.13	0.6–1.5	0.6–1.5
X_q	1.66	1.93	1.35	1.82	0.6	0.6
	1.63–1.69	1.71–2.14	1.17–1.52	1.56–2.07	0.4–0.8	0.4–0.8
X'_d	0.21	0.33	0.26	0.41	0.32	0.32
	0.18–0.23	0.264–0.387	0.25–0.27	0.35–0.467	0.25–0.5	0.25–0.5
X''_d	0.13	0.28	0.19	0.29	0.2	0.30
	0.11–0.14	0.23–0.323	0.184–0.197	0.269–0.32	0.13–0.32	0.2–0.5
X_2	$= X''_d$	$= X''_d$	$= X''_d$	$= X''_d$	0.2	0.40
					0.13–0.32	0.30–0.45
X_0§						

† Data furnished by Westinghouse Electric Corporation.
‡ Reactances of older machines will generally be close to minimum values.
§ X_0 varies so critically with armature winding pitch that an average value can hardly be given. Variation is from 0.1 to 0.7 of X''_d.

Table A.5 Typical range of transformer reactances†
Power transformers 25,000 kVA and larger

Nominal system voltage, kV	Forced-air-cooled, %	Forced-oil-cooled, %
34.5	5–8	9–14
69	6–10	10–16
115	6–11	10–20
138	6–13	10–22
161	6–14	11–25
230	7–16	12–27
345	8–17	13–28
500	10–20	16–34
700	11–21	19–35

† Percent on rated kilovoltampere base. Typical transformers are now designed for the minimum reactance value shown. Distribution transformers have considerably lower reactance. Resistances of transformers are usually lower than 1%.

Table A.6 *ABCD* constants for various networks

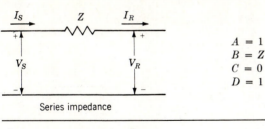

Series impedance

$A = 1$
$B = Z$
$C = 0$
$D = 1$

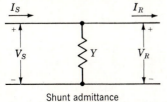

Shunt admittance

$A = 1$
$B = 0$
$C = Y$
$D = 1$

Unsymmetrical T

$A = 1 + YZ_1$
$B = Z_1 + Z_2 + YZ_1Z_2$
$C = Y$
$D = 1 + YZ_2$

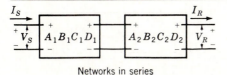

Unsymmetrical π

$A = 1 + Y_2Z$
$B = Z$
$C = Y_1 + Y_2 + ZY_1Y_2$
$D = 1 + Y_1Z$

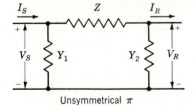

Networks in series

$A = A_1A_2 + B_1C_2$
$B = A_1B_2 + B_1D_2$
$C = A_2C_1 + C_2D_1$
$D = B_2C_1 + D_1D_2$

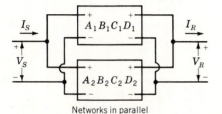

Networks in parallel

$A = (A_1B_2 + A_2B_1)/(B_1 + B_2)$
$B = B_1B_2/(B_1 + B_2)$
$C = C_1 + C_2 + (A_1 - A_2)(D_2 - D_1)/(B_1 + B_2)$
$D = (B_2D_1 + B_1D_2)/(B_1 + B_2)$

INDEX